READINGS IN HEREDITY AND DEVELOPMENT

Readings in Heredity and Development

JOHN A. MOORE

PROFESSOR OF BIOLOGY
UNIVERSITY OF CALIFORNIA, RIVERSIDE

NEW YORK
OXFORD UNIVERSITY PRESS
LONDON 1972 TORONTO

Copyright © 1972 by Oxford University Press, Inc.
Library of Congress Catalogue Card Number: 70-170264
Printed in the United States of America

To
Bentley Glass

in friendship and admiration

Preface

This anthology of great papers traces the development of ideas in the sister sciences of genetics and developmental biology. The collection can stand alone but it is designed to be used with its companion volume, *Heredity and Development* (Oxford University Press, 1972). The relation of these two volumes should be explained to facilitate their use.

Heredity and Development discusses in detail the key hypotheses, experiments, and concepts of genetics and developmental biology. The method of presentation is roughly historical.

Readings in Heredity and Development does not include the papers describing the original discoveries (these are covered in *Heredity and Development*). Instead, I have chosen some of the papers that represent the great synthetic treatments of the moment. For example, *H & D* discusses Schwann's observations on cells but *Readings* presents Virchow's famous lecture on cell theory; *H & D* discusses Wilson's discoveries on sex chromosomes but *Readings* contains his great lecture to the Royal Society, 'The Bearing of Cytological Research on Heredity.' I believe that this method of selection better serves the student interested in the intellectual development of science. He can see how giants of the field viewed the science they were making. Thus, *Readings* differs fundamentally from other anthologies covering these same fields. These other anthologies nearly always deal with the key discoveries, and are designed to provide the student with a wealth of material that only the better libraries have in the original.

Following this Preface there is a listing which correlates the major topics in *Heredity and Development* not only with the relevant articles in *Readings* but also with articles in eight other readily available anthologies.

Bibliographies. At the end of each chapter there are numerous references, which should provide the student with a broad and historically important selection. A much more selective list is given in *Heredity and Development*.

Riverside, California
October 1971

JOHN A. MOORE

Contents

For the convenience of the student there follows a listing which correlates the major topics in *Heredity and Development* not only with the relevant articles in *Readings* but also with articles in eight other readily available anthologies. Papers that appear in *Readings* are indicated by boldface type, and those appearing in the selected anthologies are identified by the last name of the anthology's compiler; when several people are involved in the anthology only the name of the first person cited in the listing is used.

SELECTED ANTHOLOGIES

(A) CARLSON, ELOF AXEL. 1967. *Gene Theory*. Belmont, Calif.: Dickenson Publishing Co.

(B) CARLSON, ELOF AXEL. 1967. *Modern Biology. Its Conceptual Foundations*. New York: George Braziller.

GABRIEL, MORDECAI L., and SEYMOUR FOGEL. 1955. *Great Experiments in Biology*. Englewood Cliffs, N.J.: Prentice-Hall.

FLICKINGER, REED A. 1966. *Developmental Biology. A Book of Readings*. Dubuque, Iowa: Wm. C. Brown.

PETERS, JAMES A. 1959. *Classic Papers in Genetics*. Englewood Cliffs, N.J.: Prentice-Hall.

STERN, CURT, and EVA R. SHERWOOD. 1966. *The Origin of Genetics. A Mendel Source Book*. San Francisco: W. H. Freeman.

VOELLER, BRUCE R. 1968. *The Chromosome Theory of Inheritance. Classical Papers in Development and Heredity*. New York: Appleton-Century-Crofts.

WILLIER, BENJAMIN H., and JANE M. OPPENHEIMER. 1964. *Foundations of Experimental Embryology*. Englewood Cliffs, N.J.: Prentice-Hall.

Major topics in **HEREDITY AND DEVELOPMENT**	Related papers in this book and in Selected Anthologies
1. DARWIN'S THEORY OF PANGENESIS Darwin's theory of pangenesis. 1868.	1. EARLY THEORIES OF INHERITANCE **Hippocrates. 'Airs, Waters, Places.'** **Aristotle. 'Generation of Animals.'** **Galton. 'Experiments in Pangenesis. . . ' 1871.** **Darwin. 'Pangenesis.' 1871.** **Galton. 'Pangenesis.' 1871.**

Kölreuter. 'Preliminary Report of Experiments and Observations concerning Some Aspects of Sexuality of Plants.' 1761–66. (Voeller)

Focke. 'Plant Hybrids.' 1881. (Stern)

2. STUDENTS OF THE CELL

Grew. 'Anatomy of Plants...' 1682.
Roget. 'Animal and Vegetable Physiology...' 1836.
Virchow. 'Cellular Pathology...' 1858.
Weismann. 'The Continuity of the Germ-plasm...' 1885.
Weismann. 'On the Number of Polar Bodies...' 1887.
Weismann. 'Amphimixis...' 1891.
Wilson. 'The Cell...' 1900.
Hooke. '...Texture of Cork...' 1665. (Gabriel; Carlson B)
Dutrochet. 'The Structural Elements of Plants.' 1824. (Gabriel)
Brown. '...Mode of Fecundation in Orchideae.' 1833. (Gabriel)
Schwann. 'Microscopical Researches.' 1839. (Gabriel)
Newport. '...Impregnation of the Ovum...' 1854. (Gabriel)
Hertwig. '...Formation, Fertilization and Division of the Animal Egg.' 1876. (Voeller)
Fol. '...Beginnings of Ontogeny...' 1877. (Voeller)
Flemming. '...Knowledge of the Cell...' 1879. (Gabriel; Voeller)
Strasburger. '...Fertilization in Phanerogams.' 1884. (Voeller)
van Beneden. '...Maturation of the Egg...' 1883. (Gabriel; Voeller)
Hertwig. 'The Problem of Fertilization...' 1885. (Voeller)

2. THE CELLULAR BASIS OF INHERITANCE

Hooke. Cells. 1665.
Schwann. Cell theory. 1839.
Schneider. Mitosis. 1873.
Flemming. Mitosis. 1882.
Hertwig. Fertilization. 1876.
Boveri. Fertilization. 1888.
Boveri. Meiosis. 1887.
Brauer. Meiosis. 1893.

3. MENDELISM
Mendel. 1866.
de Vries. 1900.
Correns. 1900.
Bateson. 1902——.

4. THE CHROMOSOMES AND INHERITANCE
Boveri. Chromosomes and inheritance. 1902.
Sutton. Chromosomes and inheritance. 1902, 1903.
Henking. Sex chromosomes. 1891.
McClung. Sex chromosomes. 1901.
Wilson and Stevens. Sex chromosomes. 1905.

3. MENDELISM
Bateson and Saunders. 'Reports to the Evolution Committee.' 1902.
Morgan. 'What Are Factors in Mendelian Explanations?' 1909.
Mendel. 'Experiments on Plant Hybrids.' 1866. (Stern; Peters; Voeller)
Mendel. 'On Hieracium-Hybrids...' 1869. (Stern)
Mendel. Letters to Nägeli. 1866–73. (Stern; Carlson B; Gabriel; Voeller)
de Vries. 'The Law of Segregation of Hybrids.' 1900. (Stern)
Correns. 'G. Mendel's Law...' 1900. (Stern)
de Vries and Correns. Letters. (Stern)
Fisher. 'Has Mendel's Work Been Rediscovered?' 1936. (Stern)
Wright. 'Mendel's Ratios.' 1966. (Stern)
Castle and Phillips. '...Ovarian Transplantation...' 1909. (Gabriel)
Johannsen. 'Heredity in...Pure Lines.' 1903. (Peters)
Bateson. 'Problems of Heredity...' 1900. (Carlson B)
Bateson & Punnett. '...Physiology of Heredity.' 1905–8. (Peters)

4. THE CHROMOSOMES AND INHERITANCE
Wilson. 'Mendel's Principles...' 1902.
Wilson. 'Croonian Lecture...' 1914.
Wilson 'Appreciation of Sutton.' 1917.
Montgomery. 'A Study of the Chromosomes...' 1901. (Voeller)
Boveri. 'On Multipolar Mitoses...' 1902. (Voeller)
McClung. 'The Accessory Chromosome.' 1902. (Voeller)
Sutton. '...Chromosome Group in Brachystola.' 1902. (Voeller)
Sutton. 'The Chromosomes in Heredity.' 1903. (Voeller; Gabriel; Peters)

Wilson. 'Chromosomes in Relation to . . . Sex. . .' 1905. (Gabriel; Voeller)

Stevens. '. . . the Accessory Chromosome.' 1905–6. (Voeller)

Carothers. 1913. 'Mendelian Ratios . . . in Chromosomes.' 1913. (Voeller)

5. THOMAS HUNT MORGAN AND DROSOPHILA
Morgan. 'On the Mechanism of Heredity.' 1922.
Morgan. 'The Rise of Genetics.' 1932.
Bateson and Punnett. '. . . Physiology of Heredity.' 1905–8. (Peters)

Morgan. 'Sex Limited Inheritance. . .' 1910. (Peters; Voeller)

Morgan. '. . . Segregation Versus Coupling. . .' 1911. (Gabriel; Voeller; Peters)

Sturtevant. '. . . Linear Arrangement of . . . Factors. . .' 1913. (Peters; Voeller)

Bridges. 'Non-disjunction. . .' 1916. (Voeller)

Wright. 'Color Inheritance in Mammals.' 1917. (Peters)

Dunn. 'Unit Character Variation in Rodents.' 1921. (Peters)

Muller. '. . . Change in the Individual Gene.' 1922. (Peters; Carlson A. B)

Blakeslee. 'Variations . . . in Chromosome Number.' 1922. (Voeller)

Muller. 'Mutation.' 1923. (Carlson B)

Bridges. 'Sex in Relation to Chromosomes. . .' 1925. (Peters)

Sturtevant. '. . . Crossing Over at the Bar Locus. . .' 1925. (Peters)

Muller. 'Artificial Transmutation of the Gene.' 1927. (Peters; Gabriel)

Dobzhansky. '. . . Cytological Proof of Translocation.' 1929. (Voeller)

5. MORGAN AND DROSOPHILA
de Vries. Mutation. 1901.
Morgan. Sex linked inheritance. 1910, 1911.
Bateson and Punnett. Linkage. 1906.
Janssens. Crossing over. 1909.
Morgan. Linkage groups. 1915.
Stern. Crossing over. 1931.
Sturtevant. Genetic maps. 1913.
Bridges. Non-disjunction. 1914, 1916.
Bridges. Sex determination.
Landsteiner. ABO blood groups. 1901.
Muller. Induced mutations. 1927.
Painter, Bridges. Salivary chromosomes. 1934.
Demerec and Hoover. Gene locus. 1936.

Creighton and McClintock '... Cytological and Genetical Crossing-over ...' 1931. (Peters)
Painter. '... Chromosome ... Maps.' 1933. (Peters)
Bridges. 'The Bar "Gene"...' 1936. (Peters)

6. GENETICS—OLD AND NEW
Muller. 'The Gene.' 1946.
Stadler. 'The Gene.' 1954. (Carlson A; Peters)
Goldschmidt. '...Philosophies of Genetics.' (Carlson A)
Beadle and Tatum. '...Control of Biochemical Reactions....' 1941. (Gabriel; Peters; Carlson B)
Horowitz and Leupold. '...One Gene–One Enzyme Hypothesis.' 1951. (Peters)

7. THE SUBSTANCE OF INHERITANCE
Avery. Letter to his brother. 1943.
Stent. 'DNA,' 1970.
Avery et al. '...Substance Inducing Transformation ...' 1944. (Peters)
Fraenkel-Conrat and Williams. '... Reconstitution of ... Virus....' 1955. (Peters)

7. THE SUBSTANCE OF INHERITANCE
Articles by Avery and Stent, listed above.
Watson and Crick. '...Structure of Nucleic Acids.' 1953. (Peters)
Watson and Crick. '...Genetical Implications of the Structure of Deoxyribonucleic Acid.' 1953. (Carlson A and B)
Benzer. 'Fine Structure of a Genetic Region....' 1955. (Peters; Carlson A and B)

6. GENETICS—OLD AND NEW
Garrod. Biochemical mutations. 1902.
Beadle and Tatum. Biochemical genetics. 1941.

7. THE SUBSTANCE OF INHERITANCE
Griffith. Transformation. 1928.
Dawson. Transformation. 1930.
Avery. DNA and transformation. 1944.
Hershey and Chase. DNA and inheritance. 1952.

8. DNA—STRUCTURE AND FUNCTION
Watson and Crick. Structure of DNA. 1953.
Pauling et al. Sickle-cell hemoglobin. 1949.
Ingram. Hemoglobin. 1957.
Volkin and Astrachan. Messenger RNA. 1957.
Nirenberg and Matthaei. Genetic code. 1961.
Jacob and Monod. Operon hypothesis. 1961.

Pauling *et al.* 'Sickle Cell Anemia...' 1949. (Carlson B)

Ingram. '...Normal and Sickle Cell Hemoglobin.' 1957. (Carlson B)

Allison. '...Sickle-Cell Trait... (and) Malaria...' 1954. (Carlson B)

Crick. 'On Protein Synthesis.' 1957. (Carlson B)

Nirenberg and Matthaei. '...Protein Synthesis...' 1961. (Carlson A and B)

Jacob and Monod. '...Biosynthesis of Proteins.' 1960. (Carlson A and B)

Yanofsky *et al.* '...Colinearity of Gene... and Protein...' 1964. (Carlson A)

8. GENETICS OF MAN

Lederberg, 'Experimental Genetics and Human Evolution.' 1966.

Lederberg, 'Genetic Engineering...' 1971.

Sturtevant. 'Social Implications of Genetics of Man.' 1954. (Peters)

9. DIFFERENTIATION

Wilson. 'The Mosaic Theory of Development.' 1894.

Spemann. 'Organizers in Animal Development.' 1927.

Roux. '...Half Embryos...' 1888. (Willier)

Roux. '...Significance of Nuclear Division...' 1895. (Voeller)

Driesch. '...Potency of the First Two Cleavage Cells...' 1892. (Willier; Gabriel)

Wilson. 'Cell-Lineage...' 1898. (Willier)

Spemann and Mangold. 'Induction of Embryonic Primordia...' 1924. (Willier)

9. GENETICS OF MAN
(General account)

12. DIFFERENTIATION

Malpighi. Development of chick. 1686.

Roux. Mechanics of development. 1880's.

Spemann. Organizer. 1910's, 1920's.

Holtfreter. Exogastrulation. 1933.

Mangold. '...Development of the Balancer...' 1931. (Gabriel)

Spemann. 'Embryonic Development and Induction.' 1938. (Carlson B)

Niu and Twitty. 'Differentiation of Gastrula Ectoderm...' 1953. (Flickinger)

9. DIFFERENTIATION

Spemann. '...Development...with Delayed Nuclear Supply.' 1958. (Gabriel)

King and Briggs. '...Transplantation of...Nuclei.' 1956. (Flickinger)

Gurdon. '...Developmental Capacity of Nuclei...' 1962. (Flickinger)

Whittaker. '...Polarity in *Fucus*...' 1938. (Flickinger)

13. DEVELOPMENTAL CONTROL OF GENETIC SYSTEMS

Briggs and King. Potentiality of nuclei. 1952.

Gurdon. Potentiality of nuclei. 1962.

Wilson. Development of *Dentalium*. 1904.

Whittaker. Cytodifferentiation of *Fucus*. 1938.

READINGS IN HEREDITY AND DEVELOPMENT

1 / Early Theories of Inheritance

The easily observed fact that offspring resemble their parents in general but rarely in all particulars, has been a profound mystery until rather recently. A mystery, that is, for those who care to speculate—most men have always been willing to accept—'that is just the way things are.' The origins of speculations of this sort are traditionally placed with the Greeks. It is appropriate, therefore, that we begin with two of them, Hippocrates and Aristotle. They differed markedly in the explanatory hypotheses they offered to explain inheritance, and their polar views were not resolved for two and a half millennia.

HIPPOCRATES

The earliest recorded theory of inheritance is to be found in *Airs, Waters, Places* attributed to Hippocrates. This renowned physician lived during the last half of the fifth century B.C. (that is, in the period 450–400) on the Greek island of Cos, which is off the southwest coast of modern Turkey. By tradition, Hippocrates is the Father of Medicine and, in addition, we can regard him as the Father of Genetics. Scholars believe that the books of Hippocrates were written by many individuals; collectively they are known as the Hippocratic Corpus. His Oath is repeated by all young physicians when they receive their medical degrees.

Airs, Waters, Places discusses the influence of climate on health and, in addition, has some lively tales about some not-so-well-known races of man. The Hippocratic theory of inheritance appears in a discussion of the Macrocephali.

1

First the Macrocephali; no other race has heads like theirs. The chief cause of the length of their heads was first found to be in their customs, but nowadays nature collaborates with tradition and they consider those with the longest heads the most nobly born. The custom was to mould the head of the newly-born children with their hands and to force it to increase in length by the application of bandages and other devices which destroy the spherical shape of the head and produce elongation instead. The characteristic was thus acquired at first by artificial means, but, as time passed, it became an inherited characteristic and the practice was no longer necessary. The seed comes from all parts of the body, healthy from the healthy parts and sickly from the sickly. If therefore bald parents usually have bald children, grey-eyed parents grey-eyed children, if squinting parents have squinting children, why should not long-headed parents have long-headed children? [p. 103]

From Hippocrates, *The Medical Works of Hippocrates. A New Translation from the Original Greek made especially for English Readers by the Collaboration of John Chadwick and W. N. Mann*. Blackwell Scientific Publications, Oxford. 1950. Reprinted by permission.

Here we find the beginnings of a theory of inheritance that was to last about 2300 years. It is pangenesis, that is, inheritance is thought to be based on minute particles produced by all parts of the body. Hippocrates called them seeds, Darwin was to call them gemmules. The amazing persistence of this theory probably indicates that it is a common sense way of looking at inheritance. Even today thoughtful individuals who are ignorant of biology may offer a similar explanation for inheritance.

ARISTOTLE

Aristotle was born 384 B.C. in Stagira, an ancient city in Greek Macedonia near modern Arnaia. He was a student of Plato, a teacher of Alexander, and the head of his own school in Athens, in the Lyceum. He died in Chalcis, near Athens, in 322 B.C. Aristotle was the greatest biologist of ancient times, and his views on inheritance were studied for centuries. Most of what he had to say on the subject is to be found in *Generation of Animals*.

First he considers the prevailing ideas:

It is generally held that all things are formed and come to be out of semen, and semen comes from the parents. And so one and the same inquiry will include the two questions: (1) Do both the male and the female discharge semen, or only one of them? and (2) Is the semen drawn from the whole of the parent's body or not?— since it is reasonable to hold that if it is not drawn from the whole of the body it is not drawn from both the parents either. There are some who assert that the semen is drawn from the whole of the body, and so we must consider the facts about this first of all. There are really four lines of argument which may be used to prove that the semen is drawn from each of the parts of the body. The first is, the intensity of the pleasure involved; it is argued that any emotion, when its scope is widened, is more pleasant than the same emotion when its scope is less wide; and obviously an emotion which affects all the parts of the body has a wider scope than one which affects a single part of a few parts only. The second argument is thát mutilated parents produce mutilated offspring, and it is alleged that because the parent is deficient in some one part no semen comes from that part, and that the part from which no semen comes does not get formed in the offspring. The third argument is the resemblances shown by the young to their parents: the offspring which are produced are like their parents not only in respect of their body as a whole, but part for part too; hence, if the reason for the resemblance of the whole is that the semen is drawn from the whole, then the reason for the resemblance of the parts is surely that something is drawn from each of the parts. Fourthly, it would seem reasonable to hold that just as there is some original thing out of which the whole creature is formed, so also it is with each of the parts: and hence if there is a semen which gives rise to the whole, there must be a special semen which gives rise to each of the parts. And these opinions derive plausibility from such evidence as the following: Children are born which resemble their parents in respect not only of congenital characteristics but also of acquired ones; for instance, there have been cases of children which have had the outline of a scar in the same places where their parents had scars, and there was a case at Chalcedon of a man who was branded on his arm, and the same letter, though somewhat confused and indistinct, appeared marked on his child. These are the main pieces of evidence which give some people ground for believing that the semen is drawn from the whole of the body.

Upon examination of the subject, however, the opposite seems more likely to be true; indeed, it is not difficult to refute these arguments, and besides that, they involve making further assertions which are impossible. First of all, then, resemblance is no proof that the semen is drawn from the whole of the body, because children resemble their parents in voice, nails, and hair and even in the way they move; but nothing whatever is drawn from these things; and there are some characteristics which a parent does not yet possess at the time when the child is generated, such as

From Aristotle, *Generation of Animals*. Translated by A. L. Peck. Loeb Classical Library. Harvard University Press, Cambridge, and William Heinemann, Ltd., London. 1943. Reprinted by permission.

grey hair or beard. Further, children resemble their remoter ancestors, from whom nothing has been drawn for the semen. Resemblances of this sort recur after many generations, as the following instance shows. There was at Elis a woman who had intercourse with a blackamoor; her daughter was not a black, but that daughter's son was. And the same argument will hold for plants. We should have to say that the seed was drawn from the whole of the plant, just as in animals. But many plants lack certain parts; you can if you wish pull some of the parts off, and some parts grow on afterwards. Further, nothing is drawn from the pericarp to contribute to the seed, yet pericarp is formed in the new plant and it has the same fashion as that in the old one. [pp. 49–53] . . .

Further, if the parts of the body are scattered about within the semen, how do they live? If on the other hand they are connected with each other, then surely they would be a tiny animal. And what about the generative organs? because that which comes from the male will be different from that which comes from the female.

Further, if the semen is drawn from all the parts of both parents alike, we shall have two animals formed, for the semen will contain all the parts of each of them. [pp. 57–59] . . .

Further, among the parts, some are distinguished by some faculty they possess, others by having certain physical qualities: thus, the non-uniform parts (such as the tongue or the hand) are distinguished by possessing the faculty to perform certain actions, the uniform parts by hardness or softness or other such qualities. Unless, there-

fore, it possesses certain special qualities, a substance is not blood or flesh; and hence it is plain that the substance which is drawn from the various parts of the parent has no right to the same name as those parts—we may not call that "blood" which is drawn from the parents' blood, and the same with flesh. This means that the offspring's blood is formed out of something which is other than blood, and if so, then the cause of its resemblance will not be due to the semen's being drawn from all the parts of the parent's body, as the supporters of this theory assert—because if blood is formed from something that is not blood, the semen need only be drawn from one part, there being no reason why all the other constituents as well as blood should not be formed out of the one substance. This theory seems to be identical with that of Anaxagoras, in asserting that none of the uniform substances comes into being; the only difference is that whereas he applied the theory universally, these people apply it to the generation of animals. Again, how are these parts which were drawn from the whole of the parent's body going to grow? Anaxagoras gives a reasonable answer; he says that the flesh already present is joined by flesh that comes from the nourishment. Those people however, who do not follow Anaxagoras in the statement just quoted, yet hold that the semen is drawn from the whole body, are faced with this question: how is the embryo to grow bigger by the addition of different substance to it unless the substance that is added changes? If however it is admitted that this added substance can change, why not admit straight away that the semen at the outset is such that out of it blood and flesh can be formed, instead of maintaining that the semen is itself both blood and flesh? [pp. 61–65] . . .

Here is another objection. Suppose it is true that the differentiation between male and female takes place during conception, as Empedocles says:

Into clean vessels were they pourèd forth;
Some spring up to be women, if so be They meet with cold.

Anyway, both men and women are observed to change: not only do the infertile become fertile, but also those who have borne females bear males; which suggests that the cause is not that the semen is or is not drawn from the whole of the parents, but depends upon whether or not that which is drawn from the man and from the woman stand in the right proportional relation to each other. Or else it is due to some other cause of this sort. Thus, if we are to assume this as true, viz., that the same semen is able to be formed into either male or female (implying that the sexual part is not present in the semen), it is clear that it is not the semen's being drawn from some one part which causes the offspring to be female, nor, in consequence, is it responsible for the special physical part which is peculiar to the two sexes. And what can be asserted about the sexual part can equally well be asserted about the other parts; since if no semen comes even from the uterus, the same will surely hold good of the other parts as well.

Further, some animals are formed neither from creatures of the same kind as themselves nor from creatures of a different kind; examples are: flies and the various kinds of fleas as they are called. Animals are formed from these, it is true, but in these cases they are not similar in character to their parents; instead we get a class of larvae. Thus in these creatures which differ in kind from their parents we clearly have animals which are *not* formed out of semen drawn from every part of the body, for if resemblance is held to be a sure sign that this has occurred, then they would resemble their parents.

Further, even among the animals there are some which generate numerous offspring from one act of coition, a phenomenon which is, indeed, universal with plants; these, as is manifest, produce a whole season's fruit as the result of one single movement. Now how is this possible on the supposition that the semen is secreted from the whole body? One act of coition, and one effort of segregation, ought necessarily to give rise to one secretion and no more. That it should get divided up in the uterus is impossible, for by that time the division would be made as it were from a new plant or animal, not of semen.

Further, transplanted cuttings bear seed—derived, of course, from themselves; which is proof positive that the fruit they bore before they were transplanted was derived from that identical amount of the plant which is now the cutting, and that the seed was not drawn from the whole of the plant. [pp. 65–69] . . .

As for mutilated offspring being produced by mutilated parents, the cause is the same as that which makes offspring resemble their parents. And anyway, not all offspring of mutilated parents are mutilated, any more than all offspring resemble their parents. The cause of these things we must consider later; the problem in both cases is the same. [p. 71] . . .

The goal of any theory of pangenesis is to offer a rational explanation for the inheritance of the specific parts of the body. If the parents have arms and the offspring also have arms, what can be the relation? It could be that the arms produce something that is transmitted to the next generation. Aristotle's reasons for rejecting such a notion of pangenesis are most interesting and, at times, remarkably prophetic. For example: '. . . why not admit straight away that the semen at the onset is such that out of it blood and flesh can be formed, instead of maintaining that the semen is itself both blood and flesh?' (p. 65). If one replaces 'semen' with 'hereditary influences,' which is the intended meaning, one is asking a very modern question and suggesting a very modern answer.

Aristotle's success in opposing a theory of pangenesis is not matched by his ability to offer a satisfying alternative hypothesis of inheritance. He believed that the parents contribute something to the offspring, menstrual secretions (or the equivalent) by the female and semen (or the equivalent) by the male. These two secretions interact, in an almost philosophical manner, to produce the child.

Now it is impossible that any creature should produce two seminal secretions at once, and as the secretion in females which answers to semen in males is the menstrual fluid, it obviously follows that the female does not contribute any semen to generation; for if there were semen, there would be no menstrual fluid; but as menstrual fluid is in fact formed, therefore there is no semen. [p. 97] . . .

By now it is plain that the contribution which the female makes to generation is the *matter* used therein, that this is to be found in the substance constituting the menstrual fluid, and finally that the menstrual fluid is a residue.

There are some who think that the female contributes semen during coition because women sometimes derive pleasure from it comparable to that of the male and also produce a fluid secretion. This fluid, however, is not seminal; it is peculiar to the part from which it comes in each several individual; there is a discharge from the uterus, which though it happens in some women does not in others. Speaking generally, this happens in fair-skinned women who are typically feminine, and not in dark women of a masculine appearance. [p. 101] . . .

. . . many animals may be formed from one semen. . . . We have as a proof of this those animals which are able to produce more offspring than one at a time, where more than one are formed as the result of one act of coitus. This shows also that the semen is not drawn from the whole body; because we cannot suppose (*a*) that at the moment of discharge it contains a number of *separate* portions from one and the same part of the body; nor (*b*) that these portions all enter the uterus *together* and separate themselves out when they have got tnere. No; what happens is what one would expect to happen. The male provides the 'form' and the 'principle of the movement,' the female provides the body, in other words, the material. Compare the coagulation of milk. Here, the milk is the body, and the fig-juice or the rennet contains the principle which causes it to set. The semen of the male acts in the same way as it gets divided up into portions within the female. (Another part of the treatise will explain the *Cause* why in some cases it gets divided into many portions, in others into few, while in others it is not divided up at all.) But as this semen which gets divided up exhibits no difference in kind, all that is required in order to produce numerous offspring is that there should be the right amount of it to suit the material available—neither so little that it fails to concoct it or even to set it, nor so much that it dries it up. If on the other hand this semen which causes the original setting remains single and undivided, then one single offspring only is formed from it.

The foregoing discussion will have made it clear that the female, though it does not contribute any semen to generation, yet contributes something, viz., the substance constituting the menstrual fluid (or the corresponding substance in bloodless animals). But the same is apparent if we consider the matter generally, from the theoretical standpoint. Thus: there must be that which generates, and that one of which it generates; and even if

these two be united in one, at any rate they must differ in kind, and in that the *logos* of each of them is distinct. In those animals in which these two faculties are separate, the body—that is to say the physical nature—of the active partner and of the passive must be different. Thus, if the male is the active partner, the one which originates the movement, and the female *qua* female is the passive one, surely what the female contributes to the semen of the male will be not semen but material. And this is in fact what we find happening; for the natural substance of the menstrual fluid is to be classed as 'prime matter.'

These then are the lines upon which that subject should be treated. And what we have said indicates plainly at the same time how we are to answer the questions which we next have to consider, viz., how it is that the male makes its contribution to generation, and how the semen produced by the male is the cause of the offspring; that is to say, Is the semen inside the offspring to start with, from the outset a part of the body which is formed, and mingling with the material provided by the female; or does the physical part of the semen have no share nor lot in the business, only the *dynamis* and movement contained in it? This, anyway, is the active and efficient ingredient; whereas the ingredient which gets set and given shape is the remnant of the residue in the female animal. The second suggestion is clearly the right one, as is shown both by reasoning and by observed fact. (*a*) If we consider the matter on general grounds, we see that when some one thing ·is formed from the conjunction of an active partner with a passive one, the active partner is not

situated within the thing which is being formed; and we may generalize
this still further by substituting 'moving' and 'moved' for 'active' and
'passive.' Now of course the female,
qua female, is passive, and the male,
qua male, is active—it is that whence
the principle of movement comes.
Taking, then, the widest formulation
of each of these two opposites, viz.,
regarding the male *qua* active and
causing movement, and the female *qua*
passive and being set in movement, we
see that the one thing which is formed
is formed *from them* only in the sense
in which a bedstead is formed from
the carpenter and the wood, or a ball
from the wax and the form. It is
plain, then, that there is no necessity
for any substance to pass from the
male; and if any does pass, this does
not mean that the offspring is formed
from it as from something situated
within itself during the process, but as
from that which has imparted movement to it, or that which is its 'form.'
The relationship is the same as that of
the patient who has been healed to
the medical art. (*b*) This piece of
reasoning is entirely borne out by the
facts. It explains why certain of those
males which copulate with the females
are observed to introduce no part at
all into the female, but on the contrary the female introduces a part into
the male. This occurs in certain insects. In those cases where the male
introduces some part, it is the semen
which produces the effect inside the
female; but in the case of these insects,
the same effect is produced by the
heat and *dynamis* inside the (male)
animal itself when the female inserts
the part which receives the residue.
And that is why animals of this sort
take a long time over copulation, and

once they have separated the young
are soon produced: the copulation
lasts until (the *dynamis* in the male)
has 'set' (the material in the female),
just as the semen does; but once they
have separated they soon discharge
the fetation, because the offspring they
produce is imperfect; all such creatures, in fact, produce larvae.

However, it is the behaviour of
birds and the group of oviparous
fishes which provides us with our
strongest proof (*a*) that the semen is
not drawn from all the parts of the
body, and (*b*) that the male does not
emit any part such as will remain
situated within the fetus, but begets the
young animal simply by means of the
dynamis residing in the semen (just as
we said happened with those insects
where the female inserts a part into
the male). Here is the evidence. Supposing a hen bird is in process of producing wind-eggs, and then that she
is trodden by the cock while the egg
is still completely yellow and has not
yet started to whiten: the result is that
the eggs are not wind-eggs but fertile
ones. And supposing the hen has been
trodden by another cock while the egg
is still yellow, then the whole brood
of chickens when hatched out takes
after the second cock. Some breeders
who specialize in first-class strains act
upon this, and change the cock for
the second treading. The implication is
(*a*) that the semen is not situated inside the egg and mixed up with it, and
(*b*) that it is not drawn from the
whole of the body of the male: if it
were in this case, it would be drawn
from both males, so the offspring
would have every part twice over.
No; the semen of the male acts otherwise; in virtue of the *dynamis* which
it contains it causes the material and

nourishment in the female to take on a particular character; and this can be done by that semen which is introduced at a later stage, working through heating and concoction, since the egg takes in nourishment so long as it is growing.

The same thing occurs in the generation of oviparous fishes. When the female fish has laid her eggs, the male sprinkles his milt over them; the eggs which it touches become fertile, but the others are infertile, which seems to imply that the contribution which the male makes to the young has to do not with bulk but with specific character. [pp. 109–17] . . .

These instances may help us to understand how the male makes its contribution to generation; for not every male emits semen, and in the case of those which do, this semen is not a part of the fetation as it develops. In the same way, nothing passes from the carpenter into the pieces of timber, which are *his* mater-ial, and there is no part of the art of carpentry present in the object which is being fashioned: it is the shape and the form which pass from the carpenter, and they come into being by means of the movement in the material. It is his soul, wherein is the 'form,' and his knowledge, which cause his hands (or some other part of his body) to move in a particular way (different ways for different products, and always the same way for any one product); his hands move his tools and his tools move the material. In a similar way to this, Nature acting in the male of semen-emitting animals uses the semen as a tool, as something that has movement in actuality; just as when objects are being produced by any art the tools are in movement, because the movement which belongs to the art is, in a way, situated in them. Males, then, that emit semen contribute to generation in the manner described. [pp. 119, 121].

CHARLES DARWIN AND FRANCIS GALTON

So little progress was made before the closing decades of the nineteenth century in explaining inheritance, that one can skip from Aristotle (384–322 B.C.) to Charles Darwin (1809–82) and miss nothing of theoretical importance. Aristotle had rejected pangenesis but could offer no acceptable alternative. Darwin revived pangenesis but failed to convince others (his ideas are discussed in *Heredity and Development,* Chapter 1). Herbert Spencer (1820–1903), for example, was not impressed. In 1884, nineteen years after Darwin first presented his views, Spencer wrote in *The Principles of Biology* (Vol. 1, p. 253): 'A positive explanation of Heredity is not to be expected in the present state of Biology.'

Nevertheless Darwin did stimulate others. Francis Galton (1822–1911), who shared a grandfather with Charles Darwin, put the theory of pangenesis to experimental test. Darwin had suggested that inheritance is via specific types of microscopic particles, the gemmules, which circulate freely through the body. Galton reasoned that these should occur in blood and, therefore, if he transfused blood from one variety of rabbit to another, the recipient rabbits should produce offspring with some of the characteristics of the blood donors.

Galton wrote to Darwin and described his intended experiments and asked for advice. It is clear from the letters that he hoped to substantiate Darwin's theory. In the beginning there were a few hopeful results but gradually Galton realized that the rabbits bred true, that is, they showed no evidence that foreign gemmules had been introduced. He then presented his results to the Royal Society of London.

March 30, 1871
General Sir EDWARD SABINE, K.C.B., President, in the Chair.
The following communications were read:—

I. 'Experiments in Pangenesis, by Breeding from Rabbits of a pure variety, into whose circulation blood taken from other varieties had previously been largely transfused.' By FRANCIS GALTON, F.R.S. Received March 23, 1871.

Darwin's provisional theory of Pangenesis claims our belief on the ground that it is the only theory which explains, by a single law, the numerous phenomena allied to simple reproduction, such as reversion, growth, and repair of injuries. On the other hand, its postulates are hypothetical and large, so that few naturalists seem willing to grant them. To myself, as a student of Heredity, it seemed of pressing importance that these postulates should be tested. If their truth could be established, the influence of Pangenesis on the study of heredity would be immense; if otherwise the negative conclusion would still be a positive gain.

It is necessary that I should briefly recapitulate the cardinal points of Mr. Darwin's theory. They are (1) that each of the myriad cells in every living body is, to a great extent, an independent organism; (2) that before it is developed, and in all stages of its development, it throws 'gemmules' into the circulation, which live there and breed, each truly to its kind, by the process of self-division, and that, consequently, they swarm in the blood, in large numbers of each variety, and circulate freely with it; (3) that the sexual elements consist of organized groups of these gemmules; (4) that the development of certain of the gemmules in the offspring depends on their consecutive union, through their natural affinities, each attaching itself to its predecessor in a regular order of growth; (5) that gemmules of innumerable varieties may be transmitted for an enormous number of generations without being developed into cells, but always ready to become so, as shown by the almost insuperable tendency to feral reversion, in domesticated animals.

It follows from this, and from the general tenor of Mr. Darwin's reasoning and illustrations, that two animals, to outward appearance of the

From *Proceedings* of the Royal Society (Biology) 19:393–404. 1871.

same pure variety, one of which has mongrel ancestry and the other has not, differ solely in the constitution of their blood, so far as concerns those points on which outward appearance depends. The one has none but gemmules of the pure variety circulating in his veins, and will breed true to his kind; the other, although only the pure variety of skin-gemmules happens to have been developed in his own skin, has abundance of mongrel gemmules in his blood, and will be apt to breed mongrels. It also follows from this that the main stream of heredity must flow in a far smaller volume from the developed parental cells, of which there is only one of each variety, than from the free gemmules circulating with the blood, of which there is a large number of each variety. If a parental developed cell bred faster than a free gemmule, an influx of new immigrants would gradually supplant the indigenous gemmules; under which supposition, a rabbit which, at the age of six months, produced young which reverted to ancestral peculiarities, would, when five years old, breed truly to his individual peculiarities; but of this there is no evidence whatever.

Under Mr. Darwin's theory, the gemmules in each individual must therefore be looked upon as entozoa of his blood, and, so far as the problems of heredity are concerned, the body need be looked upon as little more than a case which encloses them, built up through the development of some of their number. Its influence upon them can be only such as would account for the very minute effects of use or disuse of parts, and of acquired mental habits being transmitted hereditarily.

It occurred to me, when considering these theories, that the truth of Pangenesis admitted of a direct and certain test. I knew that the operation of transfusion of blood had been frequently practised with success on men as well as animals, and that it was not a cruel operation—that not only had it been used in midwifery practice, but that large quantities of saline water had been injected into the veins of patients suffering under cholera. I therefore determined to inject alien blood into the circulation of pure varieties of animals (of course, under the influence of anæsthetics), and to breed from them, and to note whether their offspring did or did not show signs of mongrelism. If Pangenesis were true, according to the interpretation which I have put upon it, the results would be startling in their novelty, and of no small practical use; for it would become possible to modify varieties of animals, by introducing slight dashes of new blood, in ways important to breeders. Thus, supposing a small infusion of bull-dog blood was wanted in a breed of greyhounds, this, or any more complicated admixture, might be effected (possibly by operating through the umbilical cord of a newly born animal) in a single generation.

I have now made experiments of transfusion and cross circulation on a large scale in rabbits, and have arrived at definite results, negativing, in my opinion, beyond all doubt, the truth of the doctrine of Pangenesis.

The course of my experiments was as follows:—Towards the end of 1869, I wrote to Dr. Sclater, the Secretary of the Zoological Society, explaining what I proposed to do, and asking if I might be allowed to keep my rabbits in some unused part of the Gardens, because I had no accommodation for them in my own house, and I was also anxious to obtain the skilled advice of Mr. Bartlett, the Superintendent of the Gardens, as to their breed and the value of my results. I further asked to be permitted to avail myself of the services of their then Prosector, Dr. Murie, to make the operations, whose skill and long experience in minute dissection is well known. I have warmly to thank Dr. Sclater for the large assistance he has rendered to me, in granting all I asked, to the full, and more than to the full; and I have especially to express my obligations to the laborious and kind aid given to me by Dr. Murie, at real inconvenience to himself, for he had little leisure to spare. The whole of the operations of transfusion into the jugular vein were performed by him, with the help of Mr. Oscar Fraser, then Assistant Prosector, and now appointed Osteologist to the Museum at Calcutta, I doing no more than preparing the blood derived from the supply-animal, performing the actual injection, and taking notes. The final series of operations, consisting of cross-circulation between the carotid arteries of two varieties of rabbits, took place after Dr. Murie had ceased to be Prosector. They were performed by Mr. Oscar Fraser in a most skilful manner, though he and I were still further indebted, on more than one occasion, to Dr. Murie's advice and assistance. My part in this series was limited to inserting and tying the canulæ, to making the cross-connexions, to recording the quality of the pulse through the exposed arteries, and making the other necessary notes.

The breed of rabbits which I endeavoured to mongrelize was the 'Silver-grey.' I did so by infusing blood into their circulation, which I had previously drawn from other sorts of rabbits, such as I could, from time to time, most readily procure. I need hardly describe Silver-grey rabbits with minuteness. They are peculiar in appearance, owing to the intimate mixture of black and grey hairs with which they are covered. They are never blotched, except in the one peculiar way I shall shortly describe; and they have never lop ears. They are born quite black, and their hair begins to turn grey when a few weeks old. The variations to which the breed is liable, and which might at first be thought due to mongrelism, are white tips to the nose and feet, and also a thin white streak down the forehead. But these variations lead to no uncertainty, especially as the white streak lessens or disappears, and the white tips become less marked, as the animal grows up. Another variation is much more peculiar: it is the tendency of some breeds to throw 'Himalayas,' or white rabbits with black tips. From first to last I have not been troubled with white Himalayas; but in one of the two breeds which I have used, and which I keep carefully separated from each other, there is a tendency to throw 'sandy' Himalayas. One of these was born a few days after I received the animals, before any operation had been made upon them, and put me on my guard. A similar one has been born since an operation. Bearing these few well-marked exceptions in mind, the Silver-grey rabbit is excellently adapted for breeding-experiments. If it is crossed with other rabbits, the offspring betray mongrelism in the highest degree, because any blotch of white or of colour, which is not 'Himalayan,' is almost certainly due to mongrelism; and so also is any decided change in the shape of the ears.

I shall speak in this memoir of litters connected with twenty silver-grey rabbits, of which twelve are does and eight are bucks; and eighteen of them have been submitted to one or two of three sorts of operations. These consisted of:—

(1) Moderate transfusion of partially defibrinized blood. The silver-grey was bled as much as he could easily bear; that was to about an ounce, a quantity which bears the same proportion to the weight of his body (say 76 oz.) that 2 lbs. bears to the weight of the body of a man (say 154 lbs.); and the same amount of partially defibrinized blood, taken from a killed animal of another variety, was thrown in in its place. The blood was obtained from a yellow, common grey, or black and white rabbit, killed by dividing the throat, and received in a warmed basin, where it was stirred with a split stick to remove part of the fibrine. Then it was filtered through linen into a measuring-glass, and thence drawn up with a syringe, graduated into drachms; and the quantity injected was noted.

(2) The second set of operations consisted in a large transfusion of wholly defibrinized blood, which I procured by whipping it up thoroughly with a whisk of rice-straw; and, in order to procure sufficient blood, I had on one occasion to kill three rabbits. I alternately bled the silver-grey and injected, until in some cases a total of more than 3 ounces had been taken out and the same quantity, wholly defibrinized, had been thrown in. This proportion corresponds to more than 6 lbs. of blood in the case of a man.

(3) The third operation consisted in establishing a system of cross-circulation between the carotid artery of a silver-grey and that of a common rabbit. It was effected on the same principle as that described by Addison and Morgan (Essay on Operation of Poisonous Agents upon the Living Body. Longman & Co., 1829), but with more delicate apparatus and for a much longer period. The rabbits were placed breast to breast, in each other's arms, so that their throats could be brought close together. A carotid of each was then exposed; the circulation in each vessel was temporarily stopped, above and below, by spring holders; the vessels were divided, and short canulæ, whose bores were larger than the bore of the artery in its normal state, were pressed into the mechanically distended mouths of the arteries; the canulæ were connected cross-wise; the four spring holders were released, and the carotid of either animal poured its blood direct into the other. The operation was complicated, owing to the number of instruments employed; but I suspended them from strings running over notched bars, with buttons as counterpoises, and so avoided entanglement. These operations were exceedingly successful; the pulse bounded through the canulæ with full force; and though, in most cases, it began to fall off after ten minutes or so, and I was obliged to replace the holders, disconnect the canulæ, extract the clot from inside them with a miniature corkscrew, reconnect the canulæ, and reestablish the cross-flow two, three, or more times in the course of a single operation, yet on two occa-

sions the flow was uninterrupted from beginning to end. The buck rabbit, which I indicate by the letter O, was 37½ minutes in the most free cross-circulation imaginable with his 'blood-mate,' a large yellow rabbit. There is no mistaking the quality of the circulation in a bared artery; for, when the flow is perfectly free, the pulse throbs and bounds between the finger and thumb with a rush, of which the pulse at the human wrist, felt in the ordinary way, gives an imperfect conception.

These, then, are the three sorts of operations which I have performed on the rabbits; it is convenient that I should distinguish them by letters. I will therefore call the operation of simply bleeding once, and then injecting, by the letter u; that of repeated bleedings and repeated injections by the letter w; and that of cross-circulation by the letter x.

In none of these operations did I use any chemical means to determine the degree to which the blood was changed; for I did not venture to compromise my chances of success by so severe a measure; but I adopted the following method of calculation instead:—

I calculate the change of blood effected by transfusion, or by cross-circulation, upon moderate suppositions as to the three following matters:—

(1) The quantity of blood in a rabbit of known weight.

(2) The time which elapses before each unit of incoming blood is well mixed up with that already in the animal's body.

(3) The time occupied by the flow, through either carotid, of a volume of blood equal to the whole contents of the circulation.

As regards 1, the quantity of blood in an animal's body does not admit, by any known method, of being accurately determined. I am content to take the modern rough estimate, that it amounts to one-tenth of its total weight. If any should consider this too little, and prefer the largest estimate, viz. that in Valentin's 'Repertorium,' vol. iii. (1838), p. 281, where it is given for a rabbit as one part in every 6·2 of the entire weight, he will find the part of my argument which is based on transfusion to be weakened, but not overthrown, while that which relies on cross-circulation is not sensibly affected.

As regards 2, the actual conditions are exceedingly complex; but we may evade their difficulty by adopting a limiting value. It is clear that when only a brief interval elapses before each unit of newly infused blood is mixed with that already in circulation, the quality of the blood which, at the moment of infusion into one of the cut ends of the artery or vein, is flowing out of the other, will be more alienized than if the interval were longer. It follows that the blood of the two animals will intermix more slowly when the interval is brief than when it is long. Now I propose to adopt an extreme supposition, and to consider them to mix *instantaneously*. The results I shall thereby obtain will necessarily be less favourable to change than the reality, and will protect me from the charge of exaggerating the completeness of intermixture.

As regards 3, I estimate the flow of blood through either carotid to be such that the volume which passes through it in ten minutes equals the whole volume of blood in the body. This is a liberal estimate; but I could afford to make it twice or even thrice as liberal, without prejudice to my conclusions.

Upon the foregoing data the following Table has been constructed. The formulæ are:—Let the blood in the Silver-grey be called a, and let its volume be V, and let the quantity u of alien blood be thrown in at each injection, then the quantity of blood a remaining in the Silver-grey's circulation, after n injections,

$$= V\left(1 - \frac{u}{V}\right)^n.$$

If the successive injections be numerous and small, so as to be equivalent to a continuous flow, then, after w of alien blood has passed in, the formula becomes $V.e^{-\frac{w}{V}}$.

A comparison of the numerical results from these two formulæ shows that no sensible difference is made if (within practicable limits) few and large, or many and small, injections are made, the total quantity injected being the same.

In cross-circulation the general formula is this:—If V' be the volume of blood in the other rabbit, after w of alien blood has passed through

TABLE I.
(Contents of circulation of Silver-grey Rabbit = 100.)

| | | | MAXIMUM PERCENTAGE OF ORIGINAL BLOOD REMAINING AFTER | | PERIOD, IN MINUTES, DURING WHICH THE CONTINUOUS FLOW THROUGH EACH CAROTID HAS LASTED. |
| | | | | CROSS-CIRCULATION. | |
QUANTITY OF BLOOD INFUSED.	SUCCESSIVE INJECTIONS OF PURELY ALIEN BLOOD, EACH = $\frac{100}{12}$.	CONTINUOUS FLOW OF PURELY ALIEN BLOOD.	RABBITS OF EQUAL SIZE.	BLOOD-MATE $\frac{1}{10}$ LARGER THAN THE SILVER-GREY.	
	Number of injections.				
25	3	77	78	80	80	2½
50	6	59	61	68	68	5
75	9	46	47	61	60	7½
100	12	35	37	56	55	10
125	15	27	29	54	52	12½
150	18	21	22	52	51	15
175	21	16	17	51	50	17½
200	24	12	14	51	49	20
300	36	4	5	50	48	30
400	48	1	2	50	48	40
infinite	infinite	0	0	50	48	infinite.

either canula, the quantity of blood a remaining in the Silver-grey exceeds*

$$\frac{V}{V + V'}\left\{ V + V'e^{-\left(\frac{1}{V}+\frac{1}{V'}\right)w} \right\}.$$

This becomes $\dfrac{V}{2}\left\{ 1 + e^{-\frac{2w}{V}} \right\}$ when

$V = V'$; also, when V' is infinite, it gives the formula already mentioned

---

* I am indebted to Mr. George Darwin for this formula.

for injection by a continuous flow of purely alien blood.

I now give a list (Table II.) of the rabbits to which, or to whose blood-mates, I shall have to refer. Every necessary particular will be found in the Table:—the weight of the rabbits; the estimated weight of blood in their veins; the operations performed on them, whether u, w, or x; the particulars of those several operations; the estimated percentage of alien blood that was substituted for their natural blood; and lastly, the colour, size, and breed of their blood-mates.

TABLE II.

SILVER-GREY DOES.	WEIGHT OF RABBIT.	ESTIMATED WEIGHT OF BLOOD.	NATURE OF OPERA-TION*.	DRACHMS IN-FUSED, AND PE-RIOD OF CROSS-CIRCULATION.	PERCENT-AGE OF ALIENIZED BLOOD.	COLOUR &C. OF BLOOD-MATE.
	lbs. oz.	drachms.				
A	5　9	79	u	9	11	Common grey and white.
B	5　13	82	u	10	12	Yellow, large.
			x	10 min. per-fect, 15 or 20 very good.	50, or more.	Common grey.
C	5　8	78	u	9·5	12	Albino, large.
D	5　4	75	u	8·5	12	Himalaya.
E	4　9	58	u	8	14	Common grey.
			x	13 min. good, 14 poor.	50, about	Common grey.
F	4　13	61	u	7·7	10	Black and white, large.
G	4　11	60	w	25·5, in 6 in-jections.	35	Grey and black, speckled.
			x	31 min. good, total.	75	Common grey.
H	..	..	x	15 min. per-fect, 15 very good.	50	Common grey.
I†	..	..	x	16 min. per-fect, not much more.	nearly 50	Common grey and white.

TABLE II con't.

J†	..	..	x	35 min. perfect.	..	Yellow, brown mouth (? Himalaya).
S	..	..	x	too unsuccessful to be worth counting.	? any.	Angora, fawn and white.
T	..	..	None.		..	None.	..
Bucks.						
K	4 14	62	u	9	14	Yellow, brown mouth.
			w	14, in 4 injections, total.	32	Yellow and white.
L	4 13	61	u	7	11	Common grey.
M	4 0	51	u	7	14	Black and white.
			w	24·5, in 6 injections, total.	45	3 black and white in succession.
N	4 9	58	u	7·5	13	Angora, grey and white, red eyes.
			w	16·5, in 4 injections, total.	34	Yellow.
O (son of C (u) by K (u))	..	..	· x	37½ min. perfect.	50	Yellow.
P†	..	..	x	25 to 30 min. perfect.	50	Common grey.
Q†	..	..	x	15 min. perfect, 15 very good.	50	Yellow and white.
	..	..	x	25 min. pretty good.	50	Common grey and white.

* *Note* (to 4th column).—*u* means simple transfusion, by one copious bleeding, and then injecting; *w* means compound transfusion by successive bleedings and successive injections; *x* means cross-circulation.

† These rabbits belong to a breed liable to throw 'Sandy' Himalayas.

In another list (Table III.) I give particulars of all the litters I have obtained from these rabbits, classified according to the operations which the parents had previously undergone.

I will now summarize the results. In the first instance I obtained five does (A, B, C, D, and E) and three bucks (K, L, and M) which had undergone the operation which I call *u*, and which had in consequence about ⅛ of their blood alienized. I bred from

TABLE III.

Litters subsequent to first transfusion. Both parents Silver-greys. Average proportion
of alienized blood in either parent $= \frac{1}{8}$; therefore in young $\frac{1}{8}$ also.

OUT OF	BY	NUMBER AND CHARACTER OF LITTERS.
A	K	4 true Silver-greys.
A	M	5 ditto, but 1 had a white foot to above knee.
B	K	5 true Silver-greys.
C	K	6 ditto.
D	K	4 ditto.
E	L	6 ditto.
		30 all true Silver-greys, except possibly one instance.

Litters subsequent to second transfusion of buck. Both parents Silver-greys. Average
proportion of alienized blood in young about $\frac{1}{4}$.

OUT OF	BY	NUMBER AND CHARACTER OF LITTERS.
A	M	6 true Silver-greys.

Litters subsequent to cross-circulation of buck only, the does being 0 or u. Both
parents Silver-greys. Average proportion of blood in young between $\frac{1}{4}$ and $\frac{1}{3}$.

OUT OF	BY	NUMBER AND CHARACTER OF LITTERS.
S	O	5 true Silver-greys.
C	O	5 ditto.
T	O	3 ditto.
		13 all Silver-greys.

Litters subsequent to cross-circulation of both parents (Silver-greys). Average proportion of alienized blood in young fully $\frac{1}{2}$.

OUT OF	BY	NUMBER AND CHARACTER OF LITTERS.
B	O	3 true Silver-greys.
H	O	7 ditto.
H	O	7 ditto.
I*	P*	6 ditto.
J*	Q*	6 ditto, all but one, a sandy Himalaya.
J*	P*	8 true Silver-greys.
	37	36 Silver-greys, 1 Himalaya.

* These rabbits belong to a breed liable to throw 'Sandy' Himalayas.

TABLE III con't.

Litters subsequent to cross-circulation of both parents (common rabbits). Average proportion of alienized blood in young a little less than ½.

OUT OF BLOOD-MATE TO	BY BLOOD-MATE TO	NUMBER AND CHARACTER OF LITTERS.
E	R	8 none Silver-grey, all like father or mother.
E	Q*	5 ditto.
G	O	9 ditto.
I*	Q*	8 ditto.
J*	Q*	8 ditto.
		38 none Silver-greys.

* These rabbits belong to a breed liable to throw 'Sandy' Himalayas.

these†, partly to see if I had produced any effect by the little I had done, and chiefly to obtain a stock of young rabbits which would be born with ⅛ of alien gemmules in their veins, and which, when operated upon themselves, would produce descendants having nearly ¼ alienized blood (the exact proportion is $1 — (1 — ⅛)^2 = {}^{15}\!/_{64}$). I obtained thirty young ones in six litters; and they were all true silver-greys, except, possibly, in one instance (out of the doe $A(u)$ by the buck M (u)), where one, of a litter of five, had a white fore leg, the white extending to above the knee-joint. This white leg gave me great hopes that Pangenesis would turn out to be true, though it might easily be accounted for by other causes; for my stock were sickly (both those on which I had not operated and those on which I had suffering severely from a skin

disease), and it was natural under those circumstances of ill health that more white than usual should appear in the young.

Having, then, had experience in transfusion, and feeling myself capable of managing a more complicated operation without confusion, I began the series which I call w. I left my old lot of does untouched, but obtained one new doe $(G(w))$, which had undergone the last operation, and three bucks $(K(u,w), M(u,w), N(u,w))$ which had undergone both operations, u and w. On endeavouring to breed from them, the result was unexpected, they appeared to have become sterile. The bucks were as eager as possible for the does; but the latter proving indifferent, I was unable to testify to their union having taken place; so I left them in pairs, in the same hutch, for periods of three days at a time. Attempts were made in this way, to breed from them in seven instances; and five of them were utter failures. One case was quite successful; and that, fortunately, was of the same pair $(A(u)$ and $M(u,w))$ which, under

—————
† I always allowed the bucks to run for awhile with waste does before commencing the breeding-experiments, that all old reproductive material might be got rid of.

the u operation, had bred the white-footed young one. This time, the offspring (six in number) were pure silver-greys. The last case was unfortunate. The doe ($E(u)$) had been once sterile to its partner ($N(u,w)$), and she had been put again in the same hutch with him for a short period, but was thought not to have taken him. She was shortly afterwards submitted to the operation x. From this she had nearly recovered when she brought forth an aborted litter and died. I was absent from town at the time; but Mr. Fraser, who examined them, wrote to say he fully believed that some were pied; if so, it must have been under the influence of the cross-circulation. But I have little faith in the appearance of the skin of naked, immature rabbits; for I have noticed that difference of transparency, and the colour of underlying tissues, give fallacious indications.

My results thus far came to this, viz. that by injecting defibrinized blood I had produced no other effect than temporary sterility. If the sterility were due to this cause alone, my results admitted of being interpreted in a sense favourable to Pangenesis, because I had deprived the rabbits of a large part of that very component of the blood on which the restoration of tissues depends, and therefore of that part in which, according to Pangenesis, the reproductive elements might be expected to reside. I had injected alien corpuscles but not alien gemmules. The possible success of the white foot, in my first litters, was not contradicted by the absence of any thing of the sort in my second set, because the additional blood I had thrown in was completely defibrinized. It was essential to the solution of the problem, that blood in its natural state should be injected; and I thought the most convenient way of doing so was by establishing cross-circulation between the carotids. If the results were affirmative to the truth of Pangenesis, then my first experiments would not be thrown away; for (supposing them to be confirmed by larger experience) they would prove that the reproductive elements lay in the fibrine. But if cross-circulation gave a negative reply, it would be clear that the white foot was an accident of no importance to the theory of Pangenesis, and that the sterility need not be ascribed to the loss of hereditary gemmules, but to abnormal health, due to defibrinization and perhaps to other causes also.

My operations of cross-circulation (which I call x) put me in possession of three excellent silver-grey bucks, four excellent silver-grey does, and one doe whose operation was not successful enough for me to care to count it. One of my x does (B) had already undergone the operation u, and I had another of my old lot ($C(u)$), which I left untouched. There were also three common rabbits, bucks, which were blood-mates to silver-greys, and four common rabbits, does, also blood-mates of silver-greys. From this large stock I have bred eighty-eight rabbits in thirteen litters, and in no single case has there been any evidence of alteration of breed. There has been one instance of a sandy Himalaya; but the owner of this breed assures me they are liable to throw them, and, as a matter of fact, as I have already stated, one of the does he sent me, did litter and throw one a few days after she reached me. The conclusion from this large series of experiments is not to be avoided, that the doctrine of

Pangenesis, pure and simple, as I have interpreted it, is incorrect.

Let us consider what were the alternatives before us. It seems *à priori* that, if the reproductive elements do not depend on the body and blood together, they must reside either in the solid structure of the gland, whence they are set free by an ordinary process of growth, the blood merely affording nutriment to that growth, or else that they reside in the blood itself. My experiments show that they are not independent residents in the blood, in the way that Pangenesis asserts; but they prove nothing against the possibility of their being temporary inhabitants of it, given off by existing cells, either in a fully developed state or else in one so rudimentary that we could only ascertain their existence by inference. In this latter case, the transfused gemmules would have perished, just like the blood-corpuscles, long before the period had elapsed when the animals had recovered from the operations.

I trust that those who may verify my results will turn their attention to the latter possibility, and will try to get the male rabbits to couple immediately, and on successive days, after they have been operated on. This might be accomplished if there were does at hand ready to take them; because it often happens that when the rabbits are released from the operating-table, they are little, if at all, dashed in their spirits; they play, sniff about, are ready to fight, and, I have no doubt, to couple. Whether after their wounds had begun to inflame, they would still take to the does, I cannot say; but they sometimes remain so brisk, that it is probable that in those cases they would do so. If this experiment succeeded, it would partly confirm the very doubtful case of the pied young of the doe which died after an operation of cross-circulation (which, however, further implies that though the ovum was detached, it was still possible for the mother gemmules to influence it), and it would prove that the reproductive elements were drawn from the blood, but that they had only a transient existence in it, and were continually renewed by fresh arrivals derived from the framework of the body. It would be exceedingly instructive, supposing the experiment to give affirmative results, to notice the gradually waning powers of producing mongrel offspring.

Darwin reacted promptly in a rejoinder published in the British scientific weekly, *Nature:*

PANGENESIS

In a paper, read March 30, 1871, before the Royal Society, and just published in the Proceedings, Mr. Galton gives the results of his interesting experiments on the inter-transfusion of the blood of distinct varieties of rabbits. These experiments were undertaken to test whether there was any truth in my provisional hypothesis of Pangenesis. Mr. Galton, in recapitulating 'the cardinal points,' says that the gemmules are supposed 'to swarm in

From *Nature*, April 27, 1871, pp. 502–3.

the blood.' He enlarges on this head, and remarks, 'Under Mr. Darwin's theory, the gemmules in each individual must, therefore, be looked upon as entozoa of his blood,' &c. Now, in the chapter on Pangenesis in my *Variation of Animals and Plants under Domestication*, I have not said one word about the blood, or about any fluid proper to any circulating system. It is, indeed, obvious that the presence of gemmules in the blood can form no necessary part of my hypothesis; for I refer in illustration of it to the lowest animals, such as the Protozoa, which do not possess blood or any vessels; and I refer to plants in which the fluid, when present in the vessels, cannot be considered as true blood. The fundamental laws of growth, reproduction, inheritance, &c., are so closely similar throughout the whole organic kingdom, that the means by which the gemmules (assuming for the moment their existence) are diffused through the body, would probably be the same in all beings; therefore the means can hardly be diffusion through the blood. Nevertheless, when I first heard of Mr. Galton's experiments, I did not sufficiently reflect on the subject, and saw not the difficulty of believing in the presence of gemmules in the blood. I have said (*Variation*, &c., vol. ii., p. 379) that 'the gemmules in each organism must be thoroughly diffused; nor does this seem improbable, considering their minuteness, and the steady circulation of fluids throughout the body.' But when I used these latter words and other similar ones, I presume that I was thinking of the diffusion of the gemmules through the tissues, or from cell to cell, independently of the presence of vessels,—as in the remark-

able experiments by Dr. Bence Jones, in which chemical elements absorbed by the stomach were detected in the course of some minutes in the crystalline lens of the eye; or again as in the repeated loss of colour and its recovery after a few days by the hair, in the singular case of a neuralgic lady recorded by Mr. Paget. Nor can it be objected that the gemmules could not pass through tissues or cell-walls, for the contents of each pollen-grain have to pass through the coats, both of the pollen-tube and embryonic sack. I may add, with respect to the passage of fluids through membrane, that they pass from cell to cell in the absorbing hairs of the roots of living plants at a rate, as I have myself observed under the microscope, which is truly surprising.

When, therefore, Mr. Galton concludes from the fact that rabbits of one variety, with a large proportion of the blood of another variety in their veins, do not produce mongrelised offspring, that the hypothesis of Pangenesis is false, it seems to me that his conclusion is a little hasty. His words are, 'I have now made experiments of transfusion and cross circulation on a large scale in rabbits, and have arrived at definite results, negativing, in my opinion, beyond all doubt the truth of the doctrine of Pangenesis.' If Mr. Galton could have proved that the reproductive elements were contained in the blood of the higher animals, and were merely separated or collected by the reproductive glands, he would have made a most important physiological discovery. As it is, I think every one will admit that his experiments are extremely curious, and that he deserves the highest credit for his ingenuity and perseverance. But it does not appear to me that Pangenesis has, as yet, received its death blow; though, from presenting so many vulnerable points, its life is always in jeopardy; and this is my excuse for having said a few words in its defence.

CHARLES DARWIN

Galton replied promptly, also.

Pangenesis

It appears from Mr. Darwin's letter to you in last week's *Nature,* that the views contradicted by my experiments, published in the recent number of the 'Proceedings of the Royal Society,' differ from those he entertained. Nevertheless, I think they are what his published account of Pangenesis (Animals, &c., under Domestication, ii. 374, 379) are most likely to convey to the mind of a reader. The ambiguity is due to an inappropriate use of three separate words in the only two sentences which imply (for there are none which tell us anything definite about) the *habitat* of the Pangenetic gemmules; the words are 'circulate,' 'freely,' and 'diffused.' The proper meaning of circulation is evident enough—it is a re-entering movement. Nothing can justly be said to circulate which does not return, after a while, to a former position. In a circulating library, books return and are re-issued. Coin is said to circulate, because it comes back into the same hands in the interchange of business. A story circulates, when a person hears it repeated over and over again in society. Blood has an undoubted claim to be called a circulating fluid, and when that phrase is used, blood is always meant. I understood Mr. Darwin to speak of blood when he used the phrases 'circulating freely,' and 'the steady circulation of fluids,' especially as the other words 'freely' and 'diffusion' encouraged the idea. But it now seems that by circulation he meant 'dispersion,' which is a totally different conception. Probably he used the word with some allusion to the fact of the dispersion having been carried on by eddying, not necessarily circulating, currents. Next, as to the word 'freely.' Mr. Darwin says in his letter that he supposes the gemmules to pass through the solid walls of the tissues and cells; this is incompatible with the phrase 'circulate freely.' Freely means 'without retardation;' as we might say that small fish can swim freely through the larger meshes of a net; now, it is impossible to suppose gemmules to pass through solid tissue without *any* retardation: 'Freely' would be strictly applicable to gemmules drifting along with the stream of the blood, and it was in that sense I interpreted it. Lastly, I find fault with the use of the word 'diffused,' which applies to movement in or with fluids, and is inappropriate to the action I have just described of solid boring its way through solid. If Mr. Darwin had given in his work an additional paragraph or two to a description of the whereabouts of the gemmules which, I must remark, is a cardinal point of his theory, my misapprehension of his meaning could hardly have occurred without more

From *Nature,* May 4, 1871, pp. 5–6.

hesitancy than I experienced, but I certainly felt and endeavoured to express in my memoir some shade of doubt; as in the phrase, p. 404, 'that the doctrine of Pangenesis, pure and simple, as I have interpreted it, is incorrect.'

As I now understand Mr. Darwin's meaning, the first passage (ii. 374), which misled me, and which stands: '. . minute granules . . . which circulate freely throughout the system' should be understood as 'minute granules . . . which are dispersed thoroughly and are in continual movement throughout the system;' and the second passage (ii. 379), which now stands: 'The gemmules in each organism must be thoroughly diffused; nor does this seem improbable, considering . . . the steady circulation of fluids throughout the body,' should be understood as follows: 'The gemmules in each organism must be dispersed all over it, in thorough intermixture; nor does this seem improbable, considering . . . the steady circulation of the blood, the continuous movement, and the ready diffusion of other fluids, and the fact that the contents of each pollen grain have to pass through the coats, both of the pollen tube and of the embryonic sack.' (I extract these latter *addenda* from Mr. Darwin's letter.)

I do not much complain of having been sent on a false quest by ambiguous language, for I know how conscientious Mr. Darwin is in all he writes, how difficult it is to put thoughts into accurate speech, and, again, how words have conveyed false impressions on the simplest matters

from the earliest times. Nay, even in that idyllic scene which Mr. Darwin has sketched of the first invention of language, awkward blunders must of necessity have often occurred. I refer to the passage in which he supposes some unusually wise, ape-like animal to have first thought of imitating the growl of a beast of prey so as to indicate to his fellow monkeys the nature of expected danger. For my part, I feel as if I had just been assisting at such a scene. As if, having heard my trusted leader utter a cry, not particularly well articulated, but to my ears more like that of a hyena than any other animal, and seeing none of my companions stir a step, I had, like a loyal member of the flock, dashed down a path of which I had happily caught sight, into the plain below, followed by the approving nods and kindly grunts of my wise and most-respected chief. And I now feel, after returning from my hard expedition, full of information that the suspected danger was a mistake, for there was no sign of a hyena anywhere in the neighbourhood. I am given to understand for the first time that my leader's cry had no reference to a hyena down in the plain, but to a leopard somewhere up in the trees; his throat had been a little out of order—that was all. Well, my labour has not been in vain; it is something to have established the fact that there are no hyenas in the plain, and I think I see my way to a good position for a look out for leopards among the branches of the trees. In the meantime, *Vive* Pangenesis.

FRANCIS GALTON

Clearly Galton was surprised, and probably hurt, by the tone of Darwin's reply. After all he had kept Darwin fully informed of the progress of his experiments. In view of this, the last paragraph of Galton's letter is quite remarkable. The two famous men remained friends.

Spencer was correct—inheritance could not be explained with the data then available. Other sorts of observations and experiments were, however, slowly building a knowledge of cell biology that was to become a basis for understanding inheritance. Examples will be given in the next chapter.

BIBLIOGRAPHY

Refer to the Preface for a list of classical papers included in other anthologies.

BABCOCK, ERNEST B. 1949–51. The development of fundamental concepts in the science of genetics. *Portugaliae Acta Biologica*. Série A., Volume R. B. Goldschmidt. Pages 1–46.

BARTHELMESS, ALFRED. 1952. *Vererbungswissenschaft*. Freiburg: Karl Alber.

COLE, F. J. 1930. *Early Theories of Sexual Generation*. Oxford: Oxford University Press.

DARLINGTON, C. D. 1969. *Genetics and Man*. New York: Schocken Books. The breadth of approach makes this book especially appealing to non-science students.

DARWIN, CHARLES. 1868. *The Variation of Animals and Plants under Domestication*. 2 Volumes. London: John Murray. Chapter 27 is 'Provisional hypothesis of pangenesis'.

DUNN, L. C. 1965. *A Short History of Genetics. The Development of Some of the Main Lines of Thought: 1864–1939*. New York: McGraw-Hill. A fine book with which to begin.

DUNN, L. C. 1965. 'Ideas about living units, 1864–1909: a chapter in the history of genetics.' *Perspectives in Biology and Medicine. 8*: 335–346.

DUNN, L. C. 1969. 'Genetics in historical perspective.' In *Genetic Organization*. Volume 1. Edited by Ernst W. Caspari and Arnold W. Ravin. New York: Academic Press. Pages 1–90.

GALTON, FRANCIS. 1889. *Natural Inheritance*. London: Macmillan.

GASKING, ELIZABETH B. 1967. *Investigations into Generation 1651–1828*. Baltimore: Johns Hopkins University Press.

GHISELIN, MICHAEL T. 1969. *The Triumph of the Darwinian Method*. Berkeley: University of California Press. Chapter 7 discusses Darwin's theories of inheritance.

GLASS, BENTLEY. 1947. 'Maupertuis and the beginning of genetics.' *Quarterly Review of Biology 22*: 196–210.

GLASS, BENTLEY. 1959. 'Maupertuis, pioneer of genetics and evolution.' In *Forerunners of Darwin: 1745–1859*. Edited by Bentley Glass, Owsei Temkin, and William L. Straus, Jr. Baltimore: Johns Hopkins University Press. Pages 51–83.

GLASS, BENTLEY. 1959. 'Heredity and variation in the 18th century concept of species.' In *Forerunners of Darwin 1745–1859*. Edited by Bentley Glass, Owsei Temkin, and William L. Straus, Jr. Baltimore: John Hopkins University Press. Pages 144–72.

OLBY, ROBERT C. 1963. 'Charles Darwin's manuscript of pangenesis.' *British Journal History of Science 1*: 251–263.

OLBY, ROBERT C. 1966. *Origins of Mendelism*. New York: Schocken Books.

PEARSON, KARL. Editor. 1924. *The Life, Letters and Labours of Francis Galton.* 4 Volumes. London: Cambridge University Press. Contains the correspondence with Darwin on the transfusion experiments.

ROBERTS, H. F. 1929. *Plant Hybridization before Mendel*. Princeton: Princeton University Press.

SACHS, J. VON. 1890. *History of Botany (1530–1860)*. Oxford: Oxford University Press. Translated into English.

STERN, CURT. 1957. 'The problem of complete Y-linkage in man.' *American Journal of Human Genetics 9*: 147–166. This article includes a reconsideration of the porcupine man, discussed by Darwin.

STUBBE, H. 1965. *Kurze geschichte der genetik bis zur wiederentdeckung der vererbungsregeln Gregor Mendels.* Second edition. Jena: Fischer.

STURTEVANT, A. H. 1965. *A History of Genetics*. New York: Harper and Row.

VORZIMMER, P. 1963. 'Charles Darwin and blending inheritance.' *Isis 54*: 371–390.

ZIRKLE, CONWAY. 1935. *The Beginnings of Plant Hybridization*. Philadelphia: University of Pennsylvania Press.

ZIRKLE, CONWAY. 1935. 'The inheritance of acquired characters and the provisional hypothesis of pangenesis.' *American Naturalist 69*: 417–445.

ZIRKLE, CONWAY. 1936. 'Further notes on pangenesis and the inheritance of acquired characters.' *American Naturalist 70*: 529–546.

ZIRKLE, CONWAY. 1946. 'The early history of acquired characters and of pangenesis.' *Transactions American Philosophical Society. N.S. 35*: 91–151.

ZIRKLE, CONWAY. 1951. 'Gregor Mendel and his precursors.' *Isis 42*: 97–104.

ZIRKLE, CONWAY. 1951. 'The Knowledge of Heredity before 1900.' In *Genetics in the 20th Century*. Edited by L. C. Dunn. New York: Macmillan. Pages 35–57.

2/Students of the Cell

The relation of the cell, a key element in biological structure, to inheritance, a key process in biological continuity, is direct: cells are the sole material link between generations. Whatever is inherited, therefore, must be parts of cells. This important relation was not realized when cells were discovered: in fact, it took nearly two centuries.

Cells were first described by Robert Hooke (1635–1703) in 1665, who chanced upon them when examining a thin slice of cork (*Heredity and Development*, Chapter 2). Hooke was an early microscopist who set about to examine various objects in the world around him. He was one of the first members of the first scientific society, the Royal Society of London, which was founded in the 1660's. He presented his observations to the Royal Society, which supported the publication of his great work, *Micrographia* (1665).

NEHEMIAH GREW

Another microscopist of the Royal Society, and a contemporary of Hooke, was Nehemiah Grew (1641–1712). His studies of the microscopic anatomy of plants were published in the 1670's and 1680 s. They clearly show the cellular nature of roots and stems. The following plates are reproduced from: *The Anatomy of Plants with an Idea of a Philosophical History of Plants and Several other Lectures Read before the Royal Society* (Printed by W. Rawlins, for the Author, 1682).

One must not conclude that Grew's dramatic illustrations of the cellular nature of plants began an active period in cytology. No special importance was attached to cells and few others studied them. The first structures of animals, later realized to be cells, were the microorganisms observed by

27

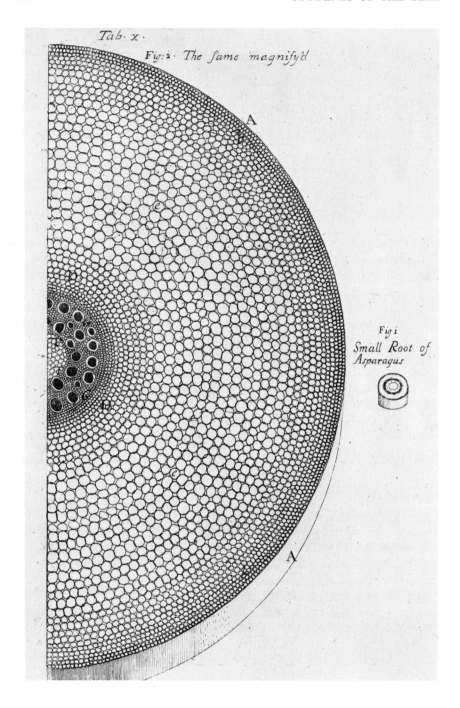

Tab. x.

Fig: 2. The same magnify'd

A

e

b

a

c

A

Fig 1
Small Root of
Asparagus

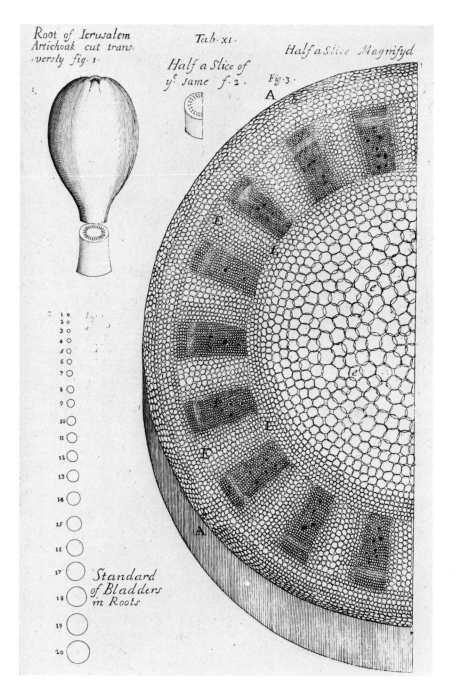

Root of Ierusalem
Artichoak cut trans-
versly fig. 1.

Tab. XI.

Half a Slice of
ye same f. 2.

Fig. 3.

Half a Slice Magnifyd

Standard
of Bladders
in Roots

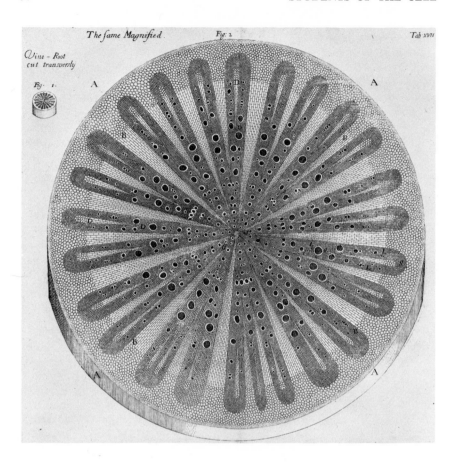

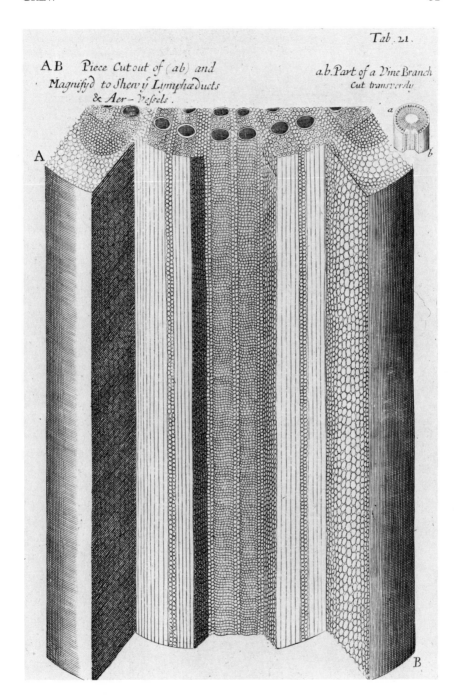

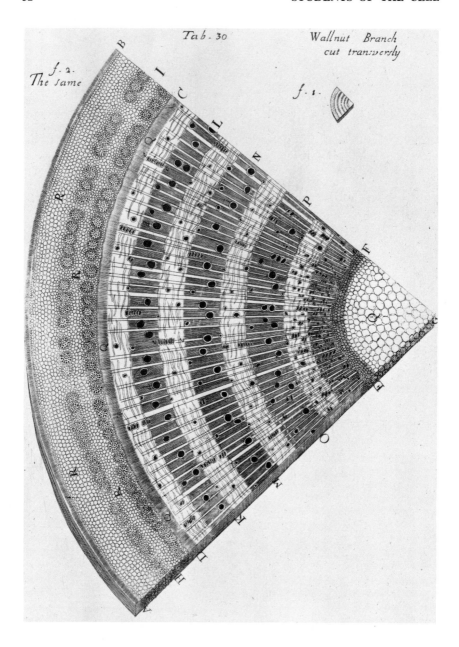

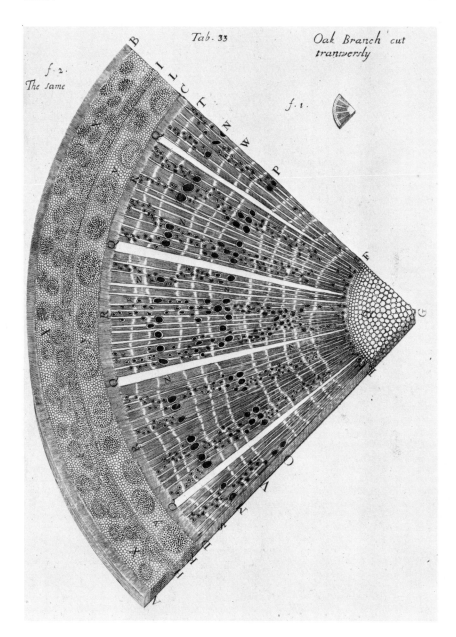

Tab. 33

Oak Branch cut transversly

f. 2.
The same

f. 1.

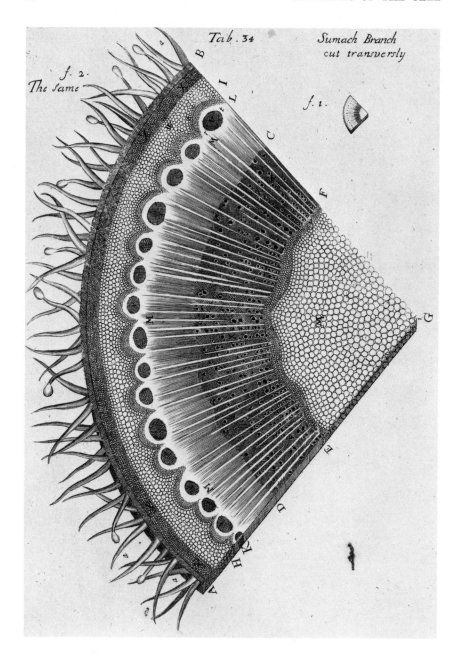

Tab. 34

Sumach Branch
cut transversly

f. 2.
The same

f. 1.

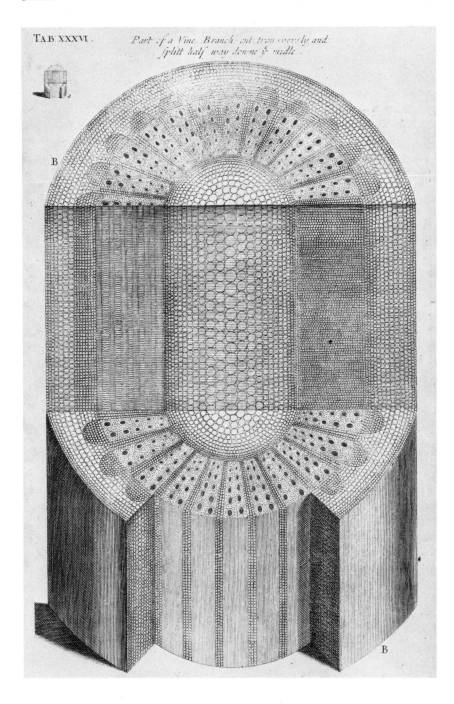

TAB. XXXVI. *Part of a Vine Branch cut transversly and split half way downe ye midle*

Antony Van Leeuwenhoek (1632–1723) of Delft, Holland. He also reported his observations to the Royal Society of London. A fine account of his work is given by Dobell, cited at the end of the chapter.

Following these observations in the late seventeenth century, the aborted science of cytology languished until the nineteenth century. Cells were observed from time to time but there was no systematic work on them. Cells seemed to be characteristic mainly of plants; the microscopic examination of animals revealed no structures corresponding to those described by Hooke and Grew. This was to be expected: the prominent cell walls of plants, which were the basis of the first descriptions of cells, are absent in animals.

PETER MARK ROGET

The next major advance came in the 1830's when Schwann and Schleiden advanced the hypothesis that cells were the basic elements of structure not only of plants but of animals as well. The following account of the basic structure of plants and animals was written just prior to this breakthrough by Peter Mark Roget, an English physician, secretary to the Royal Society, and author of the *Thesaurus of English Words and Phrases*. Roget's work, *Animal and Vegetable Physiology Considered with Reference to Natural Philosophy*, is one of the famous Bridgewater Treatises made possible by the Earl of Bridgewater who willed the sum of eight thousand pounds sterling to support the publication of books 'On the Power, Wisdom, and Goodness of God, as manifested in the Creation; illustrating such works by all reasonable arguments, as for instance the variety and formation of God's creatures in the animal, vegetable, and mineral kingdoms; the effect of digestion, and thereby of conversion; the construction of the hand of man, and an infinite variety of other arguments; as also by discoveries ancient and modern, in arts, sciences, and the whole extent of literature.' Roget was one of eight gentlemen chosen by the President of the Royal Society for this task.

Both the composition of the fluid and the texture of the solid parts of animal and vegetable bodies are infinitely varied, according to the purposes they are designed to serve in the economy. Scarcely any part is perfectly homo-

From Peter Mark Roget, *Animal and Vegetable Physiology Considered With Reference to Natural Theology*. Vol. 1. Carey, Lea, and Blanchard, Philadelphia. 1836.

geneous; that is, composed throughout of a single uniform material. Few of the fluids are entirely limpid, and none are perfectly simple in their composition; for they generally contain more or less of a gelatinous matter, which, when very abundant, imparts to them viscidity, constituting an approach to the solid state. Many fluids contain minute masses of matter, generally having a globular shape, which can be seen only by means of the microscope, and which float in the surrounding liquid, and often thicken it in a very sensible manner.* We next perceive that these globules have, in many instances, cohered, so as to form solid masses; or have united in lines, so as to constitute fibres. We find these fibres collecting and adhering together in bundles; or interwoven and agglutinated, composing various other forms of texture; sometimes resembling a loose net-work of filaments; sometimes constituting laminæ or plates; and, at other times, both plates and filaments combining to form an irregular spongy fabric. These various tissues, again, may themselves be regarded as the constituent materials of which the several organs of the body are constructed, with different degrees of complication, according to the respective functions which they are called upon to perform.

We shall now examine the several kinds of texture in relation to these functions, in the order of their increasing complexity; beginning with those of vegetables, which are apparently the simplest of all.

* Globules of this description have been found in the lymph, the saliva, and even in the aqueous humour of the eye.

2. Vegetable Organization.

Plants, being limited in their economy to the functions of nutrition and reproduction, and being fixed to the same spot, and therefore in a comparatively passive condition, require for the performance of these functions mechanical constructions of a very different kind from those which are necessary to the sentient, the active, and the locomotive animal. The organs that are essential to vegetables are those which receive and elaborate the nutritive fluids they require, those which are subservient to reproduction, and also those composing the general frame-work, which must be superadded to the whole for the purpose of giving mechanical support and protection to these finer organizations. As plants are destined to be permanently attached to the soil, and yet require the action both of air and of light; and, as they must also be defended from the injurious action of the elements, so we find these several objects provided for by three descriptions of parts: namely, first, the *Roots*, which fix plants in their situation; secondly, the *Stems*, which support them in the proper position, or raise them to the requisite height above the ground; together with the branches which are merely subdivisions of the stem; and thirdly, the *external coverings*, which correspond in their office to the teguments, or skin of animals.

The simplest and apparently the most elementary texture met with in vegetables is formed of exceedingly minute vesicles, the coats of which consist of transparent membranes of extreme tenuity. Fig. 3 is a highly magnified representation of the simplest form of these vesicles.* But they generally adhere together more closely, composing by their union a species of vegetable cellular tissue, which may be regarded as the basis or essential component material of every organ in the plant. This cellular structure is represented in figures 4 and 5, as it appears in the *Fucus vesiculosus;* the first being a horizontal, and the second a vertical

---

* These cells are well represented in the engravings which illustrate Mr. Slack's memoir on the elementary tissue of plants, contained in the 49th volume of the Transactions of the Society of Arts.

section of that plant.* The size of these cells differs considerably in different instances. Kieser states that the diameter of each individual cell varies from the 330th to the 55th part of an inch; so that from 3,000 to 100,000 cells would be contained in an extent of surface equal to a square inch. But they are occasionally met with of different sizes, from even the 1000th part of an inch to the 30th.

In their original state, these vesicles have an oval or globular form; but they are soon transformed into other shapes, either by the mutual compression which they sustain from being crowded into a limited space, or from unequal expansions in the progress of

---

* De Candolle, Organographie Végétale.

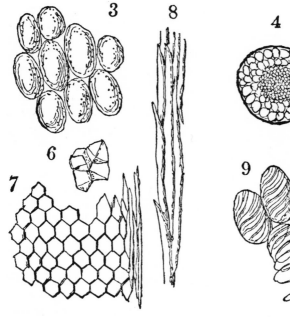

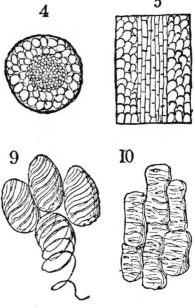

their development. From the first of these causes they often acquire angles, assuming the forms of irregular rhomboidal dodecahedrons, and often of hexagonal prisms, like the cells of a honey-comb; and by the second, they are elongated into cylinders, or slowly tapering cones, thus passing by insensible gradations into the tubular form. Figures 6, 7, and 8, are representations of some of these different states of transition from the one to the other. These various modifications of the same elementary texture have been distinguished into several classes of cells, and dignified by separate technical denominations, which I shall not stop to specify, as it does not appear that they have as yet thrown any light on vegetable physiology.

Many of the cells are fortified by the addition of elastic threads, generally disposed in a spiral course, and adhering to the inner surfaces of the membranous coats of the cells, which they keep in an expanded state. (See Fig. 9.) When the membranes are torn, the fibres; being detached, unrol themselves, and being loosely scattered among the neighbouring cells, give the appearance of fibrous connexions among these cells, which did not originally exist. Simple membranous cells, containing no internal threads, are often found intermixed with these fibrous cells. In many of the cells, again, the original spiral threads appear to have coalesced by their edges; thus presenting a more uniform surface, excepting that a few interstices are left, where the pellucid membrane, having no internal lining, presents the appearance of transverse fissures or oval perforations, (Fig. 10.) Cells of this description are said to be *reticulated* or *spotted*, and, together with

those having more regularly formed spiral threads, are very abundantly met with in plants belonging to the tribe of *Orchideæ*.

It has been much disputed whether the cells of the vegetable texture are closed on all sides, or whether they communicate with one another. Mirbel has given us delineations of what appeared to him, when he examined the coats of the cells with a microscope, to be pores and fissures. But subsequent observations have rendered it probable that these appearances arise merely from darker portions of the membranes, where opaque particles have been deposited in their substance. Fluids gain access into these cells by transuding through the membranes which form their sides, and not by any apertures capable of being detected by the highest powers of the microscope.

If all the cells consist of separate vesicles, as the concurring observations of modern botanists* appear to have satisfactorily established, the partitions which separate them, however thin and delicate, must consist of a double membrane, formed by the adhesion of the coats of the two contiguous vesicles. But as these coats can hardly be supposed to adhere in every point, we may expect to find that spaces have been left in various parts between them; and that communications exist to a certain extent between all these spaces; so as to compose what may be regarded as one large cavity. These have been denominated the *intercellular spaces;* and they have been supposed to perform, as will

---

* In particular, Treviranus, Kieser, Link, Du Petit Thouars, Pollini, Amici, Dutrochet, and De Candolle.

hereafter be seen, an important part in the functions of Nutrition.

Fluids of different kinds occupy both the cells and the intercellular spaces. The contents of some is the simple watery sap; that of others consists of peculiar liquids, the products of vegetable secretion: and very frequently they contain merely air. In many of the cells there are found small opaque and detached particles of the substance termed by chemists, *Fecula*, of which starch is the most common example. In several parts, and more especially in the leaves, and in the petals of flowers, the material which gives them their peculiar colour is contained in the cells in the form of minute globules. De Candolle has ·given it the name of *Chromule.*†

The cells of the ligneous portion of trees and shrubs are farther incrusted with particles of a more dense material, peculiar to vegetable organization, and termed *Lignine*. It is this substance which principally contributes to the density and mechanical strength of what are called the *Woody Fibres,* which consist of collections of fusiform, or tapering vessels, hereafter to be described, surrounded by assemblages of cells thus fortified, and the whole cohering in bundles, so as to present greater resistance to forces tending to displace them in the longitudinal direction than in any other.

Most of the plants which are included in the Linnean class of Cryptogamia have a structure exclusively composed of cells, as has been already shown in the *Fucus vesiculosus*. But the greater number of other plants have, in addition to these cells, numerous ducts or vessels, consisting

of membranous tubes of considerable length, interspersed throughout every part of the system. These tubes exhibit different modifications of structure, more especially with regard to the form of the fibres, or other materials, which adhere to the inner surface of their membranes; and these modifications correspond very exactly with those of the vesicles already described as constituting the simpler forms of vegetable tissue. There can be little doubt, indeed, that the vessels of plants take their origin from vesicles, which become elongated by the progress of development in one particular direction; and it is easy to conceive that where the extremities of these elongated cells meet, the partitions which separate their cavities may become obliterated at the points of junction, so as to unite them into one continuous tube with an uninterrupted interior passage. This view of the formation of the vessels of plants is confirmed by the gradation that may be traced among these various kinds of structures. Elongated cells are often met with applied to each other endwise, as if preparatory to their coalescence into tubes. Sometimes the tapering ends of fusiform cells are joined laterally (as seen in Fig. 12,) so that the partitions which divide their cavities are oblique. At other times their ends are broader, and admit of their more direct application to each other in the same line, being separated only by membranes passing transversely; in which case they present, under the microscope, the appearance of a necklace of beads (Fig. 13.) When, by the destruction of these partitions, their cavities become continuous, the tubes they form exhibit a series of contractions at cer-

---

† Organographie, Tom. 1, p. 19.

tain intervals, marking their origin from separate cells. In this state they have received the names of *monili-form, jointed* or *beaded vessels.* Traces of the membranous partitions sometimes remain where their obliteration has been only partial, leaving transverse fibres. The conical terminations occasionally observable in the vessels of plants also indicate their cellular origin.†

The membrane constituting the tube is sometimes simple, like those of the simple cells: but it frequently contains fibres, or other internal coatings, corresponding to those met with in the more compound cells. The vessels in which the internal fibres run in a spiral direction (Fig. 14,) are denominated *tracheæ*, or *spiral vessels;* or, from their being found very constantly to contain air, they are often called *air tubes.* Their diameter is generally between the 1000th and the 300th part of an inch. These spiral, or air vessels, pervade extensively the vegetable system. The threads they contain are frequently double, treble, quadruple, or even still more numerous: they are of great length, and when the external membrane of the vessel is divided, they may easily be drawn out and uncoiled, their elasticity enabling them to retain their spiral shape. The object of this structure appears to be that of keeping the cavity of the tube always pervious, by presenting resistance to any external force tending to compress and close it.‡ [pp. 59–65] . . .

In all these vegetable structures, while the objects appear to be the same, the utmost variety is displayed in the means for their accomplish-

---

* Mirbel gave them the name of *'Vaisseaux en chapelet.'*

† This theory of the derivation of vessels from cells was first advanced by Treviranus.

‡ Vessels are sometimes met with which appear to be formed simply by the coils of a spiral fibre in close juxtaposition, and unattached to any external envelope, or connecting membrane.

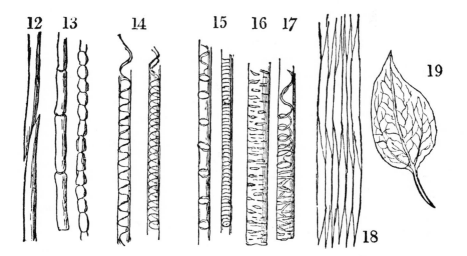

12 13 14 15 16 17   19

18

ment, in obedience, as it were, to the law of diversity which, as has been already observed, seems to be a leading principle in all the productions of nature. It is more probable, however, judging from that portion of the works of creation which we are competent to understand, that a specific design has regulated each existing variation of form, although that design may in general be utterly beyond the limited sphere of our intelligence.

4. Animal Organization.

The structures adapted to the purposes of vegetable life, which are limited to nutrition and reproduction, would be quite insufficient for the exercise of the more active functions and higher energies of animal existence. The power of locomotion, with which animals are to be invested, must alone introduce essential differences in their organization, and must require a union of strength and flexibility in the parts intended for extensive motion, and for being acted upon by powerful moving forces.

The animal, as well as the vegetable fabric is necessarily composed of a union of solid and fluid parts. Every animal texture appears to be formed from matter that was originally in a fluid state; the particles of which they are composed having been brought together and afterwards concreting by a process, which may, by a metaphor borrowed from physical science, be termed animal crystallization. Many of those animals, indeed, which occupy the lowest rank in the series, such as *Medusæ*, approach nearly to the fluid state; appearing like a soft and transparent jelly, which by spontaneous de-

composition after death, or by the application of heat, is resolved almost wholly into a limpid watery fluid.* More accurate examination, however, will show that it is in reality not homogeneous, but that it consists of a large proportion of water, retained in a kind of spongy texture, the individual fibres of which, from their extreme fineness and uniformity of distribution, can with difficulty be detected. Thus, even those animal fabrics which on a superficial view appear most simple, are in reality formed by an extremely artificial and complex arrangement of parts. The progress of development is continually tending to solidify the structure of the body. In this respect the lower orders of the animal kingdom, even when arrived at maturity, resemble the conditions of the higher classes at the earliest stages of their existence. As we rise in the scale of animals, we approximate to the condition of the more advanced states of development which are exhibited in the highest class.

Great efforts have been made by physiologists to discover the particular structure which might be considered as the simplest element of all the animal textures; the raw material, as it were, with which the whole fabric is wrought: but their labours have hitherto been fruitless. Fanciful hypotheses in abundance might be adduced on this favourite topic of

---

* Thus a Medusa, weighing twenty or thirty pounds, will, by this sort of general liquefaction, be found reduced to only a few grains of solid matter. Péron, Annales du Musée, tom. XV. p. 43. See also a memoir by *Quoy* and *Gaimard*, Annales des Sciences Naturelles, tom. I. p. 245.

speculation; but they have led to no useful or satisfactory result. Haller, who pursued the inquiry with great ardour, came to the conclusion that there existed what he calls the simple or primordial fibre, which he represents as bearing to anatomy the same relation that a line does to geometry. Chemical analysis alone is sufficient to overturn all these hypotheses of the uniformity of the proximate elementary materials of the animal organs: for they are found to be extremely diversified in their chemical composition. Neither has the microscope enabled us to resolve the problem: for although it has been alleged by many observers that the ultimate elements of every animal structure consists of minute globules, little confidence is to be placed in these results obtained by the employment of high magnifying powers, which are open to so many sources of fallacy. That globules exist in great numbers, not only in the blood, but in all animal fluids, there can be no doubt: and that these globules, by cohering, compose many of the solids, is also extremely probable. But it is very doubtful whether they are essential to the composition of other parts, such as the fibres of the muscles, the nerves, the ligaments, the tendons, and the cellular texture: for the most recent, and apparently most accurate microscopical observations tend to show that no globular structure exists in any of these textures.*

The element which we can recognise without difficulty as composing

the greater portion of animal structures, is that which is known by the name of the *cellular texture*.† Although bearing the same designation as the elementary material of the vegetable fabric, it differs widely from it, in its structure and mechanical properties. It is not, like that of plants, composed of a union of vesicles; but is formed of a congeries of extremely thin laminæ, or plates, variously connected together by fibres, and by other plates which cross them in different directions, leaving cavities or cells. (Fig. 25.) These cells, or rather intervening spaces, communicate freely with one another; and, in fact, may

[† According to Hughes (1959, p. 34) this is areolar connective tissue. (ED.)]

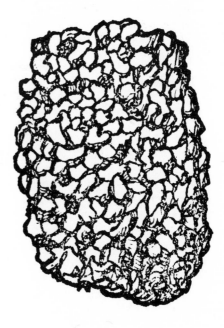

Fig. 25

* See the Appendix to Dr. Hodgkin and Dr. Fisher's translation of Edward's work on the Influence of Physical Agents on Life, p. 440.

be considered as one common cavity, subdivided by an infinite number of partitions into minute compartments. Hence the cellular texture is throughout readily permeable to fluids of all kinds, and retains these fluids in the manner, and on the same principle, as a sponge.

The cellular texture is not only the element, or essential material employed by nature in the construction of all the parts of the animal fabric; but, in its simplest form, it constitutes the general medium of connexion between adjacent organs, and also between the several parts of the same organ. Like the mortar which unites the stones of a building, the cellular texture is the universal cement employed to bind together all the solid structures. Its properties are admirably adapted to the mechanical purposes which are required in different parts of the frame: and these properties are variously modified and adjusted to suit the particular exigencies of the case. When, for instance, different parts require to be moveable upon each other, the cellular substance interposed between them has its state of condensation adapted to the degree of motion required. That which connects the muscles, or surrounds the joints, and all other parts concerned in extensive action, has a looser texture, being formed of broad and extensible plates, with few lateral adhesions, and leaving large interstices; while the more quiescent organs, the plates of the cellular substance, are thin and small, the fibres short and slender, and their intertexture closer and more condensed.

Besides being flexible and extensible, the cellular texture is also highly elastic, a property which is exceedingly advantageous in the construction of the frame. Not only the displacement of parts is resisted by their elasticity, but, when displaced, they tend to return to their natural position. This property performs a more important part in the mechanism of the animal than of the vegetable system: as might, indeed, have been anticipated from the more active and energetic movements required by the functions of the former. [pp. 79–82] . . .

RUDOLF VIRCHOW

The work of Schwann and Schleiden, which is discussed in *Heredity and Development* (Chapter 2), can be considered the first adequate statement of the Cell Theory: organisms can be thought of as composed solely of cells and cell products. This was a most important advance but, unfortunately, these two men suggested a hypothesis for the origin of cells that misled a whole generation of cytologists. They believed that cells were formed by a process similar to the cystallization of salts from a saturated solution. Only slowly did it come to be believed that cells arise only from pre-existing cells. One of the more prominent men to hold this view was Rudolf Virchow (1821–1902), a German physician, anthropologist, and statesman. His *Cellular Pathology* includes a summary of the field of cytology in the middle of the nineteenth century.

LECTURE I.
FEBRUARY 10, 1858.
CELLS AND THE CELLULAR THEORY.

Gentlemen,—Whilst bidding you heartily welcome to benches which must have long since ceased to be familiar to you, I must begin by reminding you, that it is not my want of modesty which has summoned you hither, but that I have only yielded to the repeatedly manifested wishes of many among you. Nor should I have ventured either to offer you lectures after the same fashion in which I am accustomed to deliver them in my regular courses. On the contrary, I will make the attempt to lay before you in a more succinct manner the development which I myself, and, I think, medical science also, have passed through in the course of the last fifteen years. In my announcement of these lectures, I described the subject of them in such a way as to couple histology with pathology; and for this reason, that I thought I must take it for granted that many busily occupied physicians were not quite familiar with the most recent histological changes, and did not enjoy sufficiently frequent opportunities of examining microscopical objects for themselves. Inasmuch as, however, it is upon such examinations that the most important conclusions are grounded which we now draw, you will pardon me if, disregarding those among you who have a perfect acquaintance with the subject, I behave just as if you all were not completely familiar with the requisite preliminary knowledge.

The present reform in medicine, of which you have all been witnesses, essentially had its rise in new anatomical observations, and the exposition also, which I have to make to you, will therefore principally be based upon anatomical demonstrations. But for me it would not be sufficient to take, as has been the custom during the last ten years, pathological anatomy alone as the groundwork of my views; we must add thereto those facts of general anatomy also, to which the actual state of medical science is due. The history of medicine teaches us, if we will only take a somewhat comprehensive survey of it, that at all times permanent advances have been marked by anatomical innovations, and that every more important epoch has been directly ushered in by a series of important discoveries concerning the structure of the body. So it was in those old times, when the observations of the Alexandrian school, based for the first time upon the anatomy of man, prepared the way for the system of Galen; so it was, too, in the Middle Ages, when Vesalius laid the foundations of anatomy, and therewith began the real reformation of medicine; so, lastly, was it at the commencement of this century, when Bichat developed the principles of general anatomy. What

From Rudolf Virchow. 1863. *Cellular Pathology as Based Upon Physiological and Pathological Histology. Twenty Lectures Delivered in the Pathological Institute of Berlin during the Months of February, March and April, 1858.* Translated from the Second Edition by Frank Chance. Robert M. DeWitt, New York. Reprinted by Dover, New York, 1971.

Schwann, however, has done for histology, has as yet been but in a very slight degree built up and developed for pathology, and it may be said that nothing has penetrated less deeply into the minds of all than the cell-theory in its intimate connection with pathology.

If we consider the extraordinary influence which Bichat in his time exercised upon the state of medical opinion, it is indeed astonishing that such a relatively long period should have elapsed since Schwann made his great discoveries, without the real importance of the new facts having been duly appreciated. This has certainly been essentially due to the great incompleteness of our knowledge with regard to the intimate structure of our tissues which has continued to exist until quite recently, and, as we are sorry to be obliged to confess, still even now prevails with regard to many points of histology to such a degree, that we scarcely know in favour of what view to decide.

Especial difficulty has been found in answering the question, from what parts of the body action really proceeds—what parts are active, what passive; and yet it is already quite possible to come to a definitive conclusion upon this point, even in the case of parts the structure of which is still disputed. The chief point in this application of histology to pathology is to obtain a recognition of the fact, that the cell is really the ultimate morphological element in which there is any manifestation of life, and that we must not transfer the seat of real action to any point beyond the cell. Before you, I shall have no particular reason to justify myself, if in this respect I make quite a special reservation in favour of life. In the course of these lectures you will be able to convince yourselves that it is almost impossible for any one to entertain more mechanical ideas in particular instances than I am wont to do, when called upon to interpret the individual processes of life. But I think that we must look upon this as certain, that, however much of the more delicate interchange of matter, which takes place within a cell, may not concern the material structure as a whole, yet the real action does proceed from the structure as such, and that the living element only maintains its activity as long as it really presents itself to us as an independent whole.

In this question it is of primary importance (and you will excuse my dwelling a little upon this point, as it is one which is still a matter of dispute) that we should determine what is really to be understood by the term cell. Quite at the beginning of the latest phase of histological development, great difficulties sprang up in crowds with regard to this matter. Schwann, as you no doubt recollect, following immediately in the footsteps of Schleiden, interpreted his observations according to botanical standards, so that all the doctrines of vegetable physiology were invoked, in a greater or less degree, to decide questions relating to the physiology of animal bodies. Vegetable cells, however, in the light in which they were at that time universally, and as they are even now also frequently regarded, are structures, whose identity with what we call animal cells cannot be admitted without reserve.

When we speak of ordinary vegetable cellular tissue, we generally understand thereby a tissue, which, in its most simple and regular form is, in a transverse section, seen to be composed of nothing but four- or six-sided, or, if somewhat looser in texture, of roundish or polygonal bodies, in which a tolerably thick, tough wall (*membrane*) is always to be distinguished. If now a single one of these bodies be isolated, a cavity is found, enclosed by this tough, angular, or round wall, in the interior of which very different substances, varying according to circumstances, may be deposited, *e. g.* fat, starch, pigment, albumen (*cell-contents*). But also, quite independently of these local varieties in the contents, we are enabled, by means of chemical investigation, to detect the presence of several different substances in the essential constituents of the cells.

The substance which forms the external membrane, and is known under the name of cellulose, is generally found to be destitute of nitrogen, and yields, on the adddition of iodine and sulphuric acid, a peculiar, very characteristic, beautiful blue tint. Iodine alone produces no colour; sulphuric acid by itself chars. The contents of simple cells, on the other hand, do not turn blue; when the cell is quite a simple one, there appears, on the contrary, after the addition of iodine and sulphuric acid, a brownish or yellowish mass, isolated in the interior of the cell-cavity as a special body (*protoplasma*), around which can be recognised a special, plicated, frequently shrivelled membrane (*primordial utricle*) (fig. 1, *c*). Even rough chemical analysis generally detects in the simplest cells, in addition to the non-nitrogenized (external) substance, a nitrogenized internal mass;

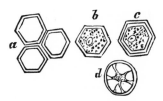

FIG. 1. Vegetable cells from the centre of the young shoot of a tuber of *Solanum tuberosum*. *a.* The ordinary appearance of the regularly polygonal, thick-walled cellular tissue. *b.* An isolated cell with finely granular-looking cavity, in which a nucleus with nucleolus is to be seen. *c.* The same cell after the addition of water; the contents (protoplasma) have receded from the wall (membrane, capsule). Investing them a peculiar, delicate membrane (primordial utricle) has become visible. *d.* The same cell after a more lengthened exposure to the action of water; the interior cell (protoplasma with the primordial utricle and nucleus) has become quite contracted, and remains attached to the cell-wall (capsule) merely by the means of fine, some of them branching, threads.

and vegetable physiology seems, therefore, to have been justified in concluding, that what really constitutes a cell is the presence within a non-nitrogenized membrane of nitrogenized contents differing from it.

It had indeed already long been known, that other things besides existed in the interior of cells, and it was one of the most fruitful of discoveries when Robert Brown detected the *nucleus* in the vegetable cell. But this body was considered to have a more important share in the formation than in the maintenance of cells, because in very many vegetable cells the nucleus becomes extremely indistinct, and in many altogether disappears,

whilst the form of the cell is preserved.

These observations were then
applied to the consideration of animal tissues, the correspondence of
which with those of vegetables
Schwann endeavoured to demonstrate.
The interpretation, which we have
just mentioned as having been put
upon the ordinary forms of vegetable
cells, served as the starting-point. In
this, however, as after-experience
proved, an error was committed. Vegetable cells cannot, viewed in their
entirety, be compared with all animal
cells. In animal cells, we find no such
distinctions between nitrogenized and
non-nitrogenized layers; in all the
essential constituents of the cells nitrogenized matters are met with. But
there are undoubtedly certain forms
in the animal body which immediately
recall these forms of vegetable cells,
and among them there are none so
characteristic as the cells of cartilage,
which is, in all its features, extremely
different from the other tissues of the
animal body, and which, especially
on account of its non-vascularity,
occupies quite a peculiar position.
Cartilage in every respect stands in the
closest relation to vegetable tissue. In
a well-developed cartilage-cell we can
distinguish a relatively thick external
layer, within which, upon very close
inspection, a delicate membrane, contents, and a nucleus are also to be
found. Here, therefore, we have a
structure which entirely corresponds
with a vegetable cell.

It has, however, been customary
with authors, when describing cartilage, to call the whole of the structure
of which I have just given you a
sketch (fig. 2, *a—d*) a cartilage-
corpuscle, and in consequence of this

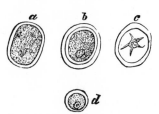

FIG. 2. Cartilage-cells as they occur at
the margin of ossification in growing
cartilage, quite analogous to vegetable
cells (cf. the explanation to fig. 1). *a—c.*
In a more advanced stage of development. *d.* Younger form.

having been viewed as analogous to
the cells in other parts of animals,
difficulties have arisen by which the
knowledge of the true state of the
case has been exceedingly obscured. A
cartilage-corpuscle, namely, is not,
as a whole, a cell, but the external
layer, the *capsule,* is the product of a
later development (secretion, excre-
tion). In young cartilage it is very
thin, whilst the cell also is generally
smaller. If we trace the development
still farther back, we find in cartilage,
also, nothing but simple cells, identical in structure with those which are
seen in other animal tissues, and not
yet possessing that external secreted
layer.

You see from this, gentlemen, that
the comparison between animal and
vegetable cells, which we certainly
cannot avoid making, is in general
inadmissible, because in most animal
tissues no formed elements are found
which can be considered as the full
equivalents of vegetable cells in the
old signification of the word; and because in particular, the cellulose membrane of vegetable cells does not correspond to the membrane of animal
ones, and between this, as containing

nitrogen, and the former, as destitute of it, no typical distinction is presented. On the contrary, in both cases we meet with a body essentially of a nitrogenous nature, and, on the whole, similar in composition. The so-called membrane of the vegetable cell is only met with a few animal tissues, as, for example, in cartilage; the ordinary membrane of the animal cell corresponds, as I showed as far back as 1847, to the primordial utricle of the vegetable cell. It is only when we adhere to this view of the matter, when we separate from the cell all that has been added to it by an after-development, that we obtain a simple, homogeneous, extremely monotonous structure, recurring with extraordinary constancy in living organisms. But just this very constancy forms the best criterion of our having before us in this structure one of those really elementary bodies, to be built up of which is eminently characteristic of every living thing—without the pre-existence of which no living forms arise, and to which the continuance and the maintenance of life is intimately attached. Only since our idea of a cell has assumed this severe form —and I am somewhat proud of having always, in spite of the reproach of pedantry, firmly adhered to it—only since that time can it be said that a simple form has been obtained which we can everywhere again expect to find, and which, though different in size and external shape, is yet always identical in its essential constituents.

In such a simple cell we can distinguish dissimilar constituents, and it is important that we should accurately define their nature also.

In the first place, we expect to find a *nucleus* within the cell; and with

regard to this nucleus, which has usually a round or oval form, we know that, particularly in the case of young cells, it offers greater resistance to the action of chemical agents than do the external parts of the cell, and that, in spite of the greatest variations in the external form of the cell, it generally maintains its form. The nucleus is accordingly, in cells of all shapes, that part which is the most constantly found unchanged. There are indeed isolated cases, which lie scattered throughout the whole series of facts in comparative anatomy and pathology, in which the nucleus also has a stellate or angular appearance; but these are extremely rare exceptions, and dependent upon peculiar changes which the element has undergone. Generally, it may be said that, as long as the life of the cell has not been brought to a close, as long as cells behave as elements still endowed with vital power, the nucleus maintains a very nearly constant form.

The nucleus, in its turn, in completely developed cells, very constantly encloses another structure within itself—the so-called *nucleolus*. With regard to the question of vital form, it cannot be said of the nucleolus that

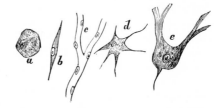

FIG. 3. *a.* Hepatic cell. *b.* Spindle-shaped cell from connective tissue. *c.* Capillary vessel. *d.* Somewhat large stellate cell from a lymphatic gland. *e.* Ganglion-cell from the cerebellum. The nuclei in every instance similar.

it appears to be an absolute requisite; and, in a considerable number of young cells, it has as yet escaped detection. On the other hand, we regularly meet with it in fully developed, older forms; and it, therefore, seems to mark a higher degree of development in the cell. According to the view which was put forward in the first instance by Schleiden, and accepted by Schwann, the connection between the three coexistent cell-constituents was long thought to be on this wise: that the nucleolus was the first to shew itself in the development of tissues, by separating out of a formative fluid (*blastema, cytoblastema*), that it quickly attained a certain size, that then fine granules were precipitated out of the blastema and settled around it, and that about these there condensed a membrane. That in this way a nucleus was completed, about which new matter gradually gathered, and in due time produced a little membrane (the celebrated watch-glass form, fig. 4, *d'*). This description of the first development of cells out of free blastema, according to which the nucleus was regarded as preceding the formation of the cell, and playing the part of a real cell-former (*cytoblast*), is the one which is usually concisely designated by the name of the *cell-theory* (more accurately, theory of *free* cell-formation), —a theory of development which has now been almost entirely abandoned, and in support of the correctness of which not one single fact can with certainty be adduced. With respect to the nucleolus, all that we can for the present regard as certain, is, that where we have to deal with large and fully developed cells, we almost constantly see a nucleolus in them; but

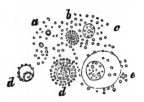

FIG. 4. From Schleiden, 'Grundzüge der wiss. Botanik,' I, fig. 1. "Contents of the embryo-sac of *Vicia faba* soon after impregnation. In the clear fluid, consisting of gum and sugar, granules of protein-compounds are seen swimming about (*a*), among which a few larger ones are strikingly conspicuous. Around these latter the former are seen conglomerated into the form of a small disc (*b, c*). Around other discs a clear, sharply defined border may be distinguished, which gradually recedes farther and farther from the disc (the cytoblast), and finally, can be distinctly recognised to be a young cell (*d, e*)."

that, on the contrary, in the case of many young cells it is wanting.

You will hereafter be made acquainted with a series of facts in the history of pathological and physiological development, which render it in a high degree probable that the nucleus plays an extremely important part within the cell—a part, I will here at once remark, less connected with the function and specific office of the cell, than with its maintenance and multiplication as a living part. The specific (in a narrower sense, animal) function is most distinctly manifested in muscles, nerves, and gland-cells; the peculiar actions of which—contraction, sensation, and secretion—appear to be connected in no direct manner with the nuclei. But that, whilst fulfilling all its functions, the element remains an element, that it is not annihilated nor destroyed by its continual activity—this seems essentially to depend upon the action of the nucleus. All those cellular formations which lose their nucleus, have a more transitory existence; they perish, they disappear, they die away or break up. A human blood corpuscle, for example, is a cell without a nucleus; it possesses an external membrane and red contents; but herewith the tale of its constituents, so far as we can make them out, is told, and whatever has been recounted concerning a nucleus in blood-cells, has had its foundation in delusive appearances, which certainly very easily can be, and frequently are, occasioned by the production of little irregularities upon the surface. We should not be able to say, therefore, that blood-corpuscles were cells, if we did not know that there is a certain period during which human blood-corpuscles also have nuclei; the period, namely, embraced by the first months of intra-uterine life. Then circulate also in the human body nucleated blood-cells, like those which we see in frogs, birds, and fish throughout the whole of their lives. In mammalia, however, this is restricted to a certain period of their development, so that at a later stage the red blood-cells no longer exhibit all the characteristics of a cell, but have lost an important constituent in their composition. But we are also all agreed upon this point, that the blood is one of those changeable constituents of the body, whose cellular elements possess no durability, and with regard to which everybody assumes that they perish, and are replaced by new ones, which in their turn are doomed to annihilation, and everywhere (like the uppermost cells in the cuticle, in which we also can discover no nuclei, as soon as they begin to desquamate) have already reached a stage in their development, when they no longer require that durability in their more intimate composition for which we must regard the nucleus as the guarantee.

On the other hand, notwithstanding the manifold investigations to which the tissues are at present subjected, we are acquainted with no part which grows or multiplies, either in a physiological or pathological manner, in which nucleated elements cannot invariably be demonstrated as the starting-points of the change, and in which the first decisive alterations which display themselves, do not involve the nucleus itself, so that we often can determine from its condition what would possibly have become of the elements.

You see from this description that, at least, two different things are of necessity required for the composition of a cellular element; the membrane, whether round, jagged or stellate, and the nucleus, which from the outset differs in chemical constitution from the membrane. Herewith, however, we are far from having enumerated all the essential constituents of the cell, for, in addition to the nucleus, it is filled with a relatively greater or less quantity of *contents,* as is likewise commonly, it seems, the nucleus itself, the contents of which are also wont to differ from those of the cell. Within the cell, for example, we see pigment, without the nucleus containing any. Within a smooth muscular fibre-cell, the contractile substance is deposited, which appears to be the seat of the contractile force of muscle; the nucleus, however, remains a nucleus. The cell may develop itself into a nerve-fibre, but the nucleus remains, lying on the outside of the medullary [white[1]] substance, a constant constituent. Hence it follows, that the special peculiarities which individual cells exhibit in particular places, under particular circumstances, are in general dependent upon the varying properties of the cell-contents, and that it is not the constituents which we have hitherto considered (membrane and nucleus), but the contents (or else the masses of matter deposited without the cell, *intercellular*), which give rise to the functional (physiological) differences of tissues. For us it is essen-

---

[1] All words included in square brackets have been inserted by the Translator, and are intended to be explanatory.

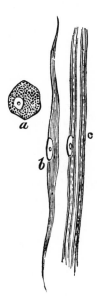

FIG. 5. *a.* Pigment-cell from the choroid membrane of the eye. *b.* Smooth muscular fibre-cell from the intestines. *c.* Portion of a nerve-fibre with a double contour, axis-cylinder, medullary sheath and parietal, nucleolated nucleus.

tial to know that in the most various tissues these constituents, which, in some measure, represent the cell in its abstract form, the nucleus and membrane, recur with great constancy, and that by their combination a simple element is obtained, which, throughout the whole series of living vegetable and animal forms, however different they may be externally, however much their internal composition may be subjected to change, presents us with a structure of quite a peculiar conformation, as a definite basis for all the phenomena of life.

According to my ideas, this is the only possible starting-point for all biological doctrines. If a definite correspondence in elementary form pervades the whole series of all living things, and if in this series something else which might be placed in the stead of the cell be in vain sought for, then must every more highly developed organism, whether vegetable or animal, necessarily, above all, be regarded as a progressive total, made up of larger or smaller number of similar or dissimilar cells. Just as a tree constitutes a mass arranged in a definite manner, in which, in every single part, in the leaves as in the root, in the trunk as in the blossom, cells are discovered to be the ultimate elements, so is it also with the forms of animal life. *Every animal presents itself as a sum of vital unities,* every one of which manifests all the characteristics of life. The characteristics and unity of life cannot be limited to any one particular spot in a highly developed organism (for example, to the brain of man), but are to be found only in the definite, constantly recurring structure, which every individual element displays. Hence it follows that the structural composition of a body of considerable size, a so-called individual, always represents a kind of social arrangement of parts, an arrangement of a social kind, in which a number of individual existences are mutually dependent, but in such a way, that every element has its own special action, and, even though it derive its stimulus to activity from other parts, yet alone effects the actual performance of its duties.

I have therefore considered it necessary, and I believe you will derive benefit from the conception, to portion out the body into *cell-territories* (Zellenterritorien). I say territories, because we find in the organization of animals a peculiarity which in vegetables in scarcely at all to be witnessed, namely, the development of large masses of so-called *intercellular substance*. Whilst vegetable cells are usually in immediate contact with one another by their external secreted layers, although in such a manner that the old boundaries can still always be distinguished, we find in animal tissues that this species of arrangement is the more rare one. In the often very abundant mass of matter which lies between the cells (*intermediate, intercellular substance*), we are seldom able to perceive at a glance, how far a given part of it belongs to one or another cell; it presents the aspect of a homogeneous intermediate substance.

According to Schwann, the intercellular substance was the cytoblastema, destined for the development of new cells. This I do not consider to be correct, but, on the contrary, I have, by means of a series of pathological observations, arrived at the conclusion that the intercellular substance is dependent in a certain defi-

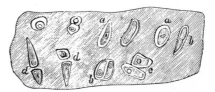

FIG. 6. Cartilage from the epiphysis of the lower end of the humerus of a child. The object was treated first with chromate of potash, and then with acetic acid. In the homogeneous mass (intercellular substance) are seen, at *a*, cartilage-cavities (Knorpelhöhlen) with walls still thin (capsules), from which the cartilage-cells, provided with a nucleus and nucleolus, are separated by a distinct limiting membrane. *b*. Capsules (cavities) with two cells produced by the division of previously simple ones. *c*. Division of the capsules following the division of the cells. *d*. Separation of the divided capsules by the deposition between them of intercellular substance—Growth of cartilage.

nite manner upon the cells, and that it is necessary to draw boundaries in it also, so that certain districts belong to one cell, and certain others to another. [pages 27–42] . . .

LECTURE II.
FEBRUARY 17, 1858.
PHYSIOLOGICAL TISSUES.

In my first lecture, gentlemen, I laid before you the general points to be noted with regard to the nature and origin of cells and their constituents. Allow me now to preface our further considerations with a review of the animal tissues in general, and this both in their physiological and pathological relations.

The most important obstacles which, until quite recently, existed in this quarter, were by no means chiefly of a pathological nature. I am convinced that pathological conditions would have been mastered with far less difficulty if it had not, until quite lately, been utterly impossible to give a simple and comprehensive sketch of the physiological tissues. The old views, which have in part come down to us from the last century, have exercised such a preponderating influence upon that part of histology which is, in a pathological point of view, the most important, that not even yet has unanimity been arrived at, and you will therefore be constrained, after you have inspected the preparations I shall lay before you, to come to your own conclusions as to how far that which I have to communicate to you is founded upon real observation.

If you read the 'Elementa Physiologiæ' of Haller, you will find, where the elements of the body are treated of, the most prominent position in the whole work assigned to *fibres,* the very characteristic expression being there made use of, that the fibre (fibra) is to the physiologist what the line is to the geometrician.

This conception was soon still further expanded, and the doctrine that fibres serve as the groundwork of nearly all the parts of the body, and that the most various tissues are reducible to fibres as their ultimate constituents, was longest maintained in the case of the very tissue in which, as it has turned out, the pathological difficulties were the greatest—in the so-called cellular tissue.

In the course of the last ten years of the last century there arose, however, a certain degree of reaction against this fibre-theory, and in the school of natural philosophers another element soon attained to honour, though it had its origin in far more speculative views than the former, namely, the *globule.* Whilst some still clung to their fibres, others, as in more recent times Milne Edwards, thought fit to go so far as to suppose the fibres, in their turn, to be made up of globules ranged in lines. This view was in part attributable to optical illusions in microscopical observation. The objectionable method which prevailed during the whole of the last and a part of the present century—of making observations (with but indifferent instruments) in the full glare of the sun—caused a certain amount of dispersion of light in nearly all microscopical objects, and the impression communicated to the observer was, that he saw nothing else than globules. On the other hand, however, this view corresponded with the ideas common amongst natural philosophers as to the primary origin of everything endowed with form.

These globules (granules, molecules) have, curiously enough, maintained their ground, even in modern histology, and there are but few histological works which do not begin with the consideration of elementary granules. In a few instances, these views as to the globular nature of elementary

FIG. 12. Diagram of the globular theory. *a.* Fibre composed of elementary granules (molecular granules) drawn up in a line. *b.* Cell with nucleus and spherically arranged granules.

FIG. 13. Diagram of the investment-(cluster-) theory. *a.* Separate elementary granules. *b.* Heap of granules (cluster). *c.* Granule-cell, with membrane and nucleus.

parts have, even not very long ago, acquired such ascendancy, that the composition, both of the primary tissues in the embryo and also of the later ones, was based upon them. A cell was considered to be produced by the globules arranging themselves in a spherical form, so as to constitute a membrane, within which other globules remained, and formed the contents. In this way did even Baumgärtner and Arnold contend against the cell theory.

This view has, in a certain manner, found support even in the history of development—in the so-called *investment-theory* (Umhüllungstheorie)—a doctrine which for a time occupied a very prominent position. The upholders of this theory imagined, that originally a number of elementary globules existed scattered through a fluid, but that, under certain circumstances, they gathered together, not in the form of vesicular membranes, but so as to constitute a compact heap, a globe (mass, cluster—Klümpchen), and that this globe was the starting point of all further development, a membrane being formed outside and a nucleus inside, by the differentiation of the mass, by apposition, or intussusception.

At the present time, neither fibres, nor globules, nor elementary granules, can be looked upon as histological

starting-points. As long as living elements were conceived to be produced out of parts previously destitute of shape, · such as formative fluids, or matters (*plastic matter, blastema, cytoblastema*), any one of the above views could of course be entertained, but it is in this very particular that the revolution which the last few years have brought with them has been the most marked. Even in pathology we can now go so far as to establish, as a general principle, *that no development of any kind begins de novo, and consequently as to reject the theory of equivocal* [spontaneous] *generation just as much in the history of the development of individual parts as we do in that of entire organisms.* Just as little as we can now admit that a tænia can arise out of saburral mucus, or that out of the residue of the decomposition of animal or vegetable matter an infusorial animalcule, a fungus, or an alga, can be formed, equally little are we disposed to concede either in physiological or pathological histology, that a new cell can build itself up out of any non-cellular substance. Where a cell arises, there a cell must have previously existed (*omnis cellula e cellula*), just as an animal can spring only from an animal, a plant only from a plant. In this manner, although there are still a few spots in the body where absolute demonstration has not yet been afforded, the principle is nevertheless established, that in the whole series of living things, whether they be entire plants or animal organisms, or essential constituents of the same, an eternal law of *continuous development* prevails. There is no discontinuity of development of such a kind that a new generation can of itself give rise to a new series of developmental forms. No developed tissues can be traced back either to any large or small simple element, unless it be unto a cell. In what manner this continuous *proliferation of cells* (Zellenwucherung), for so we may designate the process, is carried on, we will consider hereafter; to-day, my especial object only was to deter you from assuming as the groundwork of any views you might entertain with regard to the composition of the tissues, these theories of simple fibres or simple globules (elementary fibres or elementary globules). [pp. 51– 55] . . .

Thus the field of cytology was shown by Virchow to be of great medical importance: diseases often had specific pathological effects on cells and tissues and the study of these effects could be a valuable means of diagnosis. Other scientists studied cells simply because they were interested in what cells were and did, and during the 1870's and 1880's a number of exceedingly important observations were made, especially on the nucleus. It was observed that the nucleus divides, apparently in a very exact manner, when the cell divides. This process was named mitosis; during it, chromosomes appear, double in number, and are then parceled out to the daughter cells. The number of chromosomes, therefore, remains the same. Important facts were learned about fertilization: when the sperm enters an egg, part of it forms a nucleus, which then unites with a nucleus already in the egg. These and other discoveries are discussed in *Heredity and Development* (Chapter 2).

AUGUST WEISMANN

Many scientists began to speculate if these events associated with mitosis and fertilization had any meaning for inheritance. One was August Weismann (1834–1914), Professor at the University of Freiburg in Germany for nearly half a century. He made some observations of his own but his greatest contribution was an attempt to weld the data of the day into a comprehensive theory of heredity. The following excerpts are from an 1885 essay, 'The Continuity of the Germ-Plasm as the Foundation of a Theory of Heredity.'

When we see that, in the higher organisms, the smallest structural details, and the most minute peculiarities of bodily and mental disposition, are transmitted from one generation to another; when we find in all species of plants and animals a thousand characteristic peculiarities of struc-

From August Weismann, *Essays Upon Heredity and Kindred Biological Problems.* Edited by Edward B. Poulton, Selmar Schonland, and Arthur E. Shipley. Volumes 1 and 2. Oxford at the Clarendon Press. 1891–92.

ture continued unchanged through long series of generations; when we even see them in many cases unchanged throughout whole geological periods; we very naturally ask for the causes of such a striking phenomenon: and enquire how it is that such facts become possible, how it is that the individual is able to transmit its structural features to its offspring with such precision. And the immediate answer to such a question must be given in the following terms:—'A single cell out of the millions of diversely differentiated cells which compose the body, becomes specialized as a sexual cell; it is thrown off from the organism and is capable of reproducing all the peculiarities of the parent body, in the new individual which springs from it by cell-division and the complex process of differentiation.' Then the more precise question follows: 'How is it that such a single cell can reproduce the *tout ensemble* of the parent with all the faithfulness of a portrait?'

The answer is extremely difficult; and no one of the many attempts to solve the problem can be looked upon as satisfactory; no one of them can be regarded as even the beginning of a solution or as a secure foundation from which a complete solution may be expected in the future. Neither Häckel's[1] 'Perigenesis of the Plastidule,' nor Darwin's[2] 'Pangenesis,' can be regarded as such a beginning. The former hypothesis does not really treat of that part of the problem which is

here placed in the foreground, viz. the explanation of the fact that the tendencies of heredity are present in single cells, but it is rather concerned with the question as to the manner in which it is possible to conceive the transmission of a certain tendency of development into the sexual cell, and ultimately into the organism arising from it. The same may be said of the hypothesis of His[3], who, like Häckel, regards heredity as the transmission of certain kinds of motion. On the other hand, it must be conceded that Darwin's hypothesis goes to the very root of the question, but he is content to give, as it were, a provisional or purely formal solution, which, as he himself says, does not claim to afford insight into the real phenomena, but only to give us the opportunity of looking at all the facts of heredity from a common standpoint. It has achieved this end, and I believe it has unconsciously done more, in that the thoroughly logical application of its principles has shown that the real causes of heredity cannot lie in the formation of gemmules or in any allied phenomena. The improbabilities to which any such theory would lead are so great that we can affirm with certainty that its details cannot accord with existing facts. Furthermore, Brooks'[4] well-considered and brilliant attempt to modify the theory of Pangenesis, cannot escape the reproach that it is based upon possibilities, which one might certainly describe as improbabilities. But although I am of opinion that the whole foundation of

---

[1] Häckel, 'Ueber die Wellenzeugung der Lebenstheilchen, etc.,' Berlin, 1876.

[2] Darwin, 'The Variation of Animals and Plants under Domestication,' vol. ii. 1875, chap. xxvii. pp. 344–399.

[3] His, 'Unsre Körperform, etc.,' Leipzig, 1875.

[4] Brooks, 'The Law of Heredity,' Baltimore, 1883.

the theory of Pangenesis, however it may be modified, must be abandoned, I think, nevertheless, its author deserves great credit, and that its production has been one of those indirect roads along which science has been compelled to travel in order to arrive at the truth. Pangenesis is a modern revival of the oldest theory of heredity, that of Democritus, according to which the sperm is secreted from all parts of the body of both sexes during copulation, and is animated by a bodily force; according to this theory also, the sperm from each part of the body reproduces the same part[5].

If, according to the received physiological and morphological ideas of the day, it is impossible to imagine that gemmules produced by each cell of the organism are at all times to be found in all parts of the body, and

---

[5] Galton's experiments on transfusion in Rabbits have in the mean time really proved that Darwin's gemmules do not exist. Roth indeed states that Darwin has never maintained that his gemmules make use of the circulation as a medium, but while on the one hand it cannot be shown why they should fail to take the favourable opportunities afforded by such a medium, inasmuch as they are said to be constantly circulating through the body; so on the other hand we cannot understand how the gemmules could contrive to avoid the circulation. Darwin has acted very wisely in avoiding any explanation of the exact course in which his gemmules circulate. He offered his hypothesis as a formal and not as a real explanation.

Professor Meldola points out to me that Darwin did not admit that Galton's experiments disproved pangenesis ('Nature,' April 27, 1871, p. 502), and Galton also admitted this in the next number of 'Nature' (May 4, 1871, p. 5).—A. W. 1889.

furthermore that these gemmules are collected in the sexual cells, which are then able to again reproduce in a certain order each separate cell of the organism, so that each sexual cell is capable of developing into the likeness of the parent body; if all this is inconceivable, we must enquire for some other way in which we can arrive at a foundation for the true understanding of heredity. My present task is not to deal with the whole question of heredity, but only with the single although fundamental question—'How is it that a single cell of the body can contain within itself all the hereditary tendencies of the whole organism?' I am here leaving out of account the further question as to the forces and the mechanism by which these tendencies are developed in the building-up of the organism. On this account I abstain from considering at present the views of Nägeli, for as will be shown later on, they only slightly touch this fundamental question, although they may certainly claim to be of the highest importance with respect to the further question alluded to above.

Now if it is impossible for the germ-cell to be, as it were, an extract of the whole body, and for all the cells of the organism to despatch small particles to the germ-cells, from which the latter derive their power of heredity; then there remain, as it seems to me, only two other possible, physiologically conceivable, theories as to the origin of germ-cells, manifesting such powers as we know they possess. Either the substance of the parent germ-cell is capable of undergoing a series of changes which, after the building-up of a new individual, leads back again to identical germ-cells; or the germ-cells are not derived at all, as far as their essential and characteristic substance is concerned, from the body of the individual, but they are derived directly from the parent germ-cell.

I believe that the latter view is the true one: I have expounded it for a number of years, and have attempted to defend it, and to work out its further details in various publications. I propose to call it the theory of 'The Continuity of the Germ-plasm,' for it is founded upon the idea that heredity is brought about by the transference from one generation to another, of a substance with a definite chemical, and above all, molecular constitution. I have called this substance 'germ-plasm,' and have assumed that it possesses a highly complex structure, conferring upon it the power of developing into a complex organism. I have attempted to explain heredity by supposing that in each ontogeny, a part of the specific germ-plasm contained in the parent egg-cell is not used up in the construction of the body of the offspring, but is reserved unchanged for the formation of the germ-cells of the following generation.

It is clear that this view of the origin of germ-cells explains the phenomena of heredity very simply, inasmuch as heredity becomes thus a question of growth and of assimilation, —the most fundamental of all vital phenomena. If the germ-cells of successive generations are directly continuous, and thus only form, as it were, different parts of the same substance, it follows that these cells must, or at any rate may, possess the same molecular constitution, and that they would therefore pass through exactly the same stages under certain conditions of development, and would form the same final product. The hypothesis of the continuity of the germ-plasm

gives an identical starting-point to each successive generation, and thus explains how it is that an identical product arises from all of them. In other words, the hypothesis explains heredity as part of the underlying problems of assimilation and of the causes which act directly during ontogeny: it therefore builds a foundation from which the explanation of these phenomena can be attempted. [pp. 167–171] . . .

In the following pages I shall attempt to develope further the theory of which I have just given a short account, to defend it against any objections which have been brought forward, and to draw from it new conclusions which may perhaps enable us more thoroughly to appreciate facts which are known, but imperfectly understood. It seems to me that this theory of the continuity of the germ-plasm deserves at least to be examined in all its details, for it is the simplest theory upon the subject, and the one which is most obviously suggested by the facts of the case, and we shall not be justified in foresaking it for a more complex theory until proof that it can be no longer maintained is forthcoming. It does not presuppose anything except facts which can be observed at any moment, although they may not be understood,—such as assimilation, or the development of like organisms from like germs; while every other theory of heredity is founded on hypotheses which cannot be proved. It is nevertheless possible that continuity of the germ-plasm does not exist in the manner in which I imagine that it takes place, for no one can at present decide whether all the ascertained facts agree with and can be explained by it. Moreover the ceaseless activity of research brings to light new facts every day, and I am far from maintaining that my theory may not be disproved by some of these. But even if it should have to be abandoned at a later period, it seems to me that, at the present time, it is a necessary stage in the advancement of our knowledge, and one which must be brought forward and passed through, whether it prove right or wrong, in the future. In this spirit I offer the following considerations, and it is in this spirit that I should wish them to be received.

I. THE GERM-PLASM.

I must first define precisely the exact meaning of the term germplasm.

In my previous writings in which the subject has been alluded to, I have simply spoken of germ-plasm without indicating more precisely the part of the cell in which we may expect to find this substance—the bearer of the characteristic nature of the species and of the individual. In the first place such a course was sufficient for my immediate purpose, and in the second place the number of ascertained facts appeared to be insufficient to justify a more exact definition. I imagined that the germ-plasm was that part of a germ-cell of which the chemical and physical properties—including the molecular structure—enable the cell to become, under appropriate conditions, a new individual of the same species. I therefore believed it to be some such substance as Nägeli[6],

shortly afterwards, called idioplasm, and of which he attempted, in an admirable manner, to give us a clear understanding. Even at that time one might have ventured to suggest that the organized substance of the nucleus is in all probability the bearer of the phenomena of heredity, but it was impossible to speak upon this point with any degree of certainty. O. Hertwig[7] and Fol[8] had shown that the process of fertilization is attended by a conjugation of nuclei, and Hertwig had even then distinctly said that fertilization generally depends upon the fusion of two nuclei; but the possibility of the co-operation of the substance of the two germ-cells could not be excluded, for in all the observed cases the sperm-cell was very small and had the form of a spermatozoon, so that the amount of its cell-body, if there is any, coalescing with the female cell, could not be distinctly seen, nor was it possible to determine the manner in which this coalescence took place. Furthermore, it was for some time very doubtful whether the spermatozoon really contained true nuclear substance, and even in 1879 Fol was forced to the conclusion that these bodies consist of cell-substance alone. In the following year my account of the sperm-cells of *Daphnidae* followed, and this should have removed every doubt as to the cellular nature of the sperm-cells and as to their possession of an entirely normal nucleus, if only the authorities upon the subject had paid more attention

---

[6] Nägeli, 'Mechanisch-physiologische Theorie der Abstammungslehre.' München u. Leipzig. 1884.

[7] O. Hertwig, 'Beiträge zur Kenntniss der Bildung, Befruchtung und Theilung des thierischen Eies.' Leipzig, 1876.

[8] Fol, 'Recherches sur la fécondation,' etc. Genève, 1879.

to these statements[9]. In the same year (1880) Balfour summed up the facts in the following manner: 'The act of impregnation may be described as the fusion of the ovum and spermatozoon, and the most important feature in this act appears to be the fusion of a male and female nucleus[10].' It is true that Calberla had already observed in *Petromyzon,* that the tail of the spermatozoon does not penetrate into the egg, but remains in the micropyle; but on the other hand the head and part of the 'middle-piece' which effect fertilization, certainly contain a small fraction of the cell-body in addition to the nuclear substance, and although the amount of the former which thus enters the egg must be very small, it might nevertheless be amply sufficient to transmit the tendencies of heredity. Nägeli and Pflüger rightly asserted, at a later date, that the amount of the substance which forms the basis of heredity is necessarily very small, for the fact that hereditary tendencies are as strong on the paternal as on the maternal side, forces us to assume that the amount of this substance is nearly equal in both male and female germ-cells. Although I had not published anything upon the point, I was my-self inclined to ascribe considerable importance to the cell-substance in the process of fertilization; and I had been especially led to adopt this view because my investigations upon *Daphnidae* had shown that an animal produces large sperm-cells with an immense cell-body whenever the economy of its organism permits. All *Daphnidae* in which internal fertilization takes place (in which the sperm-cells are directly discharged upon the unfertilized egg), produce a small number of such large sperm-cells (*Sida, Polyphemus, Bythotrephes*); while all species with external fertilization (*Daphnidae, Lynceinae*) produce very small sperm-cells in enormous numbers, thus making up for the immense chances against any single cell being able to reach an egg. Hence the smaller the chances of any single sperm-cell being successful, the larger is the number of such cells produced, and a direct result of this increase in number is a diminution in size. But why should the sperm-cells remain or become so large in the species in which fertilization is internal? The idea suggests itself that the species in this way gains some advantage, which must be given up in the other cases; although such advantage might consist in assisting the development of the fertilized ovum and not in any increase of the true fertilizing substance. At the present time we are indeed disposed to recognize this advantage in still more unimportant matters, but at that time the ascertained facts did not justify us in the assertion that fertilization is a mere fusion of nuclei, and M. Nussbaum[11]

[9] Kölliker formerly stated, and has again repeated in his most recent publication, that the spermatozoa ('Samenfäden') are mere nuclei. At the same time he recognizes the existence of sperm-cells in certain species. But proofs of the former assertion ought to be much stronger in order to be sufficient to support so improbable a hypothesis as that the elements of fertilization may possess a varying morphological value. Compare Zeitschr. f. wiss. Zool., Bd. XLII.

[10] F. M. Balfour, 'Comparative Embryology,' vol. i. 69.

[11] Arch. f. mikr. Anat., Bd. 23. p. 182, 1884.

quite correctly expressed the state of our knowledge when he said that the act of fertilization consisted in 'the union of identical parts of two homologous cells.'

Pflüger's discovery of the 'isotropism' of the ovum was the first fact which distinctly pointed to the conclusion that the bodies of the germ-cells have no share in the transmission of hereditary tendencies. He showed that segmentation can be started in different parts of the body of the egg, if the latter be permanently removed from its natural position. This discovery constituted an important proof that the body of the egg consists of a uniform substance, and that certain parts or organs of the embryo cannot be potentially contained in certain parts of the egg, so that they can only arise from these respective parts and from no others. Pflüger was mistaken in the further interpretation, from which he concluded that the fertilized ovum has no essential relation to the organization of the animal subsequently formed by it, and that it is only the recurrence of the same external conditions which causes the germ-cell to develope always in the same manner. The force of gravity was the first factor, which, as Pflüger thought, determined the building up of the embryo: but he overlooked the fact that isotropism can only be referred to the body of the egg, and that besides this cell-body there is also a nucleus present, from which it was at least possible that regulative influences might emanate. Upon this point Born[12] first showed that the position of the nucleus is changed in eggs which are

thus placed in unnatural conditions, and he proved that the nucleus must contain a principle which in the first place directs the formation of the embryo. Roux[13] further showed that, even when the effect of gravity is compensated, the development is continued unchanged, and he therefore concluded that the fertilized egg contains within itself all the forces necessary for normal development. Finally, O. Hertwig[14] proved from observations on the eggs of sea-urchins, that at any rate in these animals, gravity has no directive influence upon segmentation, but that the position of the first nuclear spindle decides the direction which will be taken by the first divisional plane of segmentation. These observations were however still insufficient to prove that fertilization is nothing more than the fusion of nuclei.[15]

A further and more important step was taken when E. van Beneden[16] observed the process of fertilization in *Ascaris megalocephala*. Like the in-

---

[12] Born, 'Biologische Untersuchungen,' I, Arch. Mikr. Anat., Bd. XXIV.

[13] Roux, 'Beiträge zum Entwicklungsmechanismus des Embryo,' 1884.

[14] O. Hertwig, 'Welchen Einfluss übt die Schwerkraft,' etc. Jena, 1884.

[15] [Our present knowledge of the development of vegetable ova (including the position of the parts of the embryo) is also in favour of the view that it is not influenced by external causes, such as gravitation and light. It takes place in a manner characteristic of the genus or species, and essentially depends on other causes which are fixed by heredity; see Heinricher, 'Beeinflusst das Licht die Organanlage am Farnembryo?' in Mittheilungen aus dem Botanischen Institute zu Graz, II. Jena, 1888.—S. S.]

[16] E. van Beneden, 'Recherches sur la maturation de l'œuf,' etc., 1883.

vestigations of Nussbaum[17] upon the same subject, published at a rather earlier date, van Beneden's observations did not altogether exclude the possibility of the participation of the body of the sperm-cell in the real process of fertilization; still the fact that the nuclei of the egg-cell and the sperm-cell do not coalesce irregularly, but that their loops* are placed regularly opposite one another in pairs and thus form one new nucleus (the first segmentation nucleus), distinctly pointed to the conclusion that the nuclear substance is the sole bearer of hereditary tendencies—that in fact fertilization depends upon the coalescence of nuclei. Van Beneden himself did not indeed arrive at these conclusions: he was prepossessed with the idea that fertilization depends upon the union of two sexually differentiated nuclei, or rather half-nuclei—the male and female pronuclei. He considered that only in this way could a single complete nucleus be formed, a nucleus which must of course be hermaphrodite, and he believed that the essential cause of further development lies in the fact that, at each successive division of nuclei and cells, this hermaphrodite nature of the nucleus is maintained by the longitudinal division of the loops of each mother-nucleus, causing a uniform distribution of the male and female loops in both daughter-nuclei.

But van Beneden undoubtedly deserves great credit for having constructed the foundation upon which a scientific theory of heredity could be built. It was only necessary to replace the terms male and female pronuclei, by the terms nuclear substance of the male and female parents, in order to gain a starting-point from which further advance became possible. This step was taken by Strasburger, who at the same time brought forward an instance in which the nucleus only of the male germ-cell (to the exclusion of its cell-body) reaches the egg-cell. He succeeded in explaining the process of fertilization in Phanerogams, which had been for a long time involved in obscurity, for he proved that the nucleus of the sperm-cell (the pollen-tube) enters the embryo-sac and fuses with the nucleus of the egg-cell: at the same time he came to the conclusion that the body of the sperm-cell does not pass into the embryo-sac, so that in this case fertilization can only depend upon the fusion of nuclei[18].

Thus the nuclear substance must be

---

[17] M. Nussbaum, 'Ueber die Veränderung der Geschlechtsprodukte bis zur Eifurchung,' Arch. Mikr. Anat., 1884.

[* Loops=chromosomes. J.A.M. ed.]

---

[18] Eduard Strasburger, 'Neue Untersuchungen über den Befruchtungsvorgang bei den Phanerogamen als Grundlage für eine Theorie der Zeugung.' Jena, 1884.

[It is now generally admitted that, in the Vascular Cryptogams, as also in Mosses and Liverworts, the bodies of the spermatozoids are formed by the nuclei of the cells from which they arise. Only the cilia which they possess, and which obviously merely serve as locomotive organs, are said to arise from the surrounding cytoplasm. It is therefore in these plants also the nucleus of the male cell which effects the fertilization of the ovum. See Göbel, 'Outlines of Classification and Special Morphology,' translated by H. E. F. Garnsey, edited by I. B. Balfour, Oxford, 1887, p. 203, and Douglas H. Campbell, 'Zur Entwicklungsgeschichte der Spermatozoiden,' in Berichte d. deutschen bot. Gesellschaft, vol. v (1887), p. 120.—S. S.]

the sole bearer of hereditary tendencies, and the facts ascertained by van Beneden in the case of *Ascaris* plainly show that the nuclear substance must not only contain the tendencies of growth of the parents, but also those of a very large number of ancestors. Each of the two nuclei which unite in fertilization must contain the germ-nucleoplasm of both parents, and this latter nucleoplasm once contained and still contains the germ-nucleoplasm of the grandparents as well as that of all previous generations. It is obvious that the nucleoplasm of each antecedent generation must be represented in any germ nucleus in an amount which becomes less as the number of intervening generations becomes greater; and the proportion can be calculated after the manner in which breeders, when crossing races, determine the proportion of pure blood which is contained in any of the descendants. Thus while the germ-plasm of the father or mother constitutes half the nucleus of any fertilized ovum, that of a grandparent only forms a quarter, and that of the tenth generation backwards only $\frac{1}{1024}$, and so on. The latter can, nevertheless, exercise influence over the development of the offspring, for the phenomena of atavism show that the germ-plasm of very remote ancestors can occasionally make itself felt, in the sudden reappearance of long-lost characters. Although we are unable to give a detailed account of the way in which atavism happens, and of the circumstances under which it takes place, we are at least able to understand how it becomes possible; for even a very minute trace of a specific germ-plasm possesses the definite tendency to build up a certain organism,

and will develope this tendency as soon as its nutrition is, for some reason, favoured above that of the other kinds of germ-plasm present in the nucleus. Under these circumstances it will increase more rapidly than the other kinds, and it is readily conceivable that a preponderance in the quantity of one kind of nucleoplasm may determine its influence upon the cell-body. [pp. 176–182] . . .

I entirely agree with Strasburger when he says, 'The specific qualities of organisms are based upon nuclei;' and I further agree with him in many of his ideas as to the relation between the nucleus and cell-body: 'Molecular stimuli proceed from the nucleus into the surrounding cytoplasm; stimuli which, on the one hand, control the phenomena of assimilation in the cell, and, on the other hand, give to the growth of the cytoplasm, which depends upon nutrition, a certain character peculiar to the species.' 'The nutritive cytoplasm assimilates, while the nucleus controls the assimilation, and hence the substances assimilated possess a certain constitution and nourish in a certain manner the cyto-idioplasm and the nuclear idioplasm. In this way the cytoplasm takes part in the phenomena of construction, upon which the specific form of the organism depends. This constructive activity of the cyto-idioplasm depends upon the regulative influence of the nuclei.' The nuclei therefore 'determine the specific direction in which an organism developes.'

The opinion—derived from the recent study of the phenomena of fertilization—that the nucleus impresses its specific character upon the cell, has received conclusive and important confirmation in the experiments upon the regeneration of Infusoria, conducted simultaneously by M. Nussbaum[19] at Bonn, and by A. Gruber[20] at Freiburg. Nussbaum's statement that an artificially separated portion of a *Paramaecium*, which does not contain any nuclear substance, immediately dies, must not be accepted as of general application, for Gruber has kept similar fragments of other Infusoria alive for several days. Moreover, Gruber had previously shown that individual Protozoa occur, which live in a normal manner, and are yet without a nucleus, although this structure is present in other individuals of the same species. But the meaning of the nucleus is made clear by the fact, published by Gruber, that such artificially separated fragments of Infusoria are incapable of regeneration, while on the other hand those fragments which contain nuclei always regenerate. It is therefore only under the influence of the nucleus that the cell substance re-developes into the full type of the species. In adopting the view that the nucleus is the factor which determines the specific nature of the cell, we stand on a firm foundation upon which we can build with security.

If therefore the first segmentation nucleus contains, in its molecular structure, the whole of the inherited tendencies of development, it must follow that during segmentation and subsequent cell-division, the nucleoplasm will enter upon definite and varied changes which must cause the differences appearing in the cells which are produced; for identical cell-bodies depend, *ceteris paribus,* upon identical nucleoplasm, and conversely different cells depend upon differences in the nucleoplasm. The fact that the embryo grows more strongly in one direction than in another, that its cell-layers are of different nature and are ultimately differentiated into various organs and tissues,—forces us to accept the conclusion that the nuclear substance has also been changed in nature, and that such changes take place during ontogenetic development in a regular and definite manner. This view is also held by Strasburger, and it must be the opinion of all who seek to derive the development of inherited tendencies from the molecular structure of the germ-plasm, instead of from pre-formed gemmules.

We are thus led to the important question as to the forces by which the determining substance or nucleoplasm is changed, and as to the manner in which it changes during the course of ontogeny, and on the answer to this question our further conclusions must depend. The simplest hypothesis would be to suppose that, at each division of the nucleus, its specific substance divides into two halves of unequal quality, so that the cell-bodies would also be transformed; for we have seen that the character of a cell is determined by that of its nucleus. Thus in any Metazoon the first two segmentation spheres* would be transformed in such a manner that one only contained the hereditary tendencies of the endoderm and the other those of the ectoderm, and therefore, at a later stage, the cells of the endoderm would arise

---

[19] M. Nussbaum, 'Sitzungsber. der Niederrheinischen Gesellschaft für Natur- und Heilkunde.' Dec. 15, 1884.

[20] A. Gruber, 'Biologisches Centralblatt,' Bd. IV. No. 23, and V. No. 5.

[* Spheres=cells. J.A.M., ed.]

from the one and those of the ectoderm from the other; and this is
actually known to occur. In the course
of further division the nucleoplasm of
the first ectoderm cell would again
divide unequally, e.g. into the nucleoplasm containing the hereditary tendencies of the nervous system, and
into that containing the tendencies of
the external skin. But even then, the
end of the unequal division of nuclei
would not have been nearly reached;
for, in the formation of the nervous
system, the nuclear substance which
contains the hereditary tendencies of
the sense-organs would, in the course
of further cell-division, be separated
from that which contains the tendencies of the central organs, and the
same process would continue in the
formation of all single organs, and in
the final development of the most
minute histological elements. This
process would take place in a definitely
ordered course, exactly as it has taken
place throughout a very long series of
ancestors; and the determining and
directing factor is simply and solely
the nuclear substance, the nucleoplasm, which possesses such a molecular structure in the germ-cell that all
such succeeding stages of its molecular structure in future nuclei must
necessarily arise from it, as soon as
the requisite external conditions are
present. This is almost the same conception of ontogenetic development
as that which has been held by embryologists who have not accepted the
doctrine of evolution: for we have
only to transfer the primary cause of
development, from an unknown
source within the organism, into the
nuclear substance, in order to make
the views identical.

It appears at first sight that the
knowledge which has been gained by
studying the indirect division of
nuclei is opposed to such a view, for
we know that each mother-loop of the
so-called nuclear plate divides longitudinally into two exactly equal halves,
which can be stained and thus rendered visible.

In this way each resulting daughter-
nucleus receives an equal supply of
halves, and it therefore appears that
the two nuclei must be completely
identical. This at least is Strasburger's
conclusion, and he regards such identity as a fundamental fact, which cannot be shaken, and with which all
attempts at further explanation must
be brought into accord.

How then can the gradual transformation of the nuclear substance
be brought about? For such a transformation must necessarily take place
if the nuclear substance is really the
determining factor in development.
Strasburger attempts to support his
hypothesis by assuming that the inequality of the daughter-nuclei arises
from unequal nutrition; and he therefore considers that the inequality is
brought about after the division of the
nucleus and of the cell. Strasburger
has shown, in a manner which is
above all criticism, that the nucleus
derives its nutrition from the cell-
body, but then the cell-bodies of the
two *ex hypothesi* identical daughter-
nuclei must be different from the first,
if they are to influence their nuclei in
different ways. But if the nucleus
determines the nature of the cell, it
follows that two identical daughter-
nuclei which have arisen by division
within one mother-cell cannot come to
possess unequal cell-bodies. As a
matter of fact, however, the cell-bodies

of two daughter-cells often differ in size, in appearance, and in their subsequent history, and these facts are sufficient to prove that in such cases the division of the nucleus must have been unequal. It appears to me to be a necessary conclusion that, in such an instance, the mother-nucleus must have been capable of splitting into nuclear substances of differing quality. I think that, in his argument, Strasburger has over-estimated the support afforded by exact observations upon indirect nuclear division.* [pp. 187–91] . . .

There does not seem to be any objection to the view that the micro-somata of the nuclear loops—assuming that these bodies represent the idioplasm—are capable of dividing into halves, equal in form and appear-

---

[* Indirect Nuclear Division=Mitosis. (J.A.M., ed.)]

ance, but unequal in quality. We know that this very process takes place in many egg-cells; thus in the egg of the earth-worm the first two segmentation spheres are equal in size and appearance, and yet the one forms the endoderm and the other the ectoderm of the embryo.

I therefore believe that we must accept the hypothesis that, in indirect nuclear division, the formation of unequal halves may take place quite as readily as the formation of equal halves, and that the equality or inequality of the subsequently produced daughter-cells must depend upon that of the nuclei. Thus during ontogeny a gradual transformation of the nuclear substance takes place, necessarily imposed upon it, according to certain laws, by its own nature, and such transformation is accompanied by a gradual change in the character of the cell-bodies. [p. 193] . . .

Weismann continued his analysis of the problem in an 1887 essay 'On the Number of Polar Bodies and their Significance in Heredity.'

V. Conclusions with regard to Heredity.

The ideas developed in the preceding paragraphs lead to remarkable conclusions with regard to the theory of heredity,—conclusions which do not harmonize with the ideas on this subject which have been hitherto received. For if every egg expels half the number of its ancestral germ-plasms during maturation, the germ-cells of the same mother cannot contain the same hereditary tendencies, unless of course we make the supposition that cor-

responding ancestral germ-plasms are retained by all eggs—a supposition which cannot be sustained. For when we consider how numerous are the ancestral germ-plasms which must be contained in each nucleus, and further how improbable it is that they are arranged in precisely the same manner in all germ-cells, and finally how incredible it is that the nuclear thread should always be divided in exactly the same place to form corresponding loops or rods,—we are driven to the conclusion that it is quite impossible

---

From August Weismann, *Essays Upon Heredity and Kindred Biological Problems.*

for the 'reducing division' of the nucleus to take place in an identical manner in all the germ-cells of a single ovary, so that the same ancestral germ-plasms would always be removed in the polar bodies. But if one group of ancestral germ-plasms is expelled from one egg, and a different group from another egg, it follows that no two eggs can be exactly alike as regards their contained hereditary tendencies: they must all differ. In many cases the differences will only be slight, that is, when the eggs contain very similar combinations of ancestral germ-plasms. Under other circumstances the differences will be very great, viz. when the combinations of ancestral germ-plasms retained in the egg are very different. I might here mention various other considerations; but this would lead me too far from my subject, into new theories of heredity. I hope to be able at some later period to develope further the theoretical ideas which are merely indicated in the present essay. I only wish to show that the consequences which follow from my theory upon the second division of the egg-nucleus, and the formation of the second polar body, are by no means opposed to the facts of heredity, and even explain them better than has hitherto been possible.

The fact that the children of the same parents are never entirely identical could hitherto only be rendered intelligible by the vague suggestion that the hereditary tendencies of the grandfather predominate in one, and those of the grandmother in another, while the tendencies of the great-grandfather predominate in a third, and so on. Any further explanation as to why this should happen was

entirely wanting. Others even looked for an explanation to the different influences of nutrition, to which it is perfectly true that the egg is subjected in the ovary during its later development, according to its position and immediate surroundings. I had myself referred to these influences as a partial explanation[1], before I recognized clearly how extremely feeble and powerless are the influences of nourishment, as compared with hereditary tendencies. According to my theory, the differences between the children of the same parents become intelligible in a simple manner from the fact that each maternal germ-cell (I shall speak of the paternal germ-cells later on) contains a peculiar combination of ancestral germ-plasms, and thus also a peculiar combination of hereditary tendencies. These latter by their co-operation also produce a different result in each case, viz. the offspring, which are characterized by more or less pronounced individual peculiarities.

But the theory which explains individual differences by referring to the inequality of germ-cells, may be proved with a high degree of probability by an appeal to facts of an opposite kind, viz. by showing that identity between offspring only occurs when they have arisen from the same egg-cell. It is well known that occasionally some of the children of the same parents appear to be almost exactly alike, but such children are without exception twins, and there is every reason to believe that they have

---

[1] Weismann, 'Studien zur Descendenz-theorie,' ii. p. 306, Leipzig, 1876, translated by Meldola; see 'Studies in the Theory of Descent,' p. 680.

been derived from the *same* egg. In other words, the two children are exactly alike because they have arisen from the same egg-cell, which could of course only contain a single combination of ancestral germ-plasms, and therefore of hereditary tendencies[2]. The factors which by their co-operation controlled the construction of the

---

[2] [The similar conclusion that identical ova lead to the appearance of identical individuals was drawn from the same data by Francis Galton in 1875. See 'The history of the Twins, as a criterion of the relative powers of Nature and Nurture,' by Francis Galton, F.R.S., Journal of the Anthropological Institute, 1875, p. 391; also by the same author, 'Short Notes on Heredity, etc. in Twins,' in the same Journal, 1875, p. 325.

The author investigated about eighty cases of close similarity between twins, and was able to obtain instructive details in thirty-five of these. Of the latter there were no less than seven cases 'in which both twins suffered from some special ailment or had some exceptional peculiarity;' in nine cases it appeared that 'both twins are apt to sicken at the same time;' in eleven cases there was evidence for a remarkable association of ideas; in sixteen cases the tastes and dispositions were described as closely similar. These points of identity are given in addition to the more superficial indications presented by the failure of strangers or even parents to distinguish between the twins. A very interesting part of the investigation was concerned with the after-lives of the thirty-five twins. 'In some cases the resemblance of body and mind had continued unaltered up to old age, notwithstanding very different conditions of life,' in the other cases 'the parents ascribed such dissimilarity as there was, wholly, or almost wholly, to some form of illness.'

The conclusions of the author are as follows: 'Twins who closely resembled each other in childhood and early youth, and were reared under not very dissimilar conditions, either grow unlike through the development of natural characteristics which had lain dormant at first, or else they continue their lives, keeping time like two watches, hardly to be thrown out of accord except by some physical jar. Nature is far stronger than nurture within the limited range that I have been careful to assign to the latter.' And again, 'where the maladies of twins are continually alike, the clocks of their two lives move regularly on, and at the same rate, governed by their internal mechanism. Necessitarians may derive new arguments from the life histories of twins.'

The above facts and conclusions held for twins of the same sex, of which at any rate the majority are shown by Kleinwächter's observations to have been enclosed in the same embryonic membranes, and therefore presumably to have been derived from a single ovum; but in rarer cases the twins, although also invariably of the same sex, were marked by remarkable differences, greater than those which usually distinguish children of the same family. Mr. Galton met with twenty of these cases. In such twins the conditions of training, etc. had been as similar as possible, so that the evidence of the power of nature over nurture is strongly confirmed. Mr. Galton writes, 'I have not a single case in which my correspondents speak of originally dissimilar characters having become assimilated through identity of nurture. The impression that all this evidence leaves on the mind is one of wonder whether nurture can do anything at all, beyond giving instruction and professional training.'

The fact that twins produced from a single ovum seem to be invariably of the same sex is in itself extremely interesting, for it proves that the sex of the individual is predetermined in the fertilized ovum.—E. B. P.]

organism were the same, and consequently the results were also the same. Twins derived from a single egg are identical: this is a statement which, although not mathematically proved, may be looked upon as nearly certain. But there are also twins which do not possess this high degree of similarity, and these are even far commoner than the others. The explanation is to be found in the fact that the latter were derived from two egg-cells which were fertilized at the same time. In most cases, indeed, each twin is enclosed in its own embryonic membranes, while much less frequently both twins are enclosed in the same membranes. In one point only the proof is incomplete; for it has not yet been shown that identical twins are always derived from a single egg, since such an origin, together with a high degree of similarity, could only be established as occurring together in a small proportion of the cases. We therefore see that under conditions of nutriment which are as identical as possible, *two* egg-cells develope into unlike twins, *one* into identical twins; although we cannot yet affirm that the latter result invariably follows. It is conceivable that the stimulus for the production of two eggs from one may be afforded by the entrance of two spermatozoa, but these latter, as was shown above, could hardly contain identical hereditary tendencies, and thus two identical twins would not arise. It appears indeed that some cases have been observed in which differences have been exhibited by twins which were enclosed in the same embryonic membranes; but nevertheless I believe that two spermatozoa are not necessary to cause the formation of twins by a single egg. We

know, it is true, from the investigations of Fol[3], that multiple impregnation produces the simultaneous beginning of several embryos in the eggs of starfishes. But several embryos and young animals are not developed in this way, for embryonic development soon ceases, and the egg dies.

The recent observations of Born[4] upon the eggs of the frog also make it very probable that a double development is produced by the entrance of two spermatozoa into the egg, but here also only monstrosities, and not twins, were produced. On the other hand, it has been shown that in birds twins may be produced from the same egg, and there is no reason for the belief that their production is due to multiple impregnation. But if it may be assumed that human twins, when identical, have been derived from a single egg, it seems to me to be extremely probable that fertilization was also effected by a single sperm-cell. We cannot understand how such a high degree of similarity could have been produced if two sperm-cells had been made use of, for we are compelled to assume that two such cells would very rarely contain identical germ-plasms.

It is most probable that the egg-nucleus coalesces with the nucleus of a single spermatozoon, but the resulting segmentation-nucleus divides together with the cell-body itself, without the occurrence of those ontogenetic changes in the germ-plasm which

---

[3] Fol, 'Recherches sur la fécondation et le commencement de l'hénogénie.' Genève, Bâle, Lyon, 1879.

[4] Born, 'Ueber Doppelbildungen beim Frosch und deren Entstehung.' Breslauer ärztl. Zeitschrift, 1882.

normally take place. The nucleoplasm of the two daughter-cells still remains in the condition of germ-plasm, and its ontogenetic transformation begins afterwards—a transformation which must of course proceed in the same way in both cells, and must lead to the production of identical offspring. This is at least a possible explanation which we may retain until it has been either confirmed or disproved by fresh observations,—an explanation which is moreover supported by the well-known process of budding in the eggs of lower animals.

VI. RECAPITULATION.

To bring together shortly the results of this essay:—the fundamental fact upon which everything else is founded is the fact that *two* polar bodies are expelled, as a preparation for embryonic development, from all animal eggs which require fertilization, while only *one* such body is expelled from all parthenogenetic eggs.

This fact in the first place refutes every purely morphological explanation of the process. If it were physiologically valueless, such a phyletic reminiscence of the two successive divisions of the egg-nucleus must have been also retained by the parthenogenetic egg.

In my opinion the expulsion of the first polar body implies the removal of ovogenetic nucleoplasm when it has become superfluous after the maturation of the egg has been completed. The expulsion of the second polar body can only mean the removal of part of the germ-plasm itself, a removal by which the number of ancestral germ-plasms is reduced to one half. This reduction must also take place in the male germ-cells, although we are not able to associate it confidently with any of the histological processes of spermatogenesis which have been hitherto observed.

Parthenogenesis takes place when the whole of the ancestral germ-plasms, inherited from the parents, are retained in the nucleus of the egg-cell. Development by fertilization makes it necessary that half the number of these ancestral germ-plasms must be first expelled from the egg, the original quantity being again restored by the addition of the sperm-nucleus to the remaining half.

In both cases the beginning of embryogenesis depends upon the presence of a certain, and in both cases equal, quantity of germ-plasm. This certain quantity is produced by the addition of the sperm-nucleus to the egg requiring fertilization, and the beginning of embryogenesis immediately follows fertilization. The parthenogenetic egg contains within itself the necessary quantity of germ-plasm, and the latter enters upon active development as soon as the single polar body has removed the ovogenetic nucleoplasm. The question which I have raised on a previous occasion—'When is the parthenogenetic egg capable of development?'—now admits of the precise answer—'Immediately after the expulsion of the polar body.'

From the preceding facts and considerations the important conclusion results that the germ-cells of any individual do not contain the same hereditary tendencies, but are all different, in that no two of them contain exactly the same combinations of hereditary tendencies. On this fact the well-known differences between the children of the same parents depend.

But the deeper meaning of this arrangement must doubtless be sought for in the individual variability which is thus continuously kept up and is always being forced into new com- binations. Thus sexual reproduction is to be explained as an arrangement which ensures an ever-varying supply of individual differences [pp. 390–96].

Weismann's analysis of the problem led to the conclusion that the germ-plasm must be halved when the gametes were formed. He knew that this was true for eggs. Data were not available for sperm but they were soon forthcoming. He continues his analysis in 'Amphimixis or the Essential Meaning of Conjugation and Sexual Reproduction' (1891). Weismann used the term 'amphimixis' for the union of the gametes at fertilization.

Hence, after the discovery of the law of the number of polar bodies, I interpreted the first division of the nucleus as the removal of ovogenetic idioplasm from the egg, and the second as a halving of the number of ances-tral units contained in the germ-plasm. Such halving must have occurred, or the number of ancestral units would have been doubled. It necessarily fol-lowed from this view that the ancestral units contained in the spermatozoa must also have undergone a diminu-tion by half. I postulated therefore a reduction of the spermatozoa by divi-sion, and, to my mind, there was 'no doubt' that this process occurred in them 'at some time and by some means[1],' although not perhaps in the same manner as in the ova. I even said from the very first[2] that 'it is quite conceivable' that this division might occur in a manner entirely dif-ferent from that of the egg, since in the former case both daughter-cells might be of similar size and might become spermatozoa, in which case

neither of them would shrink and become polar bodies.

The Maturation of the Spermatozoon.

I have not been able to make out, by my own investigations, the facts which confirm the soundness of these views as to the spermatozoa; my impaired eyesight, which has so often put a stop to microscopic investigations, has again rendered the continuation of this research impossible. But Oscar Hertwig[3] has recently given us an account of the development of the spermatozoa of *Ascaris megaloceph-ala*, which not only proves the reduc-tion of the male germ-cells by division, but also shows that it takes place in precisely that way, which from the first I had regarded as most likely.

Since these new facts affect our conclusions with regard to many as-pects of the process of fertilization, they are here shortly abstracted. They may possibly enable us to penetrate

_____

[1] Vol. I, p. 381
[2] Vol. I. p. 385.

_____

[3] O. Hertwig, 'Ueber Ei- und Samen-bildung bei Nematoden,' Archiv f. mikr. Anat. 1890.

_____

From August Weismann, *Essays Upon Heredity and Kindred Biological Problems.*

still more deeply into the meaning and significance of the processes by which the nuclei of germ-cells are reduced in size.

Ever since Edouard van Beneden's classical researches on the process of fertilization, it has been well known that *Ascaris megalocephala* is one of the most favourable objects for the observation of the minute arrangements and changes occurring in the nuclei of germ-cells. The nuclear loops are not only relatively very large, but are also very few in number. Boveri was the first to show that, as regards this number, two varieties of the species exist, one containing two nuclear loops in the young germ-cells, the other containing four. O. Hertwig then proved, as might have been ex-

pected, that this difference in the number of loops in the youngest germ-cells exists also in the male sex. He called the variety which produces two loops *Var. univalens*, and that which produces four *Var. bivalens*. Since the development of the spermatozoa in both varieties differs only in the number of nuclear loops which are formed, I will, in the following account, deal with only one of them, the *Var. bivalens.*

The formation of the spermatozoa falls into three stages; the first is that of the 'primitive sperm-cells': these youngest male germ-cells then proceed to increase by means of successive divisions. The division of the nucleus is effected by karyokinesis after the usual manner; the four nu-

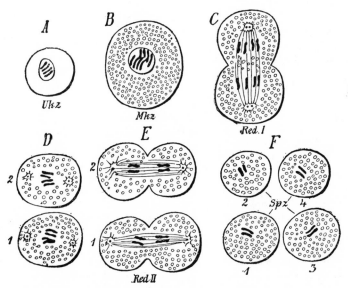

FIG. I. Formation of spermatozoa in *Ascaris megalocephala*, var. *bivalens* (modified from O. Hertwig). *A.* Primitive sperm-cells. *B.* Sperm-mother-cells. *C.* First 'reducing division.' *D.* The two daughter-cells. *E.* Second 'reducing division.' *F.* The four grand-daughter cells,—the sperm-cells.

clear loops split longitudinally and the halves form the two daughter-nuclei. After this process of multiplication has lasted for a considerable time, the cells pass into the second stage,—that of the 'mother-cells of spermatozoa.' They cease to multiply, grow considerably, and their nuclei pass into the resting condition, viz. the condition of a nuclear network into which the loops break up. When these cells have reached their full size they enter upon the preparation for fresh divisions, which are only two in number and rapidly follow each other. As soon as these are over, the whole development is complete. It is this last stage which brings about the 'reducing division' which I had predicted. The finely divided chromatin bodies contained in the nuclear network build up eight long, thin rods or threads, which afterwards shorten and form thicker rods, arranged by means of the pole-corpuscles or centrosomata, which act in such a manner that four rods are turned toward one pole and four toward the other. A division of the nucleus and of the cell now follows resulting in the formation of two daughter-cells, each of which contains as many nuclear loops as the original sperm-cells, i.e. four. This division is followed immediately by another on the same plan, but without any intervening resting stage: the number of nuclear rods is therefore again halved, so that each daughter-cell of the second order contains but **two**.

Hence the number of nuclear rods is at first increased from four to eight, and then by two consecutive divisions, this later number is first halved and then quartered, the final result being *a halving of the number of rods in the original sperm-cells.*

It is well known that precisely the same results are brought about by those divisions of the ovum which give rise to the polar bodies. In the egg the nuclear rods are first doubled and then, by two consecutive divisions, reduced to half their original number. In all essentials, the development of the ovum passes through precisely the same process as that of the spermatozoa. The first two stages, described by O. Hertwig, in the development of the spermatozoa I also find in the formation of the egg. The primitive ova correspond to the primitive sperm-cells, the mother-cell of the ova, or the mature full-sized egg, immediately before reduction by division, corresponds to the mother-cell of the spermatozoa, the only difference being that the egg in this, the second stage, has, as a rule, attained its definite shape and size and is surrounded by its membranes, and that the two last divisions, which are together spoken of as the 'reducing divisions,' generally take place after the egg has been laid or has, at any rate, left the ovary. This probably explains, as I have already maintained, why the division is so unequal, and why all the daughter-cells cannot become ova, but only the largest of them, viz. that one which alone contains the food-material necessary for the building up of the embryo.

In other respects the formation of the polar bodies corresponds with the two divisions of the mother-cells of spermatozoa: in both cases there are two successive cell-divisions, and furthermore in the egg both daughter-cells of the first generation divide again—not only the larger one, the ovum, but also the smaller or first polar body—for it is well known that

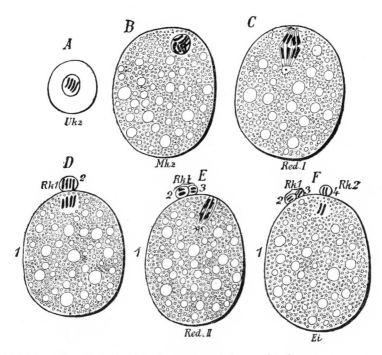

FIG. II. Formation of ova in *Ascaris megalocephala*, var. *bivalens*.

the latter body generally splits into two secondary polar bodies, and the significance of this apparently aimless division has hitherto been sought in vain. But now we see that it depends on the persistence of a phyletic stage of development, on the survival of an earlier condition, in which the original egg-cells underwent a 'reducing division,' like that of the spermatozoa, producing four cells, each of which was potentially an ovum.

Moreover in another, and obviously a decisive point, the 'reducing divisions' of ova and spermatozoa are in correspondence;—in the manner and method of the division of the nuclear rods in the daughter-nuclei. The process of karyokinesis here differs from any other mode of nuclear division, in that there is no longitudinal splitting or doubling of the nuclear rods, bringing about a contribution from each rod in the equatorial plate to both daughter-nuclei; instead of this, half the whole number of rods passes to one pole of the nuclear spindle, and half to the other. Furthermore, there is no resting-stage between the two divisions, during which the rods break up into the nuclear network, but the two divisions follow each other without any interval. If the 'reducing division,' for which I have argued, has any existence, we must look for it here; for, so far as proofs can be afforded by observation, they are forthcoming. The number of nuclear rods is reduced to half, and hence the mass of the nuclear substance is certainly halved. And if we must concede that the rods in a nucleus are not absolutely alike, but are derived from the differing germ-plasms of various ancestors (viz. that the rods consist of such different kinds of germ-plasm), it follows that a reduction of the ancestral germ-plasms is admitted.

The new facts discovered by O. Hertwig leave only one point obscure. We see indeed that, in the case of the spermatozoon as in that of the ovum, the nuclear rods are reduced to half, but we ask in vain why two successive divisions are necessary to bring about this reduction, when it seems that a single one would suffice. I had formerly concluded that since partheno-genetic eggs expel only one polar body, instead of the two which separate from all ova requiring fertilization, the first division must have a different significance from the second. I regarded the second division alone as the 'reducing division,' and this was a perfectly sound and logical conclusion, so long as it remained unknown that the mother-cells of ova contain twice as many nuclear rods as existed in the primitive egg-cells. Until this was known, the 'reducing division' was only required to effect a halving of the nuclear substance, and for this purpose one division would be sufficient. We now know that a second division is rendered necessary because the number of the rods is doubled before the process of reduction has begun. The object served by this doubling remains an obscure point upon which even the spermatogenesis of *Ascaris* does not at present en-lighten us. My previous interpretation of the first polar body as the removal of ovogenetic nucleoplasm from the egg must fall to the ground: about this there is no possible doubt, but

how can we better explain the neces-sity for two divisions? Why should the nuclear substance be doubled, only to be halved again? O. Hertwig has also propounded this question, but so far without being able to supply an an-swer. He hopes that a more accurate study of the manner and method of the arrangement of the chromatin elements in the two successive divi-sions will ultimately lead to a deeper knowledge of the essence of the whole process of maturation. I also hope the same. The processes which bring about the doubling of the chromatin rods in the resting nuclei of ova and sperm-mother-cells, contain, without doubt, the key to an understanding of the necessity for this increase in number, which at present appears to be so mysterious and superfluous.

Whether unaided observation will ever succeed in making clear the ac-cessory processes, in other words, whether morphological events can be followed in minute detail so far that we can wrest from them the secret of their meaning, we cannot say. With-out some guiding idea, it is scarcely possible that the observations of in-vestigators could be directed to the most essential part of the process, especially in this case, where differ-ences of substance are probably pres-ent—differences which might be in-visible, but are perhaps capable of being inferred by processes of reason-ing.

Thus it may be possible, on the basis of Hertwig's observations, to penetrate somewhat deeper into the meaning of the remarkable processes which attend the 'reducing divisions,' if only the subject be attacked from the point of view of the theory of ancestral germ-plasms [Vol 2. pp. 116–22] . . .

EDMUND B. WILSON

Weismann's remarkable analysis added great support to the hypothesis that the nucleus and its chromosomes are importantly involved in inheritance, a hypothesis that had wide support as the nineteenth century closed. The year 1900 was to witness a dramatic event: a paper written long before by Gregor Mendel on inheritance in the garden pea became generally known in the scientific community. It began a new era for both cytology and genetics. Just prior to this event, E. B. Wilson published the second edition of his classic *The Cell in Development and Inheritance*. These quotations from Wilson's book reveal the state of mind of students of the cell just prior to the new era.

We have thus arrived at the form in which the problems of heredity and development confront the investigator of the present day. It remains to point out more clearly how they are related to the general problems of evolution and to those post-Darwinian discussions in which Weismann has taken so active a part. All theories of evolution take the facts of variation and heredity as fundamental postulates, for it is by variation that new characters arise and by heredity that they are perpetuated. Darwin recognized two kinds of variation, both of which, being inherited and maintained through the conserving action of natural selection, might give rise to a permanent transformation of species. The first of these includes congenital or inborn variations, *i.e.* such as appear at birth or are developed 'spontaneously,' without discoverable connection with the activities of the organism itself or the direct effect of the environment upon it, though Darwin clearly recognized the fact that even such variations must indirectly be due to changed conditions acting upon the parental organism or on the germ. In a second class of variations were placed the so-called acquired characters, *i.e.* definite effects directly produced in the course of the individual life as the result of use and disuse, or of food, climate, and the like. The inheritance of congenital characters is now universally admitted, but it is otherwise with acquired characters. The inheritance of the latter, now the most debated question of biology, had been taken for granted by Lamarck a half-century before Darwin; but he made no attempt to show how such transmission is possible. Darwin, on the other hand, squarely faced the physiological requirements of the problem, recognizing that the transmission of acquired characters can only be possible under the assumption that the germ-cell definitely reacts to all other cells of the body in such wise as to register the changes taking place in them. In his ingenious and carefully elaborated theory of pangenesis,[1] Darwin framed a provisional physiological hypothesis

_____
[1] *Variation of Animals and Plants*, Chapter XXVII.

_____
From E. B. Wilson, *The Cell in Development and Inheritance*. Macmillan, New York. 1900.

of inheritance in accordance with this assumption, suggesting that the germ-cells are reservoirs of minute germs or gemmules derived from every part of the body; and on this basis he endeavoured to explain the transmission both of acquired and of congenital variations, reviewing the facts of variation and inheritance with wonderful skill, and building up a theory which, although it forms the most speculative and hypothetical portion of his writings, must always be reckoned one of his most interesting contributions to science.

In the form advocated by Darwin the theory of pangenesis has been generally abandoned in spite of the ingenious attempt to remodel it made by Brooks in 1883.[2] In the same year the whole aspect of the problem was changed, and a new period of discussion inaugurated by Weismann, who put forth a bold challenge of the entire Lamarckian principle.[3] 'I do not propose to treat of the whole problem of heredity, but only of a certain aspect of it,—the transmission of acquired characters, which has been hitherto assumed to occur. In taking this course I may say that it was impossible to avoid going back to the foundation of all phenomena of heredity, and to determine the substance with which they must be connected. In my opinion this can only be the substance of the germ-cells; and this substance transfers its hereditary tendencies from generation to generation, at first unchanged, and always unin-

---

[2] *The Law of Heredity*, Baltimore, 1883.

[3] *Ueber Vererbung*, 1883. See *Essays upon Heredity*, I., by A. Weismann, Clarendon Press, Oxford, 1889.

fluenced in any corresponding manner, by that which happens during the life of the individual which bears it. If these views be correct, all our ideas upon the transformation of species require thorough modification, for the whole principle of evolution by means of exercise (use and disuse) as professed by Lamarck, and accepted in some cases by Darwin, entirely collapses' (*l.c.*, p. 69).

It is impossible, he continues, that acquired traits should be transmitted, for it is inconceivable that definite changes in the body, or 'soma,' should so affect the protoplasm of the germ-cells as to cause corresponding changes to appear in the offspring. How, he asks, can the increased dexterity and power in the hand of a trained piano-player so affect the molecular structure of the germ-cells as to produce a corresponding development in the hand of the child? It is a physiological impossibility. If we turn to the facts, we find, Weismann affirms, that not one of the asserted cases of transmission of acquired characters will stand the test of rigid scientific scrutiny. It is a reversal of the true point of view to regard inheritance as taking place from the body of the parent to that of the child. The child inherits from the parent *germ-cell*, not from the parent-body, and the germ-cell owes its characteristics not to the body which bears it, but to its descent from a preëxisting germ-cell of the same kind. Thus the body is, as it were, an offshoot from the germ-cell (Fig. 5). As far as inheritance is concerned, the body is merely the carrier of the germ-cells, which are held in trust for coming generations. [pp. 11–13] . . .

Every discussion of inheritance and

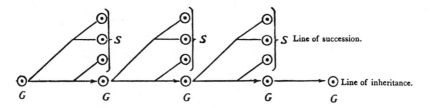

FIG. V. Diagram illustrating Weismann's theory of inheritance.

G. The germ-cell, which by division gives rise to the body or soma (*S*) and to new germ-cells (*G*) which separate from the soma and repeat the process in each successive generation.

development must take as its point of departure the fact that the germ is a single cell similar in its essential nature to any one of the tissue-cells of which the body is composed. That a cell can carry with it the sum total of the heritage of the species, that it can in the course of a few days or weeks give rise to a mollusk or a man, is the greatest marvel of biological science. In attempting to analyze the problems that it involves, we must from the outset hold fast to the fact, on which Huxley insisted, that the wonderful formative energy of the germ is not impressed upon it from without, but is inherent in the egg as a heritage from the parental life of which it was originally a part. The development of the embryo is nothing new. It involves no breach of continuity, and is but a continuation of the vital processes going on in the parental body. What gives development its marvellous character is the rapidity with which it proceeds and the diversity of the results attained in a span so brief.

But when we have grasped this cardinal fact, we have but focussed our instruments for a study of the real problem. *How* do the adult characteristics lie latent in the germ-cell; and how do they become patent as development proceeds? This is the final question that looms in the background of every investigation of the cell. In approaching it we may well make a frank confession of ignorance; for in spite of all that the microscope has revealed, we have not yet penetrated the mystery, and inheritance and development still remain in their fundamental aspects as great a riddle as they were to the Greeks. What we have gained is a tolerably precise acquaintance with the external aspects of development. The gross errors of the early preformationists have been dispelled. We know that the germ-cell contains no predelineated embryo; that development is manifested, on the one hand, by the cleavage of the egg, on the other hand, by a process of differentiation, through which the products of cleavage gradually assume diverse forms and functions, and so accomplish a physiological division of labour. We can clearly recognize the fact that these processes fall in the same category as those that take place in the tissue-cells; for the cleavage of the ovum is a form of mitotic cell-division, while, as many eminent naturalists have perceived, differentiation is nearly related to growth and

has its root in the phenomena of nutrition and metabolism. The real problem of development is *the orderly
sequence and correlation of these
phenomena toward a typical result.*
We cannot escape the conclusion that
this is the outcome of the organization of the germ-cells; but the nature
of that which, for lack of a better
term, we call "organization," is and
doubtless long will remain almost
wholly in the dark. [pp. 396–97] . . .

BIBLIOGRAPHY

Refer to the Preface for a list of classical papers appearing in other
anthologies.

ACKERKNECHT, E. H. 1953. *Rudolf Virchow, Doctor, Statesman, Anthropologist.*
Madison: University of Wisconsin Press.

BAKER, J. R. 1948–1955. 'The cell theory: a restatement, history and critique.'
Quarterly Journal of Microscopic Science. 89: 103–125; 90: 87–108, 331; 93: 157–
190; 94: 407–440; 96: 449–481.

BLUMBERG, JOE M., and many others. 1967. *The Billings Microscope Collection of
the Medical Museum Armed Forces Institute of Pathology.* Washington: American
Registry of Pathology.

BRADBURY, S. 1967. *The Evolution of the Microscope.* New York: Pergamon.

BRADBURY, S., and G. L'E. TURNER. Editors. 1967. *Historical Aspects of Microscopy.*
Cambridge: W. Heffer.

CHURCHILL, FREDERICK B. 1968. 'August Weismann and the break from tradition.'
Journal of the History of Ideas 1: 91–112.

COLEMAN, WILLIAM. 1965. 'Cell, nucleus, and inheritance: an historical study.' *Proceedings of the American Philosophical Society 109*: 124–158.

CONKLIN, EDWIN G. 1939. 'Predecessors of Schleiden and Schwann.' *American Naturalist
73*: 538–546.

CONN, H. J. 1928–1933. 'History of staining.' *Stain Technology 3*: 1–11, 110–121; 4:
37–48; 5: 3–12, 39–48; 7: 81–90; 8: 4–18.

DOBELL, CLIFFORD. (1930). *Antony van Leeuwenhoek and his 'Little Animals'.* New
York: Dover Publications. 1960 reprint.

EMBLEN, D. L. 1970. *Peter Mark Roget. The Word and the Man.* New York: Thomas
Y. Crowell.

'ESPINASSE, MARGARET. 1956. *Robert Hooke.* Berkeley: University of California Press.

GREW, NEHEMIAH. 1682. *The Anatomy of Plants, With an Idea of a Philosophical
History of Plants, and Several Other Lectures, Read before the Royal Society.*
London: W. Rawlins. Reprinted 1965 by Johnson Reprint Corporation, New York.

HERTWIG, OSCAR. 1895. *The Cell. Outlines of General Anatomy and Physiology.* New
York: Macmillan.

HOOKE, R. 1665. *Micrographia: or Some Physiological Descriptions of Minute Bodies
made by Magnifying Glasses with Observations and Inquiries Thereupon.* London:
Jo. Martyn and Ja. Allestry. Reprinted by Dover Publications, New York, in a fac-
simile edition (1961).

HUGHES, ARTHUR. 1959. *A History of Cytology.* New York: Abelard-Schuman.

HUXLEY, THOMAS HENRY. 1853. 'The cell-theory (review).' *British and Foreign Medico-
Chirurgical Review 12*: 285–314. Reprinted in *The Scientific Memoirs of Thomas*

Henry Huxley. Edited by Michael Foster and E. Ray Lankester. Volume 1. London: Macmillan. 1898. Pages 242–278.

HUXLEY, THOMAS HENRY. (1868). 'On the physical basis of life.' Reprinted in: *Lay Sermons, Addresses and Reviews* by Thomas Henry Huxley. London: Macmillan. 1891.

KARLING, JOHN S. 1939. 'Schleiden's contribution to the cell theory.' *American Naturalist 73*: 517–537.

KISCH, B. 1954. 'Forgotten leaders in modern medicine. III. Robert Remak 1815–1865.' *Transactions American Philosophical Society 44*: 227–296.

MAZZEO, JOSEPH ANTHONY. 1967. *The Design of Life. Major Themes in the Development of Biological Thought*. New York: Pantheon Books. Chapters IV and VII.

NICOLSON, MARJORIE. 1956. 'The microscope and English imagination.' In *Science and Imagination* by Marjorie Nicolson. Ithaca: Cornell University Press. Pages 155–234.

POWER, HENRY. 1664. *Experimental Philosophy, in Three Books: Containing New Experiments Microscopical, Mercurial, Magnetical*. London: T. Roycroft. The first book in English on the use of the microscope.

SCHLEIDEN, M. J. 1842. *Grundzüge der Wissenschaftlichen Botanik*. Leipzig. Translated by E. Lankaster in 1849 as *Principles of Scientific Botany*. London. See also next reference.

SCHWANN, THEODORE. 1847. *Microscopical Researches into the Accordance in the Structure and Growth of Animals and Plants*. Translated from the German by Henry Smith. London: Sydenham Society. Includes also 'Contributions to phytogenesis' by Schleiden.

SHADWELL, THOMAS. (1676). *The Virtuoso*. Edited by Marjorie Hope Nicolson and David Stuart Rodes. Lincoln: University of Nebraska Press. 1966. A satirical play poking fun at the discoveries of Robert Hooke and other members of the Royal Society.

SIRKS, M. J. 1952. 'The earliest illustrations of chromosomes.' *Genetica 26*: 65–76.

WILKIE, J. S. 1960. 'Nägeli's work on the fine structure of living matter. I. II.' *Annals of Science 16*: 11–42, 171–207.

WILSON, EDMUND B. 1896. *The Cell in Development and Inheritance*. New York: Macmillan. The second edition was published in 1900. The third, retitled *The Cell in Development and Heredity*, was published in 1925. The first edition of this classic was reprinted in 1966 by Johnson Reprint Corporation, New York.

WILSON, EDMUND B. 1899. *The Structure of Protoplasm. Biological Lectures Delivered at the Marine Biological Laboratory of Woods Holl in the Summer Session of 1897 and 1898*. Boston: Ginn & Co.

WILSON, J. WALTER. 1944. 'Cellular tissue and the dawn of the cell theory.' *Isis 35*: 168–173.

WILSON, J. WALTER. 1947. 'Dutrochet and the cell theory.' *Isis 37*: 14–21.

WILSON, J. WALTER. 1947. 'Virchow's contribution to cell theory.' *Journal of the History of Medicine 2*: 163–178.

WOODRUFF, LORANDE LOSS. 1939. 'Microscopy before the nineteenth century.' *American Naturalist 73*: 485–516.

WEISMANN, AUGUST. 1891–1892. *Essays upon Heredity and Kindred Biological Problems*. Two volumes. Edited by Edward B. Poulton and Arthur E. Shipley. Oxford: At the Clarendon Press.

WEISMANN, AUGUST. 1893. *The Germ-plasm. A Theory of Heredity*. New York: Scribners.

3 / Mendelism

Research in genetics made an abrupt change in 1900. In the last decade of the nineteenth century, many individuals were showing renewed interest in crossing varieties of plants and animals and, in contrast to earlier generations of hybridizers, more attention was paid to the inheritance of individual characters rather than to the general results. The main conclusion from the earlier days had been that the offspring were a blend of parental characters. When more attention was given to details, however, it became obvious that not all characters blended. Thus, if two variants were crossed, the offspring might be identical to one or to the other parent.

Some assumed that there existed two sorts of variation: continuous variation in which individuals of a population showed all gradations in the expression of a character—such as height. Continuous variation was associated with blended inheritance. The other type was discontinuous variation, in which the character usually existed in only two states—for example, pea seeds might be round or wrinkled. In the case of discontinuous variation, the variants maintained their distinctness in inheritance. When round and wrinkled peas were crossed, the offspring were round or wrinkled but never intermediate, that is, slightly wrinkled.

Research along these lines had progressed so far that it was becoming clear that, not only was discontinuous variation a fact, but the inheritance of these variations seemed to have a high degree of predictability. Two scientists, Hugo de Vries (1848–1935) and Carl Correns (1864–1933), working along these lines independently made enough observations to enable them to formulate a precise hypothesis to account for inheritance.

Shortly before they published their results, however, they found that they had been anticipated, both in experiment and hypothesis.

Their predecessor was an Austrian monk, Gregor Mendel (1822–84), who in 1866 had published a remarkable study of inheritance in garden peas (discussed in *Heredity and Development,* Chapter 3). His work was to become the foundation of modern genetics but it remained unknown to the general scientific community until 1900, when de Vries and Correns pointed out its great significance and confirmed its thesis.

Numerous individuals quickly saw that the precise rules of inheritance, which Mendel had found for peas, held for many other organisms. It appeared that inheritance was to become an exact science, after being a most confusing branch of biology for most of its history.

One of the most vigorous supporters of Mendel's theories was an Englishman, William Bateson (1861–1926). He was the crusader for Mendelism, to a degree reminiscent of Thomas Henry Huxley's vigorous support of the Darwinian theory of evolution. Bateson had been studying variation and inheritance for years and was, no doubt, as perplexed as all others who sought explanatory hypotheses. He realized that Mendel, and his confirmers—de Vries and Correns, had succeeded where all others had failed.

WILLIAM BATESON AND MISS E. R. SAUNDERS

In 1902, Bateson and his collaborator, Miss E. R. Saunders, presented a report to the Evolution Committee of the Royal Society of London. In it they discussed Mendel's work and presented the results of their own experiments.

EXPERIMENTAL STUDIES IN THE PHYSIOLOGY OF HEREDITY

. . . For many reasons we would rather have deferred publication until [our] work was further advanced, but we feel that with the re-discovery and confirmation of the principle which will henceforth be known as Mendel's Law, the study of heredity and the cognate problems of evolution must enter on a new phase.

At the present time the question how far Mendel's Law with its numerous corollaries is valid, to what cases and to what extent it is of general applicability, supersedes all others in significance. Consequently, our experience, however imperfect, since it bears directly on this question, has a value

From Reports to the Evolution Committee of the Royal Society. Report 1. Experiments Undertaken by W. Bateson, F. R. S., and Miss E. R. Saunders. London. Harrison & Sons, Printers in Ordinary to His Majesty. 1902. Reprinted by permission.

at this juncture, and we think it may be of use to the many investigators who will doubtless now turn their attention to the experimental study of heredity.

In the history of evolution it must ever be an astonishing fact that Mendel's discovery has so long remained unknown. The obscurity of the Brünn journal is a circumstance quite insufficient to explain the neglect of such a work.* No careful reader could doubt for a moment that he was in presence of true facts of exceptional significance. The reason may be that at the

---

* 'Abh. naturf. Ver. Brünn,' 1865, vol. 4, p. 1. The author of this remarkable paper, Gregor Johann Mendel, according to a communication made to Correns by Dr. von Schanz, was born in 1822 at Heinzendorf-bei-Odrau in Austrian Silesia. His parents were well-to-do peasants. In 1843 he entered the Königinkloster, an Augustinian house in Altbrünn and was ordained priest in 1847. From 1851–1853 he studied physics and natural sciences at Vienna. On retiring to his cloister he became a teacher in the Realschule of Brünn and afterwards Abbot. He died in 1884. His experiments were made in the garden of his cloister. Besides this paper he published a brief account of experiments with *Hieracium* in the same Journal, vol. 8, 1869, p. 26, and also various observations of a meteorological character. In 1853 and 1854 he published brief notes on *Scopolia margaritalis* and *Bruchus pisi* in the 'Verh. zool. bot. Ver. Wien,' but apparently no others dealing with natural history. Professor Correns informed us that he is said to have spent the later years of his life in the Ultramontane controversy. The Brünn Society, in 1865, was exchanging publications with most of the Academies of Europe, including the Royal and Linnean Societies.

time of Mendel's publication, the announcement of the principle of Natural Selection had almost completely distracted the minds of naturalists from the *practical* study of evolution. The labours of the hybridists were believed to have led to confusion and inconsistency, and no one heeded them any more.

It is, perhaps, even more surprising that other investigators failed to discover the same law. Naudin's conclusions came very near it.

The literature of breeding teems with facts now palpably Mendelian. Gärtner, Godron, Laxton, even Darwin himself, must have been many times of the brink of the discovery.

Looking now at such experiments as those of Rimpau with wheat, &c., of Laxton with *Pisum*, Godron with *Datura*, of Darwin with Antirrhinum and sweet peas, we can hardly understand how the conclusion was missed. In Darwin's case no doubt the theory of Pangenesis contributed to difficulty, but some part of the obscurity must have been due to the habit of regarding various species, breeds, varieties, and casual fluctuations as all comparable expressions of one phenomenon, similar in kind; and to insufficient recognition of the possibility that variation may be, in its essence, specific. We may, perhaps, attribute to this preconception the extraordinary complexity with which his experiments on poultry and pigeons were arranged. Various breeds and various crosses were mixed together, and the results are not unlike those which the early chemists would have arrived at in testing the affinities and constitution of a number of unknown elementary and compound substances mixed together at random.

Mendel's Law

In order to understand what follows it is absolutely necessary that the reader should have a general acquaintance with the present state of knowledge in regard to the Mendelian principle. The original paper is not easily accessible, but it has been reprinted in 'Flora,' 1901, and is about to appear in Ostwald's 'Klassiker der exakten Wissenschaften.' An English translation was published in the 'Journ. Roy. Hortic. Soc.,' 1901. We venture, however, to give a brief outline of the history and progress of the discovery, which is the more necessary as our own views and modes of expression do not agree wholly with those of other writers on the subject.

It may be premised that the first publication of the re-discovery was made in 1900 by de Vries,* and almost simultaneously by Correns† and Tschermak.‡ There can be no doubt that the appearance of this group of papers constitutes at length a definite advance both in the general study of the physiology of reproduction and in the particular problem of the nature of Species.

In 1865 Gregor Mendel published his discovery. His experiments are described in considerable detail, and the resulting law is stated with emphasis and precision. He at once perceived and lucidly enunciated what he regarded as the essential truth underlying the observed facts. A reader of this paper can hardly fail to recognise its masterly quality. By some strange chance it was ignored, and excepting a casual reference in Focke's 'Pflanzen-Mischlinge' (p. 110) there appears to be no allusion in literature to this remarkable performance. It is a fortunate circumstance that we need feel no hesitation in now accepting Mendel's account; for even if the original paper were such as to admit of doubt, the re-discovery comes to us with a large body of fresh evidence, the simultaneous work of three independent observers, confirming Mendel on the main points.

Mendel states that, like other investigators, he had been struck by the regularity with which offspring of certain hybrids reproduce the pure ancestral forms. But owing, as he supposes, to the complex nature of the cases studied and to want of accurate statistics, the precise facts had never been ascertained. Accordingly he set himself to work out some case from which every confusing element should as far as possible be excluded. After several trials he chose the varieties of *Pisum sativum* as best suited to his purpose. Besides other advantages, this species is well known to have the exceptional merit of being habitually self-fertilised, in N. Europe at least. From the many varieties of peas he first chose pairs of varieties, for crossing, in such a way that the members of each pair differed from each other in respect of one definite character. Of such pairs of characters he chose seven, namely: 1. Shape of seed, whether rounded *or* irregularly angular and deeply wrinkled. 2. Colour of cotyledous ['*endosperm*'], whether

---

* De Vries, 'Comptes Rendus,' March 26, 1900, and 'Ber. deut. Bot. Ges.,' xviii, 1900, p. 83; *ibid.*, p. 435; 'Rev. gén. Bot.,' 1900, p. 257.

† Correns, 'Ber. deut. Bot. Ges.,' xviii, 1900, p. 158; 'Bot. Ztg.,' 1900, p. 232.

‡ Tschermak, 'Ztschr. f. d. landw. Versuchswesen in Oesterr.,' 1900, 3, p. 465.

some shade of yellow *or* green. 3. Colour of seed-skin, whether a brownish shade *or* white (in correlation with white flowers). 4. Shape of ripe pod, whether simply inflated *or* deeply constricted between the seeds. 5. Colour of unripe pod, whether a shade of green *or* bright yellow. 6. Position of flowers, whether distributed along the stem *or* crowded near the top in a false umbel. 7. Length of stem, whether about 6–7 feet *or* about ¾–1½ feet. Between these various pairs of varieties crosses were then made, the female parent being emasculated.

As to the first two characters the result of the experiment is seen as soon as the cross-bred seeds are ripe; but to study the other five characters it is of course necessary to plant the seeds and grow the plants to maturity. On such examination it was found that in the case of each pair of characters one only was manifested ·in every cross-bred individual, to the total, or almost total, exclusion of the opposite character. The character which thus prevails is, in Mendel's terminology, the *dominant* (D), the character which is suppressed, being *recessive* (R). In the above enumeration of characters the dominant is placed first in each pair. Reciprocal crosses gave identical results. Briefly, then, D × R or R × D gave offspring which in appearance are all practically D.

The next generation is obtained by allowing the cross-bred plants DR to fertilise themselves. The result of such self-fertilisation is, according to Mendel, that the next generation instead of being uniform like their parents, breaks up into the two original forms. This takes place in such a way that there are on the average *three dominants to one recessive.*

The recessives are thenceforth not only apparently but *actually pure,* and if allowed to fertilise themselves give rise to recessives only, for any number of generations.

The dominants, on the contrary, though they may look alike, can, if allowed to fertilise themselves, be shown to consist of (*a*) individuals which are *pure* dominants, and on self-fertilisation give rise to dominants only; and (*b*) individuals which are cross-breds, and give rise on self-fertilisation to a mixture of dominants and recessives, *again in the proportion of three to one as before.* The *pure* dominants (*a*) are to the cross-bred dominants (*b*) as *one to two.*

The whole result, therefore, of self-fertilising the cross-breds is really 1DD : 2DR : 1RR, though apparently the proportion is 3D : 1R, because DR plants are not to the eye distinguishable from pure D.* The mixed offspring obtained by self-fertilising DR plants of this second generation consist again of D, DR, and R in the same proportions, and so on at each succeeding generation.

Mendel worked with considerable numbers, and it is impossible to doubt the general accuracy of his conclusions. In the case of some of the pairs of characters (notably the 5th) the evidence is much less clear than in the case of others, and it seems possible that in such cases unascertained disturbances may be at work; but on the whole the results, *as average results,* are clear. The purity of the pure domi-

---

* The attempt made by de Vries to indicate this result by an algebraical expression appears to us open to objection, and we prefer the simpler notation of Mendel.

nants and pure recessives resulting from the self-fertilisation of the cross-breds was tested in the case of characters 1 and 2 to the sixth generation, for characters 3 and 7 to the fifth generation, and for characters 4, 5, and 6 to the fourth generation.

Mendel interprets his facts as follows:—They point to the conclusion that in the cross-bred each of its pollen-grains and each of its egg-cells is *either pure dominant or pure recessive,** and that on the average there are equal numbers of each kind for each sex. If the assortment of pollen-grains and egg-cells is then supposed to take place at random, the most probable percentage result is 25D ♀ × D ♂, 25D ♀ × R ♂, 25R ♀ × D ♂, and 25R ♀ × R ♂. As D × R does not differ from R × D, we have therefore 25D : 50DR : 25R. But as cross-breds resemble dominants in appearance, the *apparent* result is 3D : 1R at each successive self-fertilisation of cross-breds.

Mendel next proceeded to cross pairs of varieties differing in respect of *two* characters; for example, a variety having seeds round and yellow with a variety having seeds angular and green; and also pairs of varieties differing in *three* characters; for example, a variety characterised by seeds round, yellow, and in grey-brown skins with a variety having seeds angular, green, and in white skins. The numerical results of these experiments are set out at length, and are too long to repeat. Briefly the result in all cases was that the dominant characters alone appeared in the first cross. When these were self-fertilised, in the case of *two* differentiating characters, *four* different kinds of seeds resulted, namely, round yellow, angular yellow, round green, and angular green. In the case of *three* differentiating characters these were similarly combined in the *eight* possible combinations. In both sets of experiments the numbers of individuals and their constitution, as tested by the seeds they were capable of producing on self-fertilisation, were consistent with the hypothesis arrived at in the case of varieties differing in respect of one character, namely, that each male and female cell of the cross-bred is pure in respect of one character of each pair of characters, and is capable of transmitting this character to the exclusion of the opposite character; that the reproductive cells are, in the cross-breds, of as many kinds as there are possible combinations of pure characters (taken two or three together, as the case may be); and, finally, that each kind is represented in the cross-breds on the average in equal numbers.

Mendel next tested the truth of the hypothesis that both male and female cells were similarly differentiated, by crossing the first crosses with pure D and pure R forms respectively, finding, as he expected, that DR × D gives the ratio 1DD : 1DR, and that DR × R gives on the average the ratio 1DR : 1RR. Both sets of germ-cells are therefore similarly differentiated.*

It must be understood that this con-

---

* In what follows such forms are spoken of as 'extracted' dominants, or 'extracted' recessives.

* As it happens, almost all our breedings of cross-breds with pure types have been in the form cross-bred ♀ × pure ♂, but reciprocal experiments are in progress.

densed statement does no justice to the lucidity and completeness of Mendel's account of his facts, and of the reasoning based on them. With regard to the numbers, it may be said that though there is a good deal of irregularity, yet taken together they plainly bear out the law as a statement of average results.

De Vries[†] working with pairs of varieties belonging to a diversity of genera and species, found that in a large number of cases one of the varietal characters was definitely dominant, prevailing in the first crosses to the exclusion of the recessive character. In several of these cases the offspring of the cross-breds fertilised *inter se* were mixed dominants and recessives in proportions fairly agreeing with Mendel's law. In the case of two colour varieties of *Papaver somniferum*, the constitution of the resulting dominants was investigated, and shown to be also according to the law. In certain cases also the purity of the recessives was tested and found to be complete.

De Vries' first announcement was followed almost immediately by the appearance of a paper by Correns,[‡] giving an account of some years' work with peas, repeating Mendel's experiments and confirming them as regards the colour of the seeds.

Correns stated that he has tested the purity of the pure dominants and pure recessives to the third generation, and he has kindly informed us that a fourth generation is equally pure.

In addition to these, Tschermak

has given an account[§] of simultaneous investigations also carried out on varieties of peas. This last paper is an elaborate memoir giving in addition valuable information on several points not directly relating to the present subject.

Tschermak worked with many varieties of peas, and though he obtained several results not wholly consistent with Mendel's law, and some actually conflicting evidence, the general tenour of his work is confirmatory. He gives, in particular, results as to the crossing of cross-breds with one or other of the pure forms.

It is an obvious corollary from Mendel's laws that the cross-bred crossed with the recessive parent should give seeds of the dominant and recessive colours in equal numbers. On the other hand, the cross-bred crossed with the dominant should give dominants only (of which half should be pure and half cross-bred in composition). Tschermak's numbers though insufficient for a thorough test, are in fair harmony with Mendel's hypothesis. He believes also that there is evidence that yellow is more decidedly dominant over green than the rounded character is over the wrinkled, and on this point further experiments are required.

Subsequently a paper by Correns[*] has appeared describing his experiments with a glabrous and a hoary form of garden stock (*Matthiola incana*), and giving results as to these varieties, tested by self-fertilisation,

---

† C.R., 1900, March 26; 'Ber. deut. Bot. Ges.,' 18, 1900, p. 83.

‡ 'Ber. deut. Bot. Ges.,' 18, 1900, p. 158.

§ 'Ztsch. f. d. landw. Versuchswesen in Oesterr.,' 3, 1900, p. 465; continued later, *ibid.*, 4, 1901, p. 641, and 'Ber. deut. Bot. Ges.,' 19, 1901, p. 35.

* 'Bot. Cblt.,' 1900, 43, p. 97.

and also by recrossing the cross-breds with the parental forms. In this, as in each of the other papers, there are many points which call for separate notice and discussion, but the facts taken together are in fairly close agreement with the expectation given by Mendel's law, though the discrepancies are decidedly greater in this case than in the others. The case happens to be one which had also formed a subject of our experiments, and the various questions raised by the facts will be discussed in connection with the statement of our results.

Lastly, Professor Correns has also published† an elaborate and important memoir on the results obtained in crossing varieties of maize. Some of these results are of a complex character, but the essential fact of the truth of Mendel's law in its application to many of these cases was fully established. From the striking differences between these varieties in several characters, notably in the constitution of the endosperm, whether round starchy (dominant) or wrinkled and sugar-containing (recessive), this case is a most attractive subject for such experiments.

In the characters of the endosperm in maize de Vries‡ also demonstrated the truth of Mendel's law. As regards the colour varieties, Correns showed many complications to exist owing to the fact that the general appearance of the seed depends partly on maternal and partly on embryonic elements; and in some of the characteristics there are other confusing factors owing to fluctuations in the intensity of the dominance.

The above is a brief sketch of a vast mast of observations all tending to the same conclusion, and the truth of the law ennunciated by Mendel is now established for a large number of cases of most dissimilar characters, beyond question.

Let us now shortly distinguish what is essential in the new discovery from what is not. The fact that in the cross-bred one character, in appearance, dominated to the exclusion of the other is not of the essence of Mendel's discovery. Tschermak, for instance, saw some exceptions to this rule in *Pisum*. Among the *Matthiola* crosses here described, and among those originally made by Trevor Clarke,* the hoary form *seems* to be not exclusively dominant. Correns saw considerable fluctuations in the dominance of the colours of maize. It is practically certain some of the poultry cases about to be described exhibit the same phenomenon. In fact, generally speaking, there are good reasons for thinking that in numerous instances purity of gametes may occur without either character exhibiting dominance.

There is also some reason for supposing that alterations in dominance may be affected by changes in conditions [see page 101], and possibly even by differences in the state of the parents, though the only considerable body of evidence relating to these points, obtained by Vernon† in

---

† 'Biblioth. Bot.,' 1901, Heft 53.

‡ 'Rev. gén. Bot.,' 12, 1900, p. 270.

---

* 'Rep. Intern. Hort. Exhib. and Bot. Congr.,' 1866, p. 142. As the sequel will show, this appearance is otherwise explicable.

† Vernon, H. M., 'Phil. Trans.,' B, 1898, vol. 164, p. 465.

Echinid crossing, still needs repetition and confirmation.

The essential part of the discovery is the evidence that *the germ-cells or gametes produced by cross-bred organisms may in respect of given characters be of the pure parental types and consequently incapable of transmitting the opposite character:* that when such pure similar gametes of opposite sexes are united together in fertilisation, the individuals so formed and their posterity are free from all taint of the cross; *that there may be, in short, perfect or almost perfect discontinuity between these germs in respect of one of each pair of opposite characters.*

Doubt might naturally be felt as to the acceptance of a proposition so far-reaching, but it is impossible to see any other possible interpretation of the facts. The pure dominant and pure recessive members of each gen-

eration are not merely like, but identical with the pure parents,‡ and their descendants obtained by self-fertilisation are similarly pure. If they are pure, surely the male and female elements of which they were composed must also be pure.

In another part of this Report we propose to discuss the bearing of these remarkable facts, and to show some of the conclusions to which they point. It will be seen that they provide satisfactory interpretations of some of those redoubtable paradoxes which have hitherto been so mysterious to all evolutionists.

We may now proceed to the description of our own experiments. [pp. 5–12] . . .

_____

‡ Correns, in maize, has found an apparent change in the degree of dominance of certain characters in the "extracted" variety.

Mr. Bateson and Miss Saunders return to general considerations after describing their experiments, most of which could be explained in Mendelian terms.

PART III.—THE FACTS OF HEREDITY IN THE LIGHT OF MENDEL'S DISCOVERY.

As was stated in the introduction to this paper, with the discovery of the Mendelian principle the problem of evolution passes into a new phase. It is scarcely possible to overrate the importance of this discovery. Every conception of biology which involves a knowledge of the physiology of reproduction must feel the influence of the new facts, and, in their light, previous ideas of heredity and variation, the nature of specific differences, and all that depends on those ideas must

be reconsidered, and in great measure modified.

If we turn to any former description of breeding experiments we generally perceive at once that the whole account must be re-stated in terms of Mendel's hypothesis, and that the discussions and arguments based on former hypotheses are now meaningless. As an illustration we may take the account which Darwin gives of his experiments with peloric Antirrhinum.* He crossed the peloric form

_____

* [Darwin], 'Animals and Plants,' vol. 2, p. 46, ed. 1885.

with the normal and *vice versâ*. The first crosses were all indistinguishable from the normal or zygomorphic form. These were allowed to fertilise themselves, and gave a crop consisting of 88 normals, 2 intermediates, and 37 perfectly peloric. He discusses these results on the hypothesis that the normal plant has a 'tendency' to become peloric, and the peloric a 'tendency' to become normal, 'so that we have two opposed latent tendencies in the same plants. Now with the crossed Antirrhinums the tendency to produce normal or irregular flowers, like those of the common Snapdragon, prevailed in the first generation; whilst the tendency to pelorism, appearing to gain strength by the intermission of a generation, prevailed to a large extent in the second set of seedlings. How it is possible for a character to gain strength by the intermission of a generation will be considered in the chapter on pangenesis.'

Now, of course, we can perceive that the zygomorphic form is dominant and the peloric recessive, and that the arguments based on other hypotheses have no longer any significance. It would be a useful task to go similarly through the literature of breeding and translate the results into Mendelian terms. Such an exercise would show that the change which must now come over the conceptions of biology can only be compared with that which in the study of physical science followed the revelations of modern chemistry.

The outcome of such a revision of current conceptions it is impossible to foresee, but we propose in the present paper to consider some of the more important questions which are immediately raised.

To denote the new conceptions some new terms are needed. Several have already been suggested by Correns, but in practice we have not found his terminology altogether convenient, or that it meets the new requirements. Correns proposes the terms 'heterodynamous' and 'homodynamous' to express that an organism is dominant or not dominant in respect of a given character. There are unfortunately objections to the use of these terms, though in some respects they are very suitable. First, they are in use by Weismann and his followers in quite different senses, as Correns states. Secondly, it is not clear whether they are to be applied to the variety, the individual, or the character. Besides these objections, it is fairly clear that dominance is a phenomenon presenting various degrees of intensity; and while the single phenomenon of dominance is well expressed by that word itself, other conditions probably consist of various phenomena which are not conveniently denoted by one word.

Correns' terms 'homöogonous' and 'schizogonous' cannot as yet be used with precision to mean more than breeding 'true' and not breeding 'true,' and, for reasons given later, the metaphor of splitting may be incorrect.

The terms also '*halb-identisch*' and '*conjugirte*' as applied to characters, are already fairly well expressed by the words in perfect or in imperfect correlation, which are already well understood. It would be confusing to introduce the metaphor of conjugation to denote these ideas.

But while doubting whether this terminology already suggested will be found adequate, we do not propose at present to substitute new terms for the same phenomena. In our view, there are other conceptions arising from the Mendelian discoveries for which brief expressions are absolutely required, and for these we suggest the following terminology.

In the introduction [p. 85] we attempted to distinguish precisely the essential fact discovered by Mendel, and to separate it from other subordinate appearances. We may now briefly recall and amplify that reasoning, showing how we propose to denote the several phenomena.

By crossing two forms exhibiting antagonistic characters, cross-breds were produced. The generative cells of these cross-breds were shown to be of two kinds, each being pure in respect of *one* of the parental characters. This purity of the germ-cells, and their inability to transmit both of the antagonistic characters, is the central fact proved by Mendel's work. We thus reach the conception of unit-characters existing in antagonistic pairs. Such characters we propose to call *allelomorphs,** and the zygote formed by the union of a pair of opposite allelomorphic gametes, we shall call a *heterozygote.* Similarly, the zygote formed by the union of gametes having similar allelomorphs, may be spoken of as a *homozygote.* Upon a wide survey, we now recognise that this first principle has an extensive application in nature. We cannot as yet determine the limits of its applicability, and it is possible that many characters may really be allelomorphic, which we now suppose to be 'transmissible' in any degree or intensity. On the other hand, it is equally possible that characters found to be allelomorphic in some cases may prove to be non-allelomorphic in others.

It will be of great interest to determine how far the purity of the germ-cells in respect of allelomorphic characters is an absolute rule, or whether there are exceptional cases in which such purity may be impaired. That such exceptions may arise is indeed almost certain from the evidence of 'mosaic' fruits in *Datura*, where it was shown . . . that the otherwise pure extracted recessives (thornless) showed exceptionally a thorny patch or segment. Unless this is an original sport on the part of the individual, such a phenomenon may be taken as indicating that the germ-cells may also have been mosaic.*

Indeed all that we know of the occurrence and distribution of variation among repeated parts, would lead us to expect such a possibility with confidence.

This is a question we can analyse no further. Were it possible to do so, it might be a real help towards getting a picture of the actual process of heredity.

But besides the strictly allelomorphic or Mendelian distribution of characters among the gametes (with or without mosaics), we can imagine three other possible arrangements. (1) There may be a substantial discontinuity, the two types of gamete being connected by a certain proportion of

---

* Correns speaks of the two opposite allelomorphs as a *'Paarling.'*

---

* Conceivably the cases of poultry having one foot with extra toe and one normal, may be of a similar nature, though for various reasons this is unlikely.

intermediates, such as are often met with in cases even of almost complete discontinuity among zygotes. (2) There may be continuous variation among the gametes, shading from gametes pure to the one type, to gametes pure to the other type, the intermediates being the most frequent. (3) There may be no differentiation among the gametes in respect of parental characters at all, each representing the heterozygote characters unresolved. This last is the homoögonous type of Correns. By a sufficiently wide survey, illustrations of each of these systems and of intermediates between them, will doubtless be found, and the classification of gametic differentiation according to these several types, in respect of various characters, in various species, will be a first step towards the construction of a general scheme of heredity.

In gametic variation we thus meet in fact the same series of possibilities with which we have been familiar in the variation of zygotic organisms.

The second fact observed by Mendel is that each heterozygote produces on an average equal numbers of gametes bearing each allelomorph of each pair. This is only enunciated as an *average* result. Unfortunately, the determinations of the results for individuals are still few, but from those that have been made, and even from the few recorded by Mendel himself . . . . we see that the fluctuations are so great, that we must suspect some special sources of disturbance. Contributing to the average result of 3:1 as between round and wrinkled peas, he mentions as extremes 43 : 2, 14 : 15; and between yellow and green 20 : 19 and 32 : 1. It is obvi-

ous that this suggests either that there has been for some cause selection among the germ-cells originally equal in numbers, or that the numbers were originally unequal, or that the assortment of male and female germs was not governed by pure chance. Probably a series of individual determinations when seriated would throw light on the nature of these remarkable fluctuations which have been observed in almost all the subjects studied. From what we already know, in respect of the output of the two kinds of gametes, it is fairly certain that fluctuations take place, corresponding probably with changes in health, age, and other conditions.

From analogy—an unsafe guide in these fields—and from what is known of discontinuous variation in general, we incline to the view that even though the figures point to a sharp discontinuity between dominant and recessive elements, we shall ultimately recognise that the discontinuity between these elements need not be *universally* absolute. We may expect to find individuals, and perhaps breeds or strains, and even individual gonads or groups of gonads, in which the discontinuity is less sharp even in respect of these very characters; similarly, for such units definite departures from statistical equality between D and R germs may be expected. In *Pisum*, for instance, we cannot be far out in considering an average of 50 per cent. D and 50 per cent. R as a close approximation to the truth for both male and female cells, but there is nothing yet which proves even here that the discontinuity *must be always and absolutely complete.*

Similarly, we are not compelled to accept the proposition that germ-cells of each allelomorph *always* exist on an average in equal numbers. The proofs of the two propositions are unfortunately as yet interdependent. The purity of the extracted recessives and dominants has been tested, and we can in such cases accept it as a fact: the *universal* purity of the gametes we cannot test. For, any dominant which gives rise to a recessive offspring, we should class as a cross-bred, because cross-breds are like dominants in appearance. Similarly, any partially impure recessive would be classed as a cross-bred. If the number of germs of each kind borne by the cross-bred is sensibly unequal, or the discontinuity between them sensibly lessened, we can perceive a result, but we shall not know to which cause to ascribe it. The statistical method unfortunately cannot distinguish between the two causes in such a case. Readers of Mendel's paper will be aware that he laid down no universal rule as to the absolute purity of gametes, but merely pointed out that his results were explicable on the hypothesis of such purity.

The statistics, however, are not so precise as to compel us to accept *both* that the germs of the cross-breds are *always* pure, and that they are *always* produced on an average in equal numbers.

The next point arising immediately out of Mendel's work concerns the characters of the heterozygote. In the *Pisum* cases the heterozygote normally exhibits only one of the allelomorphs clearly, which is therefore called the dominant. It is, however, clear from what we know of cross-breeding, that such exclusive exhibition of one allelomorph in its totality is by no means a universal phenomenon. Even in the pea it is not the case that the heterozygote always shows the dominant allelomorph as clearly and in the same intensity as the pure dominant, and speaking generally, heterozygotes, though in numerous instances readily referable to one or other of the allelomorphic types, exhibit those types in a more or less modified form.

Besides these, there are undoubtedly cases in which the heterozygote may show *either* of the allelomorphs, though one is commonly dominant. In the poultry crosses it was shown that the usually recessive foot-character (want of extra toe) may appear in the cross-bred. The want of dominance of hoariness in *Matthiola* seen in exceptional cases is a wholly different phenomenon.

From the analogy of poultry, it is scarcely doubtful that polydactylism in man is also allelomorphic to the normal, and here from the tables of heredity already recorded,* there is good evidence that both the normal and the polydactyle offspring of one polydactyle parent can transmit the polydactylism; in other words, the heterozygote may exhibit either allelomorph. Cases of the same phenomenon can indeed be multiplied. It must, however, be remembered that what is accepted as evidence of alternative inheritance, is not a proof that the dominance of either allelomorph is imperfect. This can only be known for certain when it has already been established that individuals showing either of the two allelomorphs can, when mated with an individual show-

---

* For examples see Fackenheim, 'Jen. Zt,' xxii, p. 343.

ing the same allelomorph, produce both allelomorphs among their off-spring.[†]

This leads to a point of great importance to the evolutionist. We have been in the habit of speaking of a variation as discontinuous, in proportion as between it and other forms of the species intermediates are comparatively scarce when all breed freely together. In all cases of allelomorphic characters we can now give a more precise meaning to this description. It must now be recognised that such a population consists, in respect of each pair of allelomorphs, of *three* kinds of individuals*, namely, homozygotes containing one allelomorph, homozygotes containing the other allelomorph, and heterozygotes compounded of both. The first two will thus always form discontinuous groups, and the degree to which the heterozygotes form a connecting group, will depend on whether one allelomorph regularly or chiefly dominates in the heterozygote, or the allelomorphic characters completely or partially blend in the heterozygote. *Such discontinuity will in fact primarily depend not on the*

[†] For the present, therefore, we are not entitled to assume that the numerous cases among Lepidoptera of varieties breeding together with a discontinuous mixed result are allelomorphic, probable as this conclusion is. Such cases are those of *Amphidasys betularia* and *doubledayaria; Aglia tau* and *lugens; Angerona prunaria* and *sordiata; Miana strigilis* and *œthiops*, &c. See Standfuss, 'Handb. d. pal. Gross-Schmetterl.,' 1896, p. 305, *et seq.*

[*] Four, if reciprocal heterozygotes are not identical.

blending or non-blending of the characters, as hitherto generally assumed, but on the permanent discontinuity or purity of the unfertilised germ-cells.

It will be of great interest to study the statistics of such a population in nature. If the degree of dominance can be experimentally determined, or the heterozygote recognised, and we can suppose that all forms mate together with equal freedom and fertility, and that there is no natural selection in respect of the allelomorphs, it should be possible to predict the proportions of the several components of the population with some accuracy. Conversely, departures from the calculated result would then throw no little light on the influence of disturbing factors, selection, and the like.

From the circumstance that dominance of either character is no essential accompaniment of allelomorphism, it must be determined whether the proportions of the two kinds of gametes produced by the heterozygote will vary with its individual character. Bearing on this question the experiments are very few. The determination from statistical study of zygotes must be exceedingly difficult, seeing that *both* resulting forms may be heterozygous. The ratio in which the heterozygotes are distributed in the second generation need not be the same as it was in the first, and unless this can be determined it will be almost impossible to get further with this particular inquiry.

Another difficulty will be found in the possibility that when the first cross-bred generation gives a mixture, the forms showing the usually recessive character (both in this and subsequent generations) may be *pure* recessives as regards their own gametes also (false hybrids of Millardet) [see p. 117] though heterozygous in origin. To solve these difficulties before the gametes can be microscopically differentiated may be still impossible.

We have now simple and convincing explanations of many facts hitherto paradoxical.

1. *Heterozygous Forms.*—It has long been known to breeders that certain forms cannot be fixed by selection indefinitely continued. In other words, when the most perfect examples of such forms are bred together, though they produce some offspring like themselves, they have also a large number which do not resemble them.

A case of this kind is seen in breeding crested canaries. The kind of crest desired for exhibition can, according to canary-fanciers, be produced most easily by mating crested birds and non-crested, or plain-heads as they are called. If it be supposed that the crested character is usually dominant, we have a simple explanation. When crested birds are bred together a number of birds are produced whose crests are coarse and stand up and others without crests. The latter are the recessives; the former we may suppose to be the pure dominants. What the fancier wants is a crest composed of long feathers lying evenly down over the head. These may be the heterozygotes, and consequently cannot breed true or be fixed by selection. Such birds bred together, give many plain-heads and birds with coarse crests. Fanciers hold that the plain-heads needed for crest-breeding should be themselves crest-bred, *i.e.*, from families which have had crests among them. On the view here suggested this

is probably a superstition, though one can easily see how it may have arisen.

If two crested birds are bred together it is advised that they should have imperfect crests, in all probability another form of the heterozygote.*

Another case, to which our attention was called by Mr. G. Thorne, of Broxbourne, is that of the Golden Duckwing Game Fowl. This colour can be produced by crossing Black-Reds with Silver Duckwing; but on attempting to breed the Golden Duckwing true, the colour breaks up again into its components.†

Probably the impossibility of fixing certain colours in Pigeons also illustrates the same phenomenon.

Such forms have hitherto been regarded as exhibiting 'instability.' Of this instability there is now a satisfactory account.

A more complex instance of this may be the Andalusian fowl. The colour is a blue-grey mixed with dull black. The breed will not continue true to color. Though a considerable proportion of Andalusians are produced, a number will be hatched which are too dark or too light in various ways and proportions. Selecting the best Andalusians effects nothing, and the constancy does not increase. There is, therefore, a strong

---

* An account of these facts is given in Blakiston, Swaysland, and Wiener's 'Canaries and Cage Birds,' p. 128. When birds with good crests are bred together the recessive 'plain-head' is often produced, a fact which has been exaggerated by various writers into the statement that the offspring of cresteds are *always* plain-heads, or even always *bald*.

† See also Lewis Wright, 'Book of Poultry,' 1886, pp. 289 and 356.

probability that the Andalusian is a heterozygote, though, doubtless, of a complex nature [*cf.* p. 108]. Its gametes do not fully correspond to it, and its colour must be produced by a combination of dissimilar allelomorphs.

A point of great practical and theoretical importance would be the determination whether the increased vigour so commonly observed in the offspring of some crosses is or is not correlated with the union of dissimilar allelomorphs. Hitherto we have spoken of all the offspring of crossing as 'crosses,' alike. We must now recognise that when heterozygotes are bred together their offspring *may not be crosses at all*. The great vigour seen in the first cross is known not rarely to decline in the next generation bred from them, and it may be possible to see whether such vigour was in reality associated with the union of any recognisably dissimilar allelomorphs.

The existence of forms which are exclusively heterozygous leads to the contemplation of another possibility. In the heterozygotes we have spoken of, both sexes of course bear gametes transmitting each allelomorph. If, however, one allelomorph were alone produced by the male and the other by the female we should have a species consisting *only* of heterozygotes.

So long as the heterozygotes bred together, the offspring in such a case would come true, but a proof that they were heterozygotes would be obtained by crossing them with another species or variety. It would then be found that reciprocal crosses would not give the same result. That this is actually the case we know in certain instances, of which the most familiar amongst animals is perhaps that of the Mule (Mare × Jackass) and the Hinny (She-ass × Stallion),* and amongst plants the hybrids of *Digitalis*.† In most treatises on crossing other cases are referred to, and though probably many of them are based on experiments insufficiently repeated, there can be no doubt many are authentic. Gärtner‡ acutely observes that the phenomenon of dissimilarity between the results of reciprocal crosses is more likely to be found among diœcious forms.

2. *Selection and the Phenomenon of Dominance.*—We have seen that the want of fixity in certain forms, though continually selected, may at once be explained by the hypothesis that they are heterozygous only, and have no gametes corresponding to them. Another illustration of the failure of selection is the constant recurrence of a particular 'rogue' in the best strains. Seed is never taken from such rogues. Every year they may be pulled up as soon as detected, but they continually reappear.

The hypothesis that such a 'rogue' is a recessive form *may* give a complete explanation of this phenomenon in many cases. Selection from *individuals* of known fertilisation would at once test the truth of this view, and might provide a means of producing

---

* A good description of the differences between these forms is given by Cornevin, 'Traité de Zootechnie,' 1891, p. 641.

† See Focke, 'Pflanzenmischlinge,' 1881, p. 322; and Gärtner, 'Bastarderzeugung,' 1849, p. 225. Other examples are given by Gärtner, *ibid.*; and by Swingle and Webber, 'Year-book Dept. Agric.,' 1897, p. 401.

‡ *Loc. cit.*, p. 228.

a pure strain once and for all from the pure dominants.

It is well known that some of the best modern beardless wheats which have been raised of late years by crossing distinct varieties will give a small proportion of bearded plants. This is, of course, called 'reversion' to a bearded ancestor used in the original cross.

From the experiments of Rimpau,* we find that when bearded and beardless varieties are crossed, beardlessness is dominant, and the bearded character is recessive. By subsequent breeding a form is produced with a desirable character, and after a few years of selection it is found to give this character with sufficient purity and it is put on the market. It may be a bearded or a beardless form, but if the latter, the chances are that it will always produce a certain proportion of bearded plants.† This may happen in every case where there has been a *promiscuous* selection of many dominant plants, for any one of these may be a heterozygote and bear in each year both dominant and recessive germs.

The fact that the hornless breeds of goats still give some horned offspring is probably referable to the same cause. The point is of course not certain, but from the analogy of cattle [see p. 105] we may anticipate that the hornless form is dominant. In the polled breeds of cattle, which are never *promiscuously* selected, the polled character has naturally been easily fixed pure, but in goats selec-

---

* 'Landw. J. B.,' 20.
† Such a variety is Garton's Red King.

tion among the *ewes* has been probably to a large extent promiscuous.

The phenomenon is without doubt occurring very widely in nature. To it we may perhaps attribute the undiminished persistence of some weakly varieties, which are unceasingly exterminated by natural or artificial selection without ever leaving offspring. Cases have only to be looked for to be found in abundance. We may note the paradox that, for anything we know to the contrary, a recessive allelomorph may even persist as a gamete *without the corresponding homozygote having ever reached maturity in the history of the species.*‡ It would be

---

‡ [In illustration of such a phenomenon we may perhaps venture to refer to the extraordinarily interesting evidence lately collected by Garrod regarding the rare condition known as 'Alkaptonuria.' In such persons the substance, alkapton, forms a regular constituent of the urine, giving it a deep brown colour which becomes black on exposure. The condition is exceedingly rare, and, though met with in several members of the same families, has only once been known to be directly transmitted from parent to offspring. Recently, however, Garrod has noticed that no fewer that five families containing alkaptonuric members, more than a quarter of the recorded cases, are the offspring of unions of *first cousins.* In only *two* other families is the parentage known, one of these being the case in which the father was alkaptonuric. In the other case the parents were *not* related. Now there may be other accounts possible, but we note that the mating of first cousins gives exactly the conditions most likely to enable a rare and usually recessive character to show itself. If the bearer of such a gamete mate with individuals not bearing it, the character

premature to trace out the deductions to which this suggestive fact points, but we see at once that it may give the true account of the phenomenon that domesticated forms constantly give rise to varieties not met with in the wild state, a fact often ascribed on insufficient grounds to the action of changed conditions in producing greater *variability*.

It will be clear—a point which may have some economic importance—that in any such case the recessive 'rogue' can be eliminated by selection from *individual* plants or animals, breeding only from those which give no recessives on being self-fertilised, if hermaphrodite. If the organism be diœcious the process will be more elaborate, for it will be first necessary to test for recessive allelomorphs by fertilising with a recessive, and afterwards to fertilise those that gave no recessive offspring, with a dominant similarly proved to be free from recessive influence. Nevertheless it is certain that by this process alone can a strain of pure dominants be readily made.

'Purity' then acquires a new and more precise meaning. An organism resulting from an original cross is not necessarily pure when it has been raised by selection from parents similar in appearance for an indefinite

number of generations. *It is only pure when it is compounded of gametes bearing identical allelomorphs, and such purity may occur in any individual raised from cross-bred organisms.*

An organism can be strictly defined as genetically pure if all its gametes when united with similar gametes reproduce the parent identically; and in practice the only way in which such purity can, by one breeding, be tested, is by crossing the organism in question with pure recessives.

There are also other classes of cases where progressive selection fails not only to fix a particular variety but to diminish the proportion of 'rogues' beyond a fairly definite limit. We may first consider how far the principle of dominance may give an acceptable account of such cases.

In his most valuable book, 'Die Mutationstheorie,' 1901, Professor de Vries devotes a chapter to the consideration of such phenomena, pointing out in a number of cases that progressive and continued selection has failed to fix a particular character. He draws the conclusion that such characters distinguish 'half-races,' as he calls them, which cannot be bred pure.

The cases taken are many-leaved clovers, a polypetalous Ranunculus, several plants with variegated foliage, and the biennial forms of certain species.

Selection in each case at first rapidly increases the proportions in which the selected form appears among the offspring, but soon a maximum effect is produced which is not surpassed.

Now in each of these examples fertilisation was left to insects, and though seed was saved from individual plants it is not in dispute that cross-

---

would hardly ever be seen; but first cousins will frequently be bearers of *similar* gametes, which may in such unions meet each other, and thus lead to the manifestation of the peculiar recessive characters in the zygote. See A. E. Garrod, 'Trans. Med. Chir. Soc.,' 1899, p. 367, and 'Lancet,' November 30, 1901.]

fertilisation between them occurred. In Mendelian terms some might be pure D, some pure R, and some DR. Supposing dominance complete, eradication of the pure R forms annually does not extinguish them, for by the breeding of the DR forms *inter se* they will be continually reproduced.

There are no doubt many overlying complications in each of these cases, as, for instance, the probability that dominance is in these instances imperfect, but these will not change the main result.

The case of the biennial plants is especially interesting, as here we have strong indications that treatment and conditions may determine which character shall appear. For example, de Vries quotes the evidence of the Sugar Beet, a plant of great economic importance, to the breeding of which much attention has been devoted.

The plant which forms the large sugar-bearing axis is a biennial and does not flower until it has made the sugar-store. But from the best seed which has for generations been saved from such plants only, there arises a small percentage of an annual form which runs to seed without making a thick root at all. After years of selection the proportion of such rogues is not diminished. Now, if it could be supposed that the annual is recessive and the biennial dominant, this is partly explained. On selection, seeds are taken from dominants only. But some of these will be pure dominants and others will be heterozygotes bearing *both* allelomorphs. The latter will each year give rise to a certain number of pure recessives, compounded of two recessive gametes. In the first years of selection, the proportion of recessives will be diminished rapidly

by choosing seed from dominants only, but further *promiscuous* selection of dominants, unless continued for an indefinite time, will not altogether remove the recessives, for they arise from the dominants themselves.*

But in these forms it is well known that several kinds of treatment, exposure of the young plants to frost, over crowding, heavy manuring, and forcing, will greatly increase the proportion of 'runners.' In the case of Œnothera de Vries has made some very convincing experiments, clearly proving this fact, and Rimpau has done the same for the Beet, showing that the number of 'runners' can thus be greatly increased. There are then some which are biennial in any case, some which are biennial or annual

---

* It is of course only a conjecture that the biennial form is dominant in these cases, but, owing to the great importance of the subject, it seems worth while to call the attention of those interested to the possibility. Among the many investigations already made on the Beet it does not appear that the simple experiment has been tried of seeing if the annual or biennial form can be bred true from *individual* plants fertilised under proper precautions. Still less has the possibility of dominance been investigated. The only evidence known to us is that of Rimpau, that when the annual *Beta vulgaris* was grown near the cultivated form it bore two seeds which proved biennial and fifty-eight which were annuals. Rimpau conjectures that the two were crosses with the cultivated form, in which, as we should now say, the latter was dominant. But *B. patula*, an annual, emasculated and fertilised by cultivated Beets promiscuously, gave annuals only. Here there is a cross with another species, and the evidence is of doubtful application.

according to treatment, and some which are in any case annual. This is strongly suggestive of the three Mendelian classes.

De Vries has also experimented by selection from the annual plants, getting of course a higher proportion of annuals. But it must be remembered that in order to prove that the annual character is recessive, and that it can, as such, be fixed by one selection, it is necessary to ascertain first that the plant chosen is not what de Vries calls a 'facultative' annual—on this hypothesis, a DR—and secondly that it has not been cross-fertilised, particulars not yet forthcoming.

But even if the hypothesis of dominance could be successfully applied to these cases, there are others at first sight similar, where it cannot be thus applied; for example, instances of varieties recessive in their differentiating character, producing annually a small but sensible number of a particular 'sport,' exhibiting a character already known to be dominant. Here we must suppose either that we meet the phenomenon of an *originating* variation—the 'mutation' of de Vries: or possibly, which appears to be de Vries' view of half-races—the output of a certain number of such aberrant gametes is normally incidental to the development of the type-gametes. An objection to the latter deduction in some cases exists in the fact that the 'sports' in question may be exceedingly rare, and therefore produced by few individuals only.*

---

* Excellent illustrations of this phenomenon in the case of high-class Peas have been lately supplied to us by Mr. Arthur Sutton. Of these we hope to give details hereafter.

3. *Skipping a generation.*—That marked individual peculiarities fail to appear in the immediate offspring, but may appear in a subsequent generation has been often observed, and the fact has taken a great hold on the popular imagination. It has not yet been shown that the distribution of any of these characters among the different generations in any line of descent is other than is to be expected on the hypothesis of pure chance. Nevertheless we have now in the phenomenon of dominance a fact which may possibly be a real element in the causation of such appearances, and those who are familiar with statistics of inheritance, in man for example, might usefully study them with the possibility in view. The absence of the character in the first generation may indicate merely that it is recessive, and its reappearance in the next generation may be due to the heterozygote having bred with another individual also bearing the recessive allelomorph.

4. *'New' characters may be dominant.*—We cannot as yet perceive any properties common to dominant as compared with recessive characters. It will be noted, however, that the view of many naturalists that the phylogenetically older character is prepotent, or, more correctly, dominant, is by no means of universal application. In poultry, for instance, both pea and rose combs are dominant against single, though the latter is almost certainly ancestral; the polydactyle foot is dominant against the normal, though a palpable sport. A point of some interest is that in both wheat and barley the beardless form is dominant, though we naturally, though perhaps incorrectly, regard it

as a state normal in the one species, but an innovation in the other.

In cattle the polled form is dominant over the horned, though the former is a character which in our cattle has certainly arisen since domestication.

5. *Prepotency.*—The conception of dominance avoids certain difficulties which are involved in the use of the term 'prepotent.' As we now know that the allelomorphs of the several characters may be quite independent, it is confusing to speak of the prepotency of an individual when all that we know is that one or more of its characters is dominant over the contrary character. Of the dominance or prepotency of the *whole* we know nothing. The diversity of the views which have been at various times expressed as to the respective powers of mother or father to confer special qualities has probably arisen from confusions thus caused. If the term prepotency is to be preserved it must be applied to characters rather than to organisms, and its use must be restricted to cases in which the character so qualified has been actually tested by combination with the contrary allelomorph in one heterozygote.

We have been accustomed to consider that a variety may be sometimes prepotent in respect of a given character and sometimes not prepotent. The whole evidence on which this view is based will in many cases now require careful verification, for, as was fully discussed in the case of poultry, such a result may really be due to an unsuspected heterozygote having been sometimes used for the other parent. The evidence, for instance, that on crossing pea comb and single comb the offspring may be sometimes pea and sometimes single would formerly

have been thought a clear proof that pea comb was not always dominant, whereas it is now certain that much fuller evidence is needed to establish this proposition.

The existence of the so-called 'false' hybrids of Millardet [see p. 117] is an even more serious difficulty besetting the conception of prepotency, for here, though the cross-breds are produced by a union of the male and female gametes of two varieties, it is quite uncertain that the characters of both parents are introduced at all. [See p. 118.]

As a rule fair uniformity prevails among the results of first crossings, and in every case in which a mixture of forms occurs the question must now be asked *whether the fact is not a proof that either or both of the parents are actually producing more than one sort of gametes.* It is, no doubt, possible to conceive of the elements contributed by the two gametes respectively as engaged in a conflict so balanced that some supervening circumstances may give dominance to either side with varying frequency; but from what we now know of the nature of heredity, the conception of dissimilar gametes borne by one or both parents is just as easy to form, and no less probable on the facts.

6. *Sex.*—It is often profitable to compare the phenomena of variation with those of sex, and if the suggestion alluded to in the last paragraph be found true, it is worth reflecting whether the determination of sex may not sometimes be a phenomenon similarly conditioned.

[Note, added March, 1902.

There is already a considerable body of evidence in favour of the view that difference of sex is primarily a phenomenon of gametic differentiation. The evidence, however, seems to point to the conclusion that the differentiation is sometimes a phenomenon of the male cells and sometimes of the female cells, sometimes perhaps of both. Our attention has been called to a note by McClung,* suggesting that the differentiation of the spermatozoa of many insects and of some other [Arthropods], according as they do or do not contain the 'accessory chromosome,' may be an indication of differentiation in regard to sex. This body has been the subject of extensive study on the part especially of the American cytological investigators, and further researches regarding it may be a most profitable field of inquiry.

The fact that in *Nematus ribesii*,† and in the Hive-bee, the unfertilised eggs produce males only, seems to prove that in those cases the female cells are carriers of the male character only, though whether there is sex-differentiation of the male cells is not yet known. On the other hand, we have more frequent cases of unfertilised eggs in other types producing females only.

But from the observations of de Buzareingues,* it appeared that there is a more or less definite distribution of the sexes among the seeds of diœcious plants, the females being more commonly derived from seeds of one region, and the males from those of another. This of course is no proof of *original* differentiation of sex among the female cells, but it is readily consistent with that hypothesis.

On the other hand, as on the whole *against* the hypothesis that sex depends chiefly on gametic differentiation, may be mentioned observations —especially those of Wichura ('Bastardbefruchtung,' p. 44)—that the statistical distribution of sex among first crosses shows great departure from the normal proportions. The same has been seen by many hybridisers using animal types. But the further fact that there is a still greater variation in the statistical relations of the sexes in the *offspring* of hybrids, is rather favourable to the hypothesis.

The frequent occurrence of hermaphrodites among *first* crosses is also difficult to explain on the present hypothesis.]

7. Reversions.—With the Mendelian conception of the heterozygote as a form with its own special '*hybrid character*,' we have a *rationale* of large numbers of 'reversions'; for we already know many cases where heterozygotes do present the characters of putative ancestors. This fact reduces to harmony several groups of results where different experimenters, believing themselves to have worked with similar organisms, have reached seemingly contradictory conclusions. For some have used pure forms and others heterozygotes appearing in their guise.

THE NATURE OF ALLELOMORPHISM
A. *Simple Allelomorphs.*

The following list enumerates the principal cases in which the phenom-

---

* 'Anat. Anz.,' November, 1901, p. 220.

† Professor Miall has given me a reference to Cameron, 'Phytoph. Hymenop., Ray Soc. Monogr.,' vol. 1, p. 26, where authorities are quoted. He tells me that the same result was obtained in experiments of his own.

* 'Ann. Sci. Nat.,' vols. 16, 24, and 30; 1829, &c.

enon of allelomorphism has either been actually proved to exist or may be safely inferred from the published records.† In each of these cases more or less definite dominance of one character has been found, and in this list the dominant character is put first:—

1. Hairiness and absence of hairs (Lychnis).
2. Hoariness and absence of hairs (Matthiola).
3. Felted ears and smooth ears (Wheat).‡
4. Prickliness and smoothness of fruits (Datura).
5. Style long and short (Œnothera).*
6. Beardless and bearded ears (Wheat and Barley).†
7. Pointed seed and rounded seed (Maize).‡
8. Round and wrinkled seed (Pisum).
9. Starch endosperm and sugar endosperm (Maize).
10. Inflated (generally hard) pods and constricted (generally soft) pods (Pisum, Phaseolus).
11. Axial distribution of flowers and terminal distribution of flowers (Pisum).
12. Tall habit and dwarf habit (Pisum, Phaseolus), to which, from experiments seen at Messrs. Sutton's, we think we may safely add tall habit and dwarf procumbent habit (known to gardeners as 'Cupids') in Sweet Peas (Lathyrus odoratus).§
13. Entire petals and laciniated petals (Chelidonium majus).||
14. Normal zygomorphic form and peloric form (Antirrhinum¶ and probably Linaria).**
15. Normal habit and waltzing habit (connected with malformation of the aural labyrinth) (Mouse).††
16. Presence and absence of extra toe (Fowl).‡‡

---

† [From the evidence of crosses kindly carried out for us by Mr. Leonard Sutton we are able to add the 'palm' leaf (palmatifid) and reddish stems of Primula sinensis as dominant characters, while the 'fern' leaf (pinnatifid) and purely green stems are recessive characters.— March, 1902.]

‡ Rimpau, 'Landw. J. B.,' 20, 1891, p. 346.

* De Vries.

† Rimpau, loc. cit., pp. 341 and 353. Since this paper was written we have received Tschermak's valuable analysis of the phenomena in regard to wheat, which considerably extends our knowledge of allelomorphism in that species (see 'Ztsch, für d. Landw. Versuchswesen in Oester.,' vol. 4, 1901, p. 1029.

‡ Correns, 'Biblioth. Bot.,' vol. 53, 1901

§ Some distinct exceptions to the rule of dominance of the tall form are already known. [See final note.]

|| De Vries.

¶ Darwin, 'An. and Plts.,' ed. 2, vol. 2, p. 45.

** In the case of Linaria, Naudin found that on crossing a peloric Linaria with a normal one a mixture of normal and peloric plants resulted. As to the origin of the peloric parent there is no information, and consequently it may have been a heterozygote. See Naudin, 'Nouv. Arch. du Mus.,' 1865, I, p. 137.

†† Von Guaita, 'Ber. Natur. f. Ges.,' vol. 10, 1898, p. 317, and vol. 11, 1899, p. 131. For reference to this interesting case we are indebted to Professor Correns.

‡‡ The allelomorphism is not yet fully proved in this case. It is the only obviously meristic character in which there is yet any evidence of allelomorphism.

17. Pea comb and single comb (Fowl).
18. Rose comb and single comb (Fowl).
19. Polled and horned breeds (Cattle and probably Goats).§§
20. White shanks and yellow shanks (Fowl).
21. White plumage and general brown coloration (Fowl).*
22. Several coloured forms of flowers and their white varieties.
23. Several colours of fruits and their xanthic varieties (Atropa, Solanum).

---

§§ It is almost certain that absence and presence of horns are allelomorphic characters. In England there are three principal polled breeds of cattle—the Aberdeen-Angus, Galloway, and the Red Polled. The first two are black, the last red. Between these and the horned breeds crosses are usually made in large numbers. This is especially the case with the Angus, from which great numbers of cross-bred cattle are annually bred for the meat market. These are usually Angus-Shorthorn crosses, but other horned breeds are also occasionally used. The cross between a pure Angus and a pure Shorthorn is almost always a blue-grey without horns. Generally the horns are represented by loose corns of horny material, sometimes imbedded in the skin and not rarely hidden by the hair. Such 'scurs,' as they are called in the north, are objected to in the pure polled breeds and are mostly absent.

Notes of the cross-breds exhibited at the Smithfield Club Cattle Shows in 1888, '89, '98–'01 give the following results. The animals are classified according to the descriptions in the Catalogue. No doubt, however, the actual purity of the parent breed or breeds was in many cases doubtful. Taken as they stand, the numbers exhibited in these six years were as follows:—

From $\left\{\begin{array}{l}\text{Polled Angus} \\ \text{Polled Galloway} \\ \text{Red Polled}\end{array}\right\} \times$ some horned breed, usually Shorthorn, and the reciprocal cross—104 polled, 13 horned.

From first cross animals bred as above, mated with a pure polled parent—23 polled, 1 horned.

From first cross animals mated with some horned parent—18 horned, 24 polled.

When allowance is made for the very rough materials out of which these figures come, it is clear that the facts cannot be very far from the Mendelian expectation. It is, however, likely that the allelomorphs concerned are not merely the horned character in its entirety, and total absence of horns. For in the offspring of (polled × horned) × polled, the horns, when they occur, are often *loose* though of fair size. If all parts were completely correlated we should expect *either* absence of horns (perhaps mere scurs) *or* ordinary horns like those of horned breeds. Probably, therefore, there is not *complete* correlation between the formation of horns and that of the bony cores which carry them, and these characters are divisible in transmission. Unfortunately the cross-breds are practically never bred together, so the valuable evidence thus attainable is wanting. It should be mentioned that in offspring of (polled × horned) × horned the coat-colour character also breaks up.

[*Note to page* [104], 'Cupid Sweet Pea.'—Mr. Hurst called our attention to a passage in 'Report of the Sweet Pea Bicentenary Celebration, 1900' (published by Mr. R. Dean), p. 26, where it is stated that Mr. Laxton, of Bedford, crossed Cupids and tall forms, producing almost all *talls* on the first cross. Inquiry from Messrs. Laxton elicited the fact that in this Report the facts were by mistake inverted.]

* [Wh. Dorking × Ind. Game crosses are this year giving exceptions to dominance of white.—1902.]

24. Several colours of seed coats.
25. Darker and lighter colours of endosperm (Maize).
26. Yellow and green cotyledons (Pisum).

With regard to seed-colours, Correns has shown that the question is a complex one, depending on several factors. In Maize, especially, the seed-skin and the several parts of the endosperm may all be independently concerned in giving the net result. Each must be considered separately, and in several cases the dominance is imperfect, and blendings may occur.†

----

† Full details given in Correns' memoir, 'Biblioth. Bot.,' 1901.

Between various simple allelomorphs correlations may of course occur. A few of these we know already. But in these cases of simple correlation the gametes may each transmit the correlated groups or the opposite allelomorph entire.

From the foregoing list it appears that allelomorphism may occur in a great diversity of characters, involving many different physiological factors.

In the plants albinism appears to be recessive, but in the case of fowls white plumage is dominant, though not completely so. It does not appear as yet that simple allelomorphism occurs between any two colours, of which neither is xanthic or albino.

Bateson now begins to list numerous crosses that could not be explained in simple Mendelian terms and you will find the next ten pages difficult to follow. In some of the crosses, the two varieties differed from one another in many genes. These situations could be resolved by additional crosses. Many of the crosses, however, could not be understood at all. Later research revealed that additional genetic principles were involved—principles unknown to Mendel and Bateson. Some of the solutions are given in Chapters 3 and 5 of *Heredity and Development*. You will not learn much genetics in the next few pages but, possibly, you will learn something more valuable: the difficulties of applying a new principle, assumed to be universal, to all existing data. Such an attempt has to be made and, if the 'universal' principle cannot be applied to all data, the principle must be modified. The history of genetics was to show that Mendelism was correct, though incomplete.

B. *Compound Allelomorphs.*

So far, in all or nearly all the cases we have considered, the dominant and recessive characters are each *simple*. In other words, when the heterozygotes breed together, they produce dominants and recessives like their parents, heterozygotes like themselves, and no other forms. The gametes therefore respectively bear characters which are the same as those of the varieties which were used to produce the heterozygotes. We have next to consider a numerous and important group of cases in which a character of one of the original parental varieties after crossing is itself split up. Of these we will give illustrations.

1. *Sweet Pea.*—By the great cour-

tesy of Messrs. Sutton and Sons we have been permitted to watch many of the experiments conducted at their nurseries. We cannot sufficiently express our indebtedness for the splendid opportunities of study in these fields thus provided. For the most part these phenomena are not dealt with in the present paper, and amongst many interesting results there witnessed we propose now to refer very briefly to the following only:—

Sweat Pea (Lathyrus odoratus).— Stanley, standard dark maroon or chocolate, with wings similar but somewhat tinged with violet, crossed with Giant White, gave *all* Giant Purple Invincible, viz., standards as in Stanley, but wings blue. These first crosses self-fertilised, gave Giant White, Giant Purple (without blue wings), Mars (a well-known red variety), Her Majesty (a full magenta, well known), and a form like Her Majesty, *but flaked with white.**

One plant of each was saved and its self-fertilised seed sown. Mars and Her Majesty came true. The Giant White was tested, and it came true also. The Her Majesty flaked with white, however, gave Whites, Her Majesty, and Her Majesty flaked white again. The Giant Purple gave Giant White, Her Majesty, Giant Purples, and two plants of a streaky cream color.

---

* It is possible that this complex result does not always occur; for in another case a Giant Rich Purple, very like Stanley, crossed with Giant White, gave seedlings *all* Giant Rich Purple. These on self-fertilisation gave a mixture of Giant White and Giant Rich Purple again. One plant of each on self-fertilisation gave only offspring like itself.

The facts point to a higher degree of complexity than we can yet realise, but we see that the first crosses are all alike, though differing from the coloured parent. The same form, or something very like it, was often observed to come in other cases where a blue or purple parent was used in crossing. Now on self-fertilisation the first cross gave a variety of forms. It therefore produced a variety of gametes, not two kinds, but several. Of these forms some, Mars, Her Majesty (Giant White also in all probability), reproduced themselves exactly. Therefore they had only one kind of gamete, and they must be supposed to have been formed by the union of similar gametes. The purples, on the contrary, produced most of the whole series again, showing that they were producing a variety of gametes like the first cross parent itself.

Her Majesty flaked with white, gave some Her Majesty, some White, some Her Majesty flaked white. Therefore the flaked plants are heterozygotes, formed by the union of a Her Majesty gamete with a white gamete.

We are then led to the conclusion that the allelomorph transmitting the coloration of Stanley is *compound,* and that it can be broken up into simpler and possibly component elements. When *similar* elements, thus extracted, combine in fertilisation, they do not split up again on the formation of gametes. The constituents of the compound allelomorphs may perhaps be spoken of as *hypallelomorphs.*

The fact that Stanley did not occur again is another indication that its colour character had been broken up into *more than two* elements.

Another fact which may point in the same direction is that the purple formed on the first cross is different from that which recurs in the next generation. In fact, this Giant Purple Invincible results from the union of the whole compound allelomorph of Stanley with that of Giant White. We may suppose that it does not come again for the reason that the compound allelomorph has been broken up among the gametes borne by the first cross, and that the union of no two of these, or of any of them with white, results in that particular heterozygote form, Giant Purple Invincible. Inasmuch, however, as Giant Purple Invincible, not yet distinguishable from that produced in this cross, is a well-known and stable form, there must *either* be gametes corresponding to it* (or its male and female gametes must be dissimilar and combine in that definite heterozygote, which is most unlikely). Till the experiment has been repeated on a large scale we must not lay much stress on the absence of Purple Invincible after the break up of the first cross, because in other experiments where White Cupid (a procumbent form) was crossed with Mme. Carnot (a blue), Purple Invincible again resulted together with White Cupid (? the result of imperfect emasculation). These Purple Invincibles, self-fertilised, gave several forms, amongst them Mme. Carnot and some Purple Invincibles again. Whether this indicates that the

---

* Similarly from other crosses seen at Messrs. Sutton's it is clear that the form called 'Painted Lady' may be another heterozygote form, though the same is one of the oldest and most familiar fixed forms. According to Mr. S. B. Dicks, there is good reason to believe the purple and the Painted Lady forms to be the oldest varieties. 'Report of Sweet Pea Conference,' 1900.

compound allelomorph is not wholly broken up, or that its character may again be synthetically reproduced, cannot yet be said. Corroborative evidence that the blue elements are definitely extracted from the 'derived' Her Majesty was seen in the fact that this variety when crossed with various pink and cream kinds gives no blues or purples.

To the whole subject of the results of crossing Sweet Peas we hope to return when our own experiments are further advanced.

The probability is that in this, as in other similar cases of compound allelomorphs, there is a heterozygote form which may be common to several combinations of dissimilar gametes, and it is characteristic of such forms that they may reproduce *in appearance* some putative ancestor. It is to this class of phenomena that Darwin's famous 'reversions on crossing' are probably to be referred.

2. *Poultry.*—Another example of the splitting up of a compound allelomorph is probably to be seen in the poultry experiments. The first cross between Indian Game and White Leghorns, for instance, is white flecked with a few black or grey feathers, sometimes barred, sometimes irregularly marked with pigment. Such first crosses bred together give some dark birds and some light, (see p. 000), the latter being sometimes pure white, sometimes flecked with black, and sometimes pile (brown and white). When White Dorkings are crossed with Brown Leghorns the result is very similar; but in each of these cases the dark birds resulting from the interbreeding of the first crosses are not simply like their dark grand-parent, but belong to several distinct types of coloration such as black, cuckoo,

silver-grey,* together with some more or less nearly reproducing the dark grand-parental type. The numbers reared are far too small to justify a comprehensive deduction, but that the types of coloration thus produced have some definiteness is quite clear. Whether any of them will breed pure must be unknown till next season. As already stated, some of these colours are already well known as characterising various breeds.

Until experiments have been carried out with the express object of proving the compound nature of allelomorphic characters and of resolving them into their constituents, we can only gather indications of such phenomena from experiments undertaken for other objects. Of these there are a considerable number which leave little doubt that further examination would disclose such a result. We may mention the observations of von Guaita on mice, from which it appeared that the first cross of albino mice with black-and-white Japanese waltzing mice, gave a grey house-mouse resembling in size, colour, and wildness the wild house-mouse.* The first crosses bred together gave albinos, grey mice, black and white, grey and white, and black mice (with the waltzing charac-

ter distributed among them in proportions closely obeying the Mendelian ratio); of these the albinos produced, with one exception, albinos only when bred together. The grey marked with white, bred together, produced no more blacks or black-and-whites; and the blacks and the black and whites bred together gave no more greys, though both descriptions may still give albinos. Facts like these strongly suggest that, with suitable mating, the classes could be shown to consist of the original albino, and a number of forms, some of which would henceforth be pure, while others would be found to be heterozygous.

3. Another case, possibly of the same nature, is that of the Himalayan rabbit, of which an account is given by Darwin.†

The literature of pigeon-fancying abounds with information pointing to a similar *rationale* of the colour phenomena there seen. Formerly the recipes given in such treatises as to the methods of mating to be followed for the production of particular colours would have seemed mere nostrums, but now we can see at least the general basis of fact whence they have been derived.

The experiments with stocks described give cases probably also analogous. Several forms crossed together all gave purple for the first cross, which on being self-fertilised gave other colours in addition to those of the pure parental forms and that of the first cross.

This conception of compound allelo-

---

* The appearance of silver-grey in the offspring of first crosses between White Dorking and Brown Leghorn may be attributed to the certainty that White Dorking were related to Silver-grey Dorkings. The colour may, nevertheless, have come from resolution of the Leghorn colour, for it is not peculiar to Dorkings, but is known in other breeds, *e.g.*, Game Duckwings.

* Haacke, crossing albinos with grey-and-white Japanese waltzing mice, usually obtained the same result, viz., grey mice, but more rarely *black* mice. The

latter result must be taken as indicating impurity in one or other parent. Vosseler, quoted by von Guaita, obtained greys only. See Haacke, 'Biol. Cblt.,' vol. 15, 1895, p. 45.

† 'An. and Plts.,' I, p. 113.

morphs is almost the same as that which Mendel himself introduces in speaking of his Phaseolus crosses.‡ His analysis does not, however, seem to be strictly correct, and the subsequent reasoning is consequently obscure and not altogether valid. He says if the colour of the red Phaseolus be made up of $A_1 + A_2 + \ldots$, then on crossing with a white form a, hybrid unions are produced, $A_1a + A_2a +$ &c.

But it is the group $A_1A_2 \ldots$ which is allelomorphic to a, and the heterozygote is $A_1A_2 \ldots a$, and not $A_1a + A_2a +$ &c. It cannot be till the crosses form their gametes that the compound allelomorph breaks up.

It is not evident how this error of expression came about. Mendel in consequence misses the point that by the breaking-up of the compound character after the cross, new fixed forms may be produced by union of the elements of the original compound allelomorph, without any admixture from the variety with which the first cross was made. Such pure forms may be represented as A_1A_1, A_2A_2, &c., and of these we have already seen instances in the case of the Sweet Peas, Mars and Her Majesty.

Of the coloured forms appearing as offspring of the first crosses inter-bred, some are compounded of colour-bearing gametes meeting similar or dissimilar colour-bearing gametes, and some (like the Sweet Pea, Her Majesty, flaked with white) of a colour-bearing gamete meeting a white-bearing gamete.

We have good reason to believe that the compound allelomorph is not in every case resolved into its ultimate

‡ *Loc. cit.*, p. 35.

constituents when the gametes of the first cross are formed, and indeed we must suppose such imperfect resolution to be present whenever, as in the case of the Sweet Pea, among the resolved forms (White, Mars, Her Majesty) there occur complex heterozygotes like Giant Purple, which can itself produce a series of forms in the next generation. Such a form may be represented as $A_2A_3 \ldots a$. It is to this class of complex heterozygotes that we conceive the Andalusian fowl to belong.

It is doubtful whether and in what sense we are entitled to regard the whole compound character as *one* allelomorph. Some justification for this conception is to be found in the fact that in the poultry crosses the light chicks bore to the *whole number* of dark chicks the proportions of $3 : 1$. On the Mendelian hypothesis this must be taken to show that the crossbreds produce on an average white-bearing gametes equal in number to the whole number of colour-bearing gametes, which may bear the colour-allelomorph in various stages of resolution.

By statistical investigation of such cases it should be possible to determine with some success how the unresolved characters are related to the elementary characters, and to make a scheme of *equivalence*.

It is, perhaps, hardly too much to suggest that in a great number of cases the familiar fact so often observed that first crosses bred together give a profusion of new forms may be capable of similar explanation. With such new forms the usual experience is that some breed true from the beginning, while some continue to give rise to other forms, of which some may have already been produced, while others again are new. The cases we have taken are those of colour-varieties, as the facts in those cases are clearer, but their nature is probably not different. It is in this sense that crossing may be truly spoken of as a 'cause' of variability, and some picture of that phenomenon is now provided.

The importance of this reasoning lies in the fact that we can now recognise that these different new forms may be, in their genetic composition, diverse. We are no longer to expect that it is a matter of chance whether each will be able to transmit any of the other forms, but we perceive that this is a question to be determined by actual observation once for all. When such determinations shall have been made on a statistical basis we shall be able to state precisely the numerical proportions which the gametes of the several classes bear to each other, and hence to determine the actual number of constituents of the compound allelomorph and their relationships to each other. This investigation is now merely a matter for precise quantitative analysis.

Remembering that we have no warrant for regarding any hereditary character as depending on a material substance for its transmission, we may, with this proviso, compare a compound character with a double salt, such as an alum, from which one or other of the metals of the base can be dissociated by suitable means, while the compound acid-radicle may be separated in its entirety, or again be decomposed into its several constituents. Though a crude metaphor, such an illustration may serve to explain the great simplification of the physiology

of heredity to which the facts now point.

A marked feature in connection with compound allelomorphism* is the frequency with which in such cases one or more of the heterozygotes present what we have reason to regard as ancestral characters. To such 'reversion' we referred in speaking of Sweet Pea crosses. The Sweet Peas produced a flower with purple standard and blue wings approaching what we may regard as a primitive Sweet Pea. Several white varieties of Stock produce a purple form; many of the crosses with the 'half-hoary' type gave fully hoary heterozygotes. The Albino and Japanese mice produce a grey house-mouse as their heterozygote. *Why* such heterozygotes should show ancestral characters we do not know; but we can now recognise that such 'reversions' are heterozygous mixtures and not constant forms. To speak of such reappearances of ancestral characters as a reappearance of the ancestral *form* is entirely misleading. These heterozygotes will not breed true, and *are* ancestral in no real sense. Not only are they heterozygous and in constitution compound, but, as in the Sweet Pea, several different compounds agree in having the same ancestral form as their specific heterozygote.

It is unfortunate that Darwin's own experiments with poultry and pigeons were so complex that it is now impossible to disentangle the results or

---

* There is no reason for supposing such reversion to be absent in all cases of heterozygotes formed by the union of *simple* allelomorphs, but the few clear cases known seem to be all cases including compound allelomorphs.

to use them for the purposes of these deductions. He records the most complicated unions of birds of different breeds, some homo-, some heterozygotes, some exhibiting simple and others compound allelomorphs, and in the statement of results the all-important distinctions between the generations and the offspring of the several individual birds are often not observed.

To sum up the phenomena of compound allelomorphism, we may say that the evidence shows that the characters of a pure form when crossed with another may be broken up into component characters or hypallelomorphs, and that the decomposition may take place in various degrees of completeness.

To the variations which thus arise by resolution of compound characters we propose to give the name *Analytical Variations*. There can be no doubt that a very large proportion of the discontinuous variations in colour, at all events, met with both in wild and domesticated species are of this nature. The fact that similar component forms are similarly related to each other and of the type, in various species, thus provides the true account of numerous phenomena of 'parallel' variation.

The facts thus grouped, suggest the following questions. Has a given organism a fixed number of unit-characters? Can we rightly conceive of the whole organism as composed of such unit-characters, or is there some residue—a basis—upon which the unit-characters are imposed? We know, of course, that we cannot isolate this residue from the unit-characters. We cannot conceive a pea, for ex-

ample, that has no height, no colour, and so on; if all these were removed there would be no living organism left. But while we know that all these characters can be interchanged, we are bound to ask is there something not thus interchangeable? And if so, what is it? We are thus brought to face the further question of the bearing of the Mendelian facts on the nature of Species. The conception of Species, however we may formulate it, can hardly be supposed to attach to allelomorphic or analytical varieties. We may be driven to conceive 'Species' as a phenomenon belonging to that 'residue' spoken of above, but on the other hand we get a clearer conception of the nature of sterility on crossing.

Though some degree of sterility on crossing is only one of the divers properties which may be associated with Specific difference, the relation of such sterility to Mendelian phenomena must be a subject for most careful inquiry. So far as we yet know, it seems to be an essential condition that in these cases the fertility of the cross-bred should be complete. We know no Mendelian case in which fertility is impaired. We may, perhaps, take this as an indication that the sterility of certain crosses is merely an indication that *they cannot divide up the characters among their gametes*. If the parental characters, however dissimilar, can be split up, the gametes can be formed, and the inability to form gametes may mean that the process of resolution cannot be carried out. In harmony with this suggestion is the well-known experience of hybridisers, that if there is any degree of fertility in the first cross, with subsequent inter-bred generations the fertility may increase.*

Such increase in fertility is generally associated with some greater approximation to one of the parental forms. In terms of our hypothesis, we may conceive this fact as denoting that offspring formed of gametes which have successfully resolved the characters of the heterozygote, and are not bearers of the irresoluble characters, can form their own gametes with less difficulty.

That the sterility of hybrids is generally connected in some way with inability to form germ-cells correctly, especially those of the male, is fairly clear, and there is in some cases actual evidence that this deformity of the pollen grains of hybrids is due to irregularity or imperfection in the processes of division from which they result. It is a common observation that the grains of hybrid pollen are too large or two small, or imperfectly divided from each other.† Such conditions are what we should expect on the hypothesis here suggested.‡

However this may be, it would be of the utmost importance to discover at which of the divisions leading to the production of the gametes, the allelomorphic characters are divided. Correns has pointed out that the evidence of maize proves that in that case the two nuclei of the pollen tube must both be transmitters of the same character, for, in the fruit of the first cross between starch and sugar varieties, those seeds which have sugar endosperms produce pure recessive (sugar) offspring. This fact proves therefore that the nucleus which fertilises the embryo and that which fertilises the endosperm, are transmitters of the same character. Therefore, the separation of the characters does not take place in this case when the two generative nuclei divide from each other.* Further evidence on this

---

* Focke, 'Pflanzen Mischlinge,' p. 483; Gärtner, 'Bastarderzeugung,' pp. 333 and 373.

† See e.g., Naudin, 'Nouv. Arch. du Mus.,' 1865, I, p. 95, and Wichura, 'Bastardbefruchtung im Pflanzenreich,' 1865, p. 37. Cases are easy to find.

‡ Remarkable observations bearing directly on this question have recently been published by Guyer ('Science,' vol. 11, 1900, p. 248), as to the spermatogenesis in hybrid Pigeons. The species used are not named, and the account is very brief. He states that in both sterile and fertile hybrids much variation in cell-division was seen, inequalities in chromatin distribution were common and multipolar spindles were abundant. In hybrid spermatogonia there were often more than

eight (the normal number) large ring chromosomes. Sometimes there were sixteen small rings. In this case they usually located in two spindles, eight to each. Frequently both large and small rings were present. Guyer suggests, though apparently in ignorance of Mendel's work, that this phenomenon may indicate a 'tendency in the chromatin of each parent species to retain its individuality.' If so, he points out that in cells with two spindles and eight chromosomes, after division, some of the new cells will have chromatin from one parent and some from the other, and the observed 'reversion' of the offspring of hybrids to parent species 'may be due to the persistence of the chromatin of only one species in one or both of the germ-cells.'

* Correns inclines to the view (based on the fact that pollen grains of crosses between forms of Epilobium differing in pollen colour do not show a mixture of the two parental colours) that the sepa-

question is wholly wanting. Several attempts are being made by others and by ourselves to determine this point by crossing varieties with recognisably different pollens; but, so far, the desired mixture of dissimilar gametes in our cross-bred has not been satisfactorily observed. As soon as some means shall have been found of making visible that differentiation which we now know must exist between the germ-cells of the same heterozygote, a vast field of research will be opened up. Till then, the microscopical appearances accompanying the segregation of the characters must remain unknown, and we are obliged to resort to the cumbrous and protracted method of deduction from the statistical study of the zygotes formed by the union of the several kinds of gametes.

Variation, especially discontinuous

---

ration of characters does not take place when the pollen-grains divide from each other, but when the generative nucleus separates from the vegetative nucleus. Such an observation is, however, surely inconclusive. The pollen-grain is not the germ-cell, but the carrier of the germ-cell, and in any case there *may* be no universal correlation between the appearance of the pollen-grain and the characters it transmits. From what we know of discontinuous variation, and especially from the analogy of that 'dichotomy' of characters seen in various parts of hybrids, we incline to the view that the separation of characters will be found to occur at various divisions in various forms. Information on these phenomena is given especially by Naudin, 'Nouv. Arch. Mus.,' I, 1865, p. 150; Focke, 'Oesterr. bot. Ztschr., 1868, p. 139; Macfarlane, 'Trans. Roy. Soc. Edin.,' 1895, vol. 37, p. 203.

variation, of zygotes is in great measure thrown back on that of the gamete. We perceive, in fact, that the production of dissimilar gametes by one zygote may be compared, to take a rough illustration, to a bud-variation, constantly recurring in each heterozygote. Whether the divisions resulting in the formation of the dissimilar gametes are symmetrical or asymmetrical we cannot yet tell; but as in most cases of discontinuous variation, by sufficient searching, occasional instances, particular individuals or strains, will probably be found where the discontinuity is imperfect. As already pointed out also, the existence of exceptional gametes of a mosaic nature must already be inferred. It is unfortunate that so long as the statistical distribution of the zygotes is the only criterion by which the nature of the gametes can be deduced, even cases of impurity in extracted recessives—the readiest form in which imperfect differentiation will be seen— will not suffice to show whether there has been in fact such imperfect differentiation, or only defective dominance.

Mendel's discovery, it will be understood, applies only to the manner of transmission of a character already existing. It makes no suggestion as to the manner in which such a character came into existence. The facts, however, leave no room for doubt that at least one character of each pair of simple allelomorphs has arisen discontinuously. The fact that the gametes of the cross transmit each member of the pair pure, is as strong an indication as can be desired of the discontinuity between them. From imperfection of the records, however, we cannot point to many cases where we know both that the origin was sudden, and that the characters obey Mendel's law, though no one practically acquainted with these subjects will feel any doubt that if those records were complete, there would be abundant evidence to this effect. A positive example, however, is that of *Chelidonium majus laciniatum,* of which the modern origin is recorded,* and the allelomorphic nature was proved by de Vries.† It is scarcely doubtful that such varieties repeatedly arise. The Cupid Sweet Pea is another [p. 104].

With regard to the compound allelomorphs, it must be determined by further investigation whether they similarly can come into existence in their entirety, or whether they are capable of synthesis. At present, though we can perceive the fact that they are capable of decomposition, we know nothing of the reverse process.

In the cases [p. 106] we have discussed, it is plainly the simple allelomorph that has discontinuously arisen.

While we can hardly doubt that, of each pair of simple allelomorphs, *one* must have come suddenly into existence, we cannot tell whether this fact means that something is *added* to the original organism, or whether, from the first, the appearance of the new character is to be regarded as a *replacement* of the corresponding character. For example, we do not know whether the greenness of the peas is due to an *addition* of something to

---

* For literature see Korschinsky, Heterogenesis, trans., 'Flora,' Ergänzungsheft, vol. 89, 1901, p. 248.

† 'Ber. deut. Bot. Ges.,' 1900, p. 87.

the whole sum of the yellow pea, or to a *substitution* of something for the yellow character. We may partly understand the physiological nature of the yellowness and the greenness, or to take a clearer case, of the relation of the starch endosperm to the sugar endosperm, but this is as yet no help in elucidating the question. If it shall appear that the process is one of addition, the conception of the characters *splitting* may prove an incorrect one, and some other metaphor must be substituted.

Of special importance in this regard will be the study of cases where three or more characters are capable of mutual replacement. All cases studied so far are examples in which the allelomorphs are in *pairs*, but we know instances where three or more alternative forms of the organism occur, and an investigation of such cases may throw light on this part of the problem.

Attention of those who propose to experiment in this direction must, however, be called to the fact that so long as we are dealing with simple allelomorphs, though there may conceivably be more than two forms of gamete (apart from 'mosaics,' &c.), in respect of each group of simple allelomorphs, yet each zygote can, variation apart, bear two only. Consequently, no zygote can be formed by the sexual process which shall be capable of bearing more than two forms of gamete of each sex. But it is not inconceivable that by grafting or some other form of union, a combination of three or more allelomorphs in one organism may be brought about.

Non-Mendelian Cases.—In the case of Matthiola and among the poultry, instances have been apparently found of definite departure from Mendel's law. It is certain that these exceptions at all events indicate the existence of other principles which we cannot yet formulate. But besides these cases there are three distinct classes of phenomena met with in breeding to which the Mendelian principles cannot be readily applied. It will be useful to consider briefly how each case departs from these principles, and whether by any modification they can be extended to such cases.

Such phenomena are—

1. The ordinary blended inheritance of continuous variations.
2. Cases in which the form resulting from the first cross breeds true.
3. The 'false hybrids' of Millardet.

1. *Blended Inheritance.*—At first sight it seems that cases of continuous variations, blending in their hereditary transmission, form a class apart from those to which Mendel's principles apply. But, though it may well be so, the question cannot be so easily disposed of. The essence of the Mendelian conception is, as we have seen, that each gamete may transmit one allelomorph pure. So long as each heterozygote can only exhibit *one* allelomorphic character, the dominant, we can from a study of the heterozygotes and their offspring demonstrate the purity of the gametes. But dominance is a distinct and subordinate phenomenon. We readily perceive that the heterozygotes may show either of the parental characters discontinuously, or various blends between them, while the gametes which composed the heterozygotes may still be pure in respect of the parental char-

acters. The degree of blending in the heterozygote has nothing to do with the purity of the gametes.

It must be recognised that in, for example, the stature of a civilised race of man, a typically continuous character, there must certainly be on any hypothesis more than one pair of possible allelomorphs. There may be many such pairs, but we have no certainty that the number of such pairs and consequently of the different kinds of gametes are altogether *unlimited* even in regard to stature. If there were even so few as, say, four or five pairs of possible allelomorphs, the various homo- and hetero-zygous combinations might, on seriation, give so near an approach to a continuous curve, that the purity of the elements would be unsuspected, and their detection practically impossible. Especially would this be the case in a character like stature, which is undoubtedly very sensitive to environmental accidents.

It is, of course, quite possible that the gametes in such cases do in fact vary as continuously as we see the zygotes do, but this cannot yet be affirmed. The great theoretical significance of this question should therefore lead us to suspend judgment for the present.

2. *First Crosses Breeding True.*— With respect to this phenomenon no experiments on a large scale have yet been made. Most examples are recorded in the form that A and B were crossed together and produced a third form, C. The C's were then bred together and some C's were again produced. We hardly ever are told that in this generation *only C's* were produced. We hardly ever are told that in this generation *only C's* were produced.

Generally, however, we do not even know so much. The cases for example given by Darwin,* are for the most part general statements that certain new and now definite forms, the Swede turnip, for instance, were produced by crossing. Any such case may, therefore, be merely one of the resolution of compound allelomorphs followed by selection of the forms produced by the union of similar component allelomorphs. This, indeed, is probably the true account of most permanent forms produced by crossing.†

There remain, however, a few cases

---

* 'An. and Plts.,' ed. 2, vol. 2, pp. 73–77.

† We cannot avoid expressing a doubt whether the wonderful series of 'mutations' which de Vries has lately recorded ('Die Mutationstheorie,' 1901) as arising from *Œnothera Lamarckiana* do not fall under suspicion that they may owe their origin to some unsuspected original cross. Nothing can take away the extraordinary interest which attaches to these experiments, but until it has been shown in the clearest way that the *Lamarckiana* which gave rise to the 'mutants' is a genuine uncrossed form we must feel hesitation in accepting the conclusion which de Vries has drawn from the facts.

This possibility is strengthened by the fact which Professor de Vries has told us, that the pollen of his *Lamarckiana* contains deformed grains, a point which is also mentioned by Pohl ('Oesterr. Bot. Ztschr.', 1895, vol. 45, p. 212) in a paper to which de Vries refers (*loc. cit.*, p. 153).

On the other hand, we can scarcely suppose crossing to be the only cause determining the production of heterogeneous gametes, or in other words, variation in sexual descent.

of which Mendel's‡ own crosses among Hieracia are a good example, in which a distinct form, produced by the first cross, has proved able to transmit its characters to its offspring. Of such cases we know very little. We may, perhaps, notice two features as apparently characteristic of these cases. First, that the results of the first cross may show no uniformity; secondly, that there is often a considerable degree of sterility.

In Correns' terminology such crosses are 'homodynamous' and 'homoögonous.' De Vries speaks of them as '*erbungleich*.' In these instances the new form is able to give off gametes, male and female, carrying its own new character. Such facts plainly indicate a degree of complexity higher than that to which the Mendelian principles can apply, and for the present we have no insight into their nature.

3. *Millardet's 'False Hybrids.'*— Some allusion must be made to the remarkable results described by Millardet,* which have been the subject of frequent discussion among practical evolutionists. Put briefly, Millardet found that when certain varieties, especially of strawberry, are crossed together, (1) the cross-breds may precisely reproduce the maternal type, without any indication of the paternal characters; (2) in other cases the cross-bred individuals may show *either* the maternal characters pure (save in one case the colour of fruits) *or* the paternal characters pure. Seeds

---

‡Mendel, 'Abh. Ver. Brünn.,' vol. 8, 1869. See also Swingle and Webber, 'Year-book Dept. Agric.,' 1897, p. 393.

* 'Mém. Soc. Sci. Bordeaux,' sér. iv, vol. 4, p. 347.

from plants thus exclusively reproducing one parental type themselves gave
plants again exclusively of that type.
To such forms he gives the name
'*faux hybrides*' or '*hybrides sans
croisement.*'

In order to estimate the significance
of these facts we ought to know of
what variations the pure forms are
capable, when bred *inter se*, without
crossing. Upon this point we have as
yet no evidence. If we assume that
each of the forms used would, if bred
pure, transmit its characters regularly
to its offspring, then we should have
established that the heterozygote produced exclusively gametes, transmitting the character which appeared as
'dominant' in itself, and a new order
of facts is thus revealed. It is difficult
to see any escape from this conclusion,
but, on the other hand, if it could be
shown that the pure-bred offspring of
the one form could themselves exhibit
the characters of the other parent used
in the cross, we should recognise that
the parent forms themselves produced
mixed gametes, and in such a case we
should expect that when similar
gametes meet in fertilisation the offspring resulting would breed true. On
the whole this explanation is very
improbable, but as yet it is not wholly
excluded in some of the cases in which
Millardet's phenomenon is alleged to
have occurred.

In our experiments with Matthiola,
cases were described which, it can
scarcely be doubted, are fully proven.
The same is true of some of de Vries'
instances,[†] notably that of *Œnothera
Lamarckiana* crossed with the *cruciata*
var. of *biennis*. Possible instances occurred in regard to the combs of

---

[†] 'Ber. deut. Bot. Ges.,' 1900, vol. 18,
p. 441.

poultry though, as [we] pointed out,
a simpler explanation is not altogether excluded in those examples.
Such phenomena may perhaps be regarded as fulfilling the conception of
Strasburger and Boveri, that fertilisation may consist of two distinct operations, the stimulus to development and
the union of characters in the zygote.

[Note, added March, 1902.

Several times in the course of these
pages, reference has been made to the
phenomenon known as the 'false
hybridism' of Millardet. We are not
aware that attempt has yet been made
to elucidate that phenomenon. In view
of the Mendelian discovery, we think
it may not be altogether premature to
suggest a possibility, which may perhaps be some guide to further experiment with this phenomenon.

In the false hybrid then, one or
more characters are contributed to
the zygote by one parent alone, to the
exclusion of the corresponding character of the other parent. This exclusive character is exhibited on the
development of the zygote; and that
the opposite character is really excluded appears from the fact that the
offspring of the 'false hybrid' do not
reproduce the excluded character.

The terms 'false hybridism' and
'false hybrid,' though they have done
good service, are clearly inconvenient
for the fuller discussion that must arise
respecting these facts, and we propose
to denote the phenomenon by use of
the term *monolepsis*, the ordinary result of fertilisation being distinguished
as *amphilepsis*.

It is not yet certain whether monolepsis is a phenomenon peculiar to
recessive characters; but while we are

fairly sure that some of the cases in which it is seen are instances of recessive characters, we know no certain example of the monoleptic transmission of a dominant character. By the nature of the case, positive evidence of such transmission must be peculiarly difficult to obtain; for the first-cross-bred generation would have to be individually tested on a considerable scale by subsequent breeding before such a possibility could be established.

Let us first consider certain features of the process of fertilisation as it may be supposed to occur between gametes bearing similar allelomorphs—for example, an R character. Each gamete bears R, the zygote exhibits it, and the gametes produced by that zygote bear it again.

But we note that we do not *know* whether the character exhibited by such zygote is really the product of the allelomorph of *both* gametes, or is due to the exclusive development of that of one gamete only. Commonly we conceive of all characters of a zygote as the product of both gametes, and in cases of true blended inheritance we must so conceive them. Such a view also accords well with all that we know of the visible processes of fertilisation. Nevertheless, the fact is not certain in the case of the union of similar gametes, and the case may— to take a rough and partially incorrect illustration—be comparable to the known fact that the faculty of speech is, in the normal case, controlled by the centre in the left hemisphere only, the corresponding structures presumed to exist in the right hemisphere not developing or at least not becoming functional. We do not *know*, in fact, whether the character in the zygote depends on, or is in any way affected

by, the fact that *both* gametes were bearers of that character.

But if we suppose that the zygote character is thus a product of the two similar allelomorphs in the normal case, we may on that hypothesis form a conception of what may be imagined to take place in the case of monolepsis. For returning to the heterozygote we perceive that on the formation of its gametes there is a resolution or separation of the two dissimilar allelomorphs which came into it at fertilisation. May we not then suppose that in the case of the homozygote a similar separation takes place? The gametes of the heterozygote DR are bearers of D and R respectively separated out of DR; may not the gametes of the homozygote, which are bearers of R and R, receive those allelomorphs by a similar separation occurring between R and R?

If this reasoning prove valid, we suggest the possibility that in the case of false hybridisation the allelomorph is passed on from the zygote to the gamete without such resolution, and that thus it is not in a state which admits of its being affected by the contrary allelomorph of the other gamete. The case may perhaps be compared with the known fact that on separating the two segmentation spheres of an egg, each half may develop into symmetrical larva.

Unproved as such a suggestion must necessarily be, it is in accord with several of the facts of crossing, of which no other account is as yet forthcoming. If, then, in a cross between D & R, an R be produced in circumstances which leave no doubt that such production is not due to mere environmental disturbance, we must suppose that the D character has never really '*met*' the R character.

Apart from examples of the appearance of a completely recessive form in the first cross, there are curious cases of the appearance of mosaic or pied forms in which the D and R characters form an irregular patchwork. In such a case Correns speaks of the characters as *pœcilodynamous*, a sufficiently expressive term. If, however, it were true that the pied condition is not really due to the dominance failing sometimes and succeeding sometimes, but to the existence in the mosaic of islands of the recessive character in the 'paired' or unresolved state, we ought not to describe the phenomenon by reference to dominance at all.

In the introduction to this paper reference was made to the case of Canary—Goldfinch mules. Here the Goldfinch colour is normally dominant. It is said that, generally speaking, 99 per cent. of mules are thus 'dark.' As was also there stated, the belief is prevalent that in-breeding the hen Canaries has an effect in increasing the proportion of 'light'—or canary-like mules. Others have disputed and denied the truth of this belief.

Nevertheless, it is generally admitted that to get 'light' mules one should begin with a strain of Canaries which, on mating with the Goldfinch, throw some pied birds. On the hypothesis here suggested, the pied character is supposed to be due to the partially unresolved character of the recessive allelomorph. On in-breeding we may conceive the process of non-resolution on formation of gametes to be carried further. We have seen that cross-breeding leads to the fuller resolution of characters—in-breeding may lead to the contrary result.

With the Canary, as the mule is almost (if not quite) universally sterile, further experiment is impossible, but other cases are available for the experimental testing of this hypothesis.

If it is correct, it should appear that when on crossing a D and R a pied form is produced, showing patches of the R character, then such a pied form on crossing with the dominant again is more likely to give pied recessive or recessive offspring than a pure normal recessive would be, for we are on the hypothesis entitled to believe the gametes of the pied mule to partake of the same character as the zygote itself.

On the older view of breeding such a fact would be paradoxical; for the pied form, inasmuch, as it already is part way to the D form, would be supposed *less* likely to show any R in its hybrid than the pure R form.

The fact that Tschermak in his crosses between the pea *Telephone* and yellow varieties obtained a considerable number of seeds greenish or patched with green, is consistent with this view; for this pea, though commonly a green or greenish pea, is liable to great variation, and is frequently mosaic or pied yellow and green.*

The remarkable series of Orchid crosses given by Hurst,† in which the female parent's characters alone appeared as the result of certain extreme crosses, seem rather to illustrate the possibility of parthenogenesis following the stimulus of fertilisation, without zygotic union.]

---

* See Weldon, 'Biometrika,' I, 1902, Pt. II.

† 'J. R. Hort. Soc.,' xxiv, 1900, pp. 104–5.

Galton's Law of Ancestral Heredity in relation to the new Facts.

Such a preliminary survey of the phenomena of heredity as we have attempted would be incomplete without some reference to this subject. We note at once that the Mendelian conception of heredity effected by *pure* gametes representing definite allelomorphs is quite irreconcilable with Galton's conception in which *every* ancestor is brought to account in reckoning the probable constitution of every descendant. With respect of each allelomorphic pair of characters we now see that only four kinds of zygotes can exist, the pure forms of each character, and the two reciprocal heterozygotes. On Galton's view the number of kinds is indefinite.

At first sight it may appear that as the two views are quite incompatible, they must relate to different classes of phenomena. In so far as Galton's law relates to continuous variations, with blended inheritance, this may be the case [see p. 116], but in some of the cases following Galton's rule, notably that of the colour of Basset hounds, the colours dealt with are discontinuous.* Let us consider what evidence there is in this case that the gametes are not pure tricolour *or* non-tricolour, as we should now expect them to be. The first question is, does either colour show dominance? If either were dominant it must clearly be the tricolour,

and in view of the fact that both tricolour × tricolour and non-tricolour × non-tricolour are said to have given mixtures, neither colour can be supposed to be exclusively dominant. In this case, therefore, as it is impossible to tell which individuals are pure and which are heterozygotes, Galton's results might possibly have occurred, *and the gametes yet be pure.* More cannot be said without reference to the actual details out of which the tables were constructed.

Attention may also be called to the fact that in cases which fully obey Mendel's ratio (and exhibit dominance), two of the commonest matings happen to give the same result as they would do on Galton's expectation, though the latter is founded on wholly different considerations. Mendel, for instance, expects

DR × DR to give 3Ds and 1R,

and that DR × R will give equal numbers of Ds and Rs. Both these results are, *cæteris paribus*, to be expected on Galton's law, so that it might need a good deal of experiment to distinguish the two classes of cases. A clear distinction would, however, at once be found by comparing the result of DR × R with that of DR × D.

Bearing this in mind, and having regard to the considerations mentioned in the paragraph on blended inheritance, it is impossible to avoid the suggestion that Galton's law may be a representation of particular groups of cases which are in fact Mendelian, in the sense, that is, that there may be purity of gametes in respect of allelomorphic characters. In any case it is now certain that Galton's law cannot be accepted as 'universally applicable to bi-sexual descent.'

---

* Pearson ('Roy. Soc. Proc.,' vol. 66, 1900, p. 142) has suggested a distinct formula for these cases of alternative inheritance, which he terms the 'Law of Reversion.' He urges that such phenomena should be treated separately from those of blended inheritance. Both laws alike are of course based on the numerical composition of the ancestry.

By any practical breeder this must have been always expected, for he knows that while he can rapidly fix some characters, some never come true at all, and others will not come true with any certainty after long selection. The expectation after simple selection is, in fact, quite different for different characters. Mendel's principle disposes of a great part of these difficulties, for we now know that any recessive character may be fixed at once by selecting recessives, and that this fixity may have nothing to do with the novelty of the character, its 'prepotency,' &c., and that the heterozygote may never come true.

Galton's law in fact does not recognise that *absolute* purity which is so common a phenomenon in breeding, as it is in nature. The breeder, in hosts of instances, is not, as a matter of fact, constantly troubled by recurrences of forms with which, even in his own practice, his strain has been crossed. Of this the full explanation is now seen; for if two similar gametes meet, their offspring will be no more likely to show the other allelomorph than if no cross had ever taken place.

Conclusion.

We have now sketched the principal deductions already attained by the study of cross-breeding, and we have pointed out some of the results now attainable by that method. The lines on which such experiments can be profitably undertaken are now clear and a wide field of research is open.

The properties of each character in each organism have, as regards heredity and variation to be separately investigated, and, for the present, generalisation in regard to those properties must be foregone. The outlook, in

fact, is not very different from that which opened in chemistry when definiteness began to be perceived in the laws of chemical combination. It is reasonable to infer that a science of Stoechiometry will now be created for living things, a science which shall provide an analysis, and an exact determination of their constituents. The units with which that science must deal, we may speak of, for the present, as character-units, the sensible manifestations of physiological units of as yet unknown nature. As the chemist studies the properties of each chemical substance, so must the properties of organisms be studied and their composition determined.

To the solution of the practical problems of heredity, and a determination of the laws of breeding both plants and animals, this is the first step. The attainment of these solutions is now only a question of time and patience.

That the same method will give the key to the nature of specific differences, we may perhaps fairly hope. Certain it is that until the several characters are thus disentangled and their variations classified, no real progress with this question can be expected.*

---

* It is absolutely necessary that in work of this description some uniform notation of generations should be adopted. Great confusion is created by the use of merely descriptive terms, such as 'first generation,' 'second generation of hybrids,' &c., and it is clear that even to the understanding of the comparatively simple cases with which Mendel dealt, the want of some such system has led to difficulty. In the present paper we have followed the usual modes of expression, but in future we propose to use a system of notation modelled on that used by

[Since this Report was written, a paper has appeared by Professor Weldon, entitled 'Mendel's Laws of Alternative Inheritance in Peas,'*

Galton in 'Hereditary Genius.' We suggest as a convenient designation for the parental generation the letter P. In crossing, the P generation are the pure forms. The offspring of the first cross are the first filial generation F. Subsequent filial generations may be denoted by F_2, F_3, &c. Similarly, starting from any subject-individual, P_2 is the grandparental, P_3 the great-grandparental generation, and

questioning the importance of Mendel's discovery. This paper will be dealt with in a separate publication by W. Bateson, entitled 'Mendel's Principles of Heredity,' with which is incorporated a translation of Mendel's papers.—March 1902.]

so on. We suggest this terminology here for the consideration of others who are working in the same field. All that is essential is to obtain uniformity, and it is quite likely that a better system may be suggested.

* 'Biometrika,' 1, 1902, Pt. 11.

Remember the dates. Bateson first learned of Mendel early in 1900; this report for the Royal Society was presented late in 1901. In less than two years it proved possible to marshal considerable evidence to confirm the data for peas. Bateson's effective presentation was important in gaining acceptance for Mendelism, but throughout the first decade of the twentieth century there was considerable opposition. Much of it centered on the question of the 'purity of the gametes,' that is, how could an **Aa** organism produce pure **A** or pure **a**, but never **Aa** gametes?

THOMAS HUNT MORGAN

In 1909, Thomas Hunt Morgan (1866–1945), published the following article revealing his skepticism of Mendelism and its explanatory hypotheses. This article is not especially important for any intrinsic merit—the arguments are not very convincing—but in a few years its author's name was to replace that of Mendel as the principal formulator of genetic theory.

WHAT ARE 'FACTORS' IN MENDELIAN EXPLANATIONS?
By Prof. T. H. Morgan.
Columbia University, New York, N. Y.

In the modern interpretation of Mendelism, facts are being transformed into factors at a rapid rate. If one factor will not explain the facts, then two are invoked; if two prove insufficient, three will sometimes work out. The superior jugglery sometimes nec-

essary to account for the results may blind us, if taken too naively, to the common-place that the results are often so excellently 'explained' because the explanation was invented to explain them. We work backwards from the facts to the factors, and then, presto! explain the facts by the very factors that we invented to account for them. I am not unappreciative of the distinct advantages that this method has in handling the facts.

From *Proceedings American Breeder's Association 5*: 365–368. 1909.

I realize how valuable it has been to us to be able to marshal our results under a few simple assumptions, yet I cannot but fear that we are rapidly developing a sort of Mendelian ritual by which to explain the extraordinary facts of alternative inheritance. So long as we do not lose sight of the purely arbitrary and formal nature of our formulae, little harm will be done; and it is only fair to state that those who are doing the actual work of progress along Mendelian lines are aware of the hypothetical nature of the factor-assumption. But those who know the results at second hand and hear the explanations given almost invariably in terms of factors, are likely to exaggerate the importance of the interpretations and to minimize the importance of the facts.

By way of illustrating empirically what I have in mind, I should like to point out certain implications in the current assumption that the factors (sometimes referred to as the actual characters themselves—unit-characters, not infrequently) are dissociated in the germ-cells of the hybrids into their allelomorphs. For instance a tall pea crossed with a dwarf pea produces in the first generation a tall hybrid. Such tall peas inbred produce three tall peas to one dwarf. Such are the surprising facts. Mendel pointed out that the numerical results could be explained if we assume that the hybrid peas produce germ-cells of two kinds, tall-producing and dwarf-producing. The simplicity of the explanation, its wide applicability and what I may call its intrinsic probability will recommend his interpretation to all who have worked with such problems of heredity. Out of this assumption the modern factor-hypothesis has emerged. The tallness of the tall pea is said to

be due to a tall-factor; the dwarfness of the dwarf-pea, to be a dwarf-factor. When they meet in the hybrid, the tall-factor gets the upper hand. So far we do little more than restate Mendel's view. But when we turn to the germ-cells of the hybrid we go a step further. We assume that the tall-factor and the dwarf-factor retire into separate cells after having lived together through countless generations of cells without having produced any influence on each other. We have come to look upon them as entities that show a curious antagonism, so that when the occasion presents itself, they turn their backs on each other and go their several ways. Here it seems to me is the point where we are in danger of over-looking other possibilities that may equally well give us the two kinds of germ-cells that the Mendelian explanation calls for.

In the first place the assumption of separation of the factors in the gametes is a purely preformation idea. The factors have become entities that may be shuffled like cards in a pack, but cannot become mixed. The whole mechanism turns on the old preformation conception of the way the characters of the adult are contained in the egg. The success of the method as a ready means of explanation does not, in my opinion, justify the procedure; for the preformation idea has always led to immediate, if temporary, successes; while the epigenetic conception, although laborious, and uncertain, has, I believe, one great advantage, it keeps open the door for further examination and re-examination. Scientific advance has most often taken place in this way.

Can we offer one or more alternative points of view that will accord with the assumption of two kinds of germ-cells in the first generation of hybrids? Until all the possibilities are exhausted it will be at least judicious to hold the segregation hypothesis, as currently interpreted—a purely formal procedure.

I think that the condition of two alternative characters may equally well be imagined as the outcome of alternative states of stability (or of conditions) that stand for the characters that make up the individual. We can conceive of tallness and dwarfness as two protoplasmic stages or states without intermediates, just as chemical compounds are alternative and separate, each with its own properties. If we mix, in the hybrid, materials that have the power to produce one, or the other condition, it is conceivable that they remain at first independent; the stronger the result in most cases, although the weaker not infrequently shows its presence also. In one and the same individual both the dominant and the recessive characters may appear. For example, chocolate and black mice give Mendelian results; but I have obtained individuals that were black in front and chocolate behind. Again black-eye and pink eye are Mendelian alternatives, yet I have three mice with one pink-eye and one black one. Local conditions, I infer, determine, in these heterozygotes, that at one time the dominant and at another the recessive characters come to the front, and I could bring forward evidence to show that the results are not due to segregations of unit-characters. The continuous series of types not infrequently shown in the second hybrid generation, shows also, I think, that segregation of characters in the germ-cells is by no means an established Mendelian rule.

There is a consideration here of capital importance. The egg need not contain the *characters* of the adult, nor need the sperm. Each contains a particular material which *in the course of the development produces* in some unknown way the character of the adult. Tallness, for instance, need not be thought of as represented by that character in the egg, but the material of the egg is such that placed in a favorable medium it continues to develop until a tall plant results. Similarly for shortness. The fertilized egg of the hybrid between these types likewise contains both possibilities. In the soma or body of these produce their characters—one kind at least. If the germ cell of the hybrid can be traced directly to the undifferentiated germ plasma of that individual, it too contains as yet not the characters but the undeveloped materials from which the characters develop. It follows that we are not justified in speaking of the materials in the germ-cells as the same thing as the adult characters until they develop. As already stated, it seems possible that at some time in the history of the germ-cells, the new materials reach in some cells one condition of equilibrium; in others the alternating condition. On the average it is possible that equal numbers of each are formed, if they are equally potent —but only on an average; for it is well known that Mendelian results are only average results, and it by no means follows that in all individuals equal numbers of the two states are present. In fact the well-recognized principle of pre-potency clearly indicates that equal numbers of the alternative conditions are not always present in each individual. In some such way, I think, it is possible to conceive of the admitted presence of two kinds of germ-cells that produce the average Mendelian proportions. The point of view lacks the sharpness of the preformed-factor assumption— which is not altogether a disadvantage when all the facts are considered! The view has two advantages of its own, one is that it shows at least the possibility of another interpretation, and the more such we have the less likely are we to become blind followers of one idea. The other advantage, to my mind, is that the assumption is more epigenetic in character than the conventional one. This, however, may not be conceded by everyone.

The crucial point for the view here presented is found in the critical stage in the germ-cells of the hybrid. It is an equally difficult point for the segregation theory. According to the latter, the characters separate at this time; according to my view, they now enter into an intimate relation with each other—for which view there is good cytological evidence—and emerge in such a condition, that, in some cells, the state that will produce one character has established itself; in other cells, the opposite state.[1] We have fairly good evidence to suppose that in the critical stage the maternal and paternal elements are brought into a new relation to each other—one that has not existed in the somatic cells. It is at this time that the changes take place that lead to the postulated differences.

In conclusion I wish to point out a parallel that we cannot afford to ignore. In the development of the egg the organs of the individual become just as sharply separated from each other as are the characters of Mendelism. Experimental embryology has made it more than probable, that the alternative nature of the characters of the different organs is not the result of disjunction of unit-characters present in the egg although the preformationists have attempted to explain development in this way. The process is far more subtle and in a sense more complicated. It may be claimed that we have here only an analogy, not a similar series of events, but I incline to think that the comparison worthy of serious consideration.

---

[1] This statement has a superficial resemblance to another view I proposed two years ago, namely of alternate dominance. In the present view alternate dominance is different in so far as the recessive need not contain the dominant latent, etc. Also in other respects.

BIBLIOGRAPHY

Refer to the Preface for a list of classical papers included in other anthologies.

ALLEN, GARLAND E. 1969. 'Hugo de Vries and the reception of the "mutation theory."' *Journal of the History of Biology* 2: 55–87.
BATESON, BEATRICE. 1928. *William Bateson, F.R.S. Naturalist.* London: Cambridge University Press.
BATESON, W. and others. 1902–1908. *Royal Society. Reports to the Evolution Committee. Reports I–V.* London: Harrison and Sons.

BATESON, W. 1902. *Mendel's Principles of Heredity. A Defence.* London: Cambridge University Press.

BATESON, W. 1908. *The Methods and Scope of Genetics. An Inaugural Lecture Delivered 23 October 1908.* London: Cambridge University Press.

BATESON, W. 1909. *Mendel's Principles of Heredity.* London: Cambridge University Press. (Third Impression with Additions, 1913)

BATESON, WILLIAM. 1928. *The Scientific Papers of William Bateson.* Edited by R. C. Punnett. 2 Volumes. London: Cambridge University Press.

BENNETT, J. H. Editor. 1965. *Experiments in Plant Hybridization. Mendel's Original Paper in English Translation with Commentary and Assessment by the Late Sir Ronald A. Fisher Together with a Reprint of W. Bateson's Biographical Notice of Mendel.* Edinburgh: Oliver and Boyd.

BOYES, B. C. 1966. 'The impact of Mendel.' *Bioscience 16*: 85–92.

BRINK, R. A. Editor. 1967. *Heritage from Mendel.* Madison: University of Wisconsin Press. Chapter 1 by C. P. Oliver, and Chapter 2 by A. H. Sturtevant.

CASTLE, W. E. 1903. 'Mendel's laws of heredity.' *Science 18*: 396.

CASTLE, WILLIAM E. 1911. *Heredity in Relation to Evolution and Animal Breeding.* New York: D. Appleton.

CASTLE, W. E. 1951. 'The beginnings of Mendelism in America.' In *Genetics in the 20th Century.* Edited by L. C. Dunn. New York: Macmillian.

CONKLIN, E. G. 1908. 'The mechanism of heredity.' *Science 27*: 89–99. A prominent biologist surveys the field and omits Mendel.

COOK, O. F. 1907. 'Mendelism and other methods of descent.' *Proceedings of the Washington Academy of Sciences 9*: 189–240. An example of an anti-Mendelian argument.

CREW, F. A. E. 1966. *The Foundation of Genetics.* New York: Pergamon.

CREW, F. A. E. 1968. 'R. C. Punnett.' *Genetics 58*: 1–7.

DARBISHIRE, A. D. 1911. *Breeding and the Mendelian Discovery.* London: Cassell.

DAVENPORT, CHARLES B. 1901. 'Mendel's law of dichotomy in hybrids.' *Biological Bulletin 2*: 307–10.

DAVENPORT, CHARLES B. 1907. 'Heredity and Mendel's law. *Proceedings of the Washington Academy of Science 9*: 179–187.

DODSON, E. O. 1955. 'Mendel and the rediscovery of his work.' *Scientific Monthly 81*: 187–195.

DORSEY, M. J. 1944. 'Appearance of Mendel's paper in American libraries.' *Science 99*: 199–200.

DUNN, L. C. 1965. 'Mendel, his work and his place in history.' *Proceedings of the American Philosophical Society 109*: 189–198.

DUNN, L. C. 1965. *A Short History of Genetics.* New York: McGraw-Hill.

DUNN, L. C. 1969. 'Genetics in historical perspective.' In *Genetic Organization.* Edited by Ernst Caspari and Arnold W. Ravin. New York: Academic Press.

EAST, E. M. 1922. 'As genetics comes of age.' *Journal of Heredity 13*: 207–214.

EAST, E. M. 1923. 'Mendel and his contemporaries.' *Scientific Monthly 16*: 225–236.

EICHLING, C. W. 1942. 'I talked with Mendel.' *Journal of Heredity 33*: 243–246.

FISHER, R. A. 1936. 'Has Mendel's work been rediscovered?' *Annals of Science 1*: 115–137. Reprinted in Stern and Sherwood. A suggestion that Mendel's data are 'too good.' See also Fong; Orel.

FONG, PETER. 1969. 'Letters to the editor.' *Perspectives in Biology and Medicine 12*: 636. Chance and the rediscovery of Mendel's paper.

GASKING, ELIZABETH B. 1959. 'Why was Mendel's work ignored?' *Journal of the History of Ideas 20*: 60–84.

GENETICS, Editors of. 1950. 'The birth of genetics. Mendel, de Vries, Correns, Tschermak in english translation.' *Genetics 35*: Supplement, 1–47.

GLASS, BENTLEY. 1953. 'The long neglect of a scientific discovery: Mendel's laws of inheritance.' In *Studies in Intellectual History*. Edited by G. Boaz. Baltimore: Johns Hopkins University Press.

GLASS, BENTLEY. 1947. 'Maupertuis and the beginnings of genetics.' *Quarterly Review of Biology 22*: 196–210.

ILTIS, HUGO. 1932. *Life of Mendel*. New York: Norton. Reprinted 1966 by Haffner, New York.

ILTIS, HUGO. 1947. 'A visit to Gregor Mendel's home.' *Journal of Heredity 38*: 163–167.

ILTIS, HUGO. 1951. 'Gregor Mendel's life and heritage.' In *Genetics in the 20th Century*. Edited by L. C. Dunn. New York: Macmillan.

KRIZENECKY, JAROSLAV. Editor. 1965. *Fundamenta Genetica (the Revised Edition of Mendel's Classic Paper with a Collection of Twenty Seven Original Papers Published During the Rediscovery Era)* Prague: Publ. House Czecho. Acad. Sci.

LOCK, R. H. 1906. *Recent Progress in the Study of Variation, Heredity, and Evolution*. New York: Dutton.

MORGAN, T. H. 1926. 'William Bateson.' *Science N. S. 63:* 531–535.

OLBY, ROBERT C. 1966. *Origins of Mendelism*. New York: Schocken Books.

OREL, VÍTEZSLAV. 1968. 'Will the story on "too good" results of Mendel's data continue?' *Bioscience 18*: 776–778.

OREL, V., and M. VÁRVA. 1968. 'Mendel's program for the hybridization of apple trees.' *Journal of the History of Biology 1*: 219–224.

PUNNETT, R. C. 1911. *Mendelism*. Third Edition. New York: Macmillan.

PUNNETT, R. C. 1950. 'Early days of genetics.' *Heredity 4*: 1–10.

ROBERTS, H. F. 1929. *Plant Hybridization Before Mendel*. Princeton: Princeton University Press.

SPILLMAN, W. J. 1902. 'Exceptions to Mendel's laws.' *Science 16*: 794–796.

STERN, CURT, and EVA R. SHERWOOD. 1966. *The Origin of Genetics. A Mendel Source Book*. San Francisco: W. H. Freeman. Papers and letters of Mendel, Focke, de Vries, Correns, Fisher, and Wright.

STOMPS, T. J. 1954. 'On the rediscovery of Mendel's work by Hugo de Vries.' *Journal of Heredity 45*: 293–294.

STUBBE, H. 1965. *Kurze Geschichte der Genetik bis zur Wiederentdeckung der Vererbungsregelen Gregor Mendels*. Jena: Fischer.

STURTEVANT, A. H. 1965. *History of Genetics*. New York: Harper and Row.

STURTEVANT, A. H. 1965. 'The early Mendelians.' *Proceedings of the American Philosophical Society 109*: 199–204.

TSCHERMAK, E. VON. 1951. 'The rediscovery of Gregor Mendel's work.' *Journal of Heredity 42*: 163–171.

WEIR, J. A. 1968. 'Agassiz, Mendel, and heredity.' *Journal of the History of Biology 1*: 179–203.

WELDON, W. F. R. 1902. 'Mendel's laws of alternative inheritance in peas.' *Biometrika 1*: 228–254. One of the first vehement attacks on Mendelism. Answered by Bateson in *Mendel's Principles of Heredity* (1902).

ZIRKLE, CONWAY. 1951. 'Gregor Mendel and his precursors.' *Isis. 42*: 97–104.

ZIRKLE, CONWAY. 1964. 'Some oddities in the delayed discovery of Mendelism.' *Journal of Heredity 55*: 65–72.

ZIRKLE, CONWAY. 1968. 'Mendel and his era.' In *Mendel Centenary: Genetics, Development and Evolution.* Edited by Roland M. Nardone. Washington: Catholic University of America Press.

ZIRKLE, CONWAY. 1968. 'The role of Liberty Hyde Bailey and Hugo de Vries in the rediscovery of Mendelism.' *Journal of the History of Biology 1*: 205–218.

4 / The Chromosomes and Inheritance

Progress in understanding scientific phenomena is most rapid when testable hypotheses are available. The hypothesis that the nucleus (and or its chromosomes) is largely responsible for inheritance was widely believed during the 1880's (Chapters 2 of this volume and *Heredity and Development*). Two generations later it was to be accepted as true beyond a reasonable doubt, yet it was not until the early years of the twentieth century that effective testing became possible.

Mendel's model for inheritance demanded that the genes behave in very specific and predictable ways. His first law, segregation, demanded that heterozygous individuals, **Aa**, produce pure gametes. That is, they must be either **A** or **a** and, furthermore, they should be formed in statistically equal numbers. His second law, independent assortment, maintained that different pairs of genes are inherited independently of one another.

If these rules are universal or nearly so, and Bateson suggested they were (Chapter 3), then a cytologist can make some specific deductions from the hypothesis, 'chromosomes are responsible for inheritance.' Thus, if Mendel's genes are responsible for inheritance, and if chromosomes are also responsible for inheritance, then the simplest deduction is that genes are parts of chromosomes. If this is so, then the behavior of genes as observed in Mendelian crosses must be based on a parallel behavior of the chromosomes. The chromosomes must behave in such a way as to provide a basis for segregation and independent assortment. The work of cytologists and geneticists suggested that these parallel events would be occurring during meiosis and fertilization.

EDMUND B. WILSON

Synthesizing the data of breeding and cytology to produce the modern theory of genetics began with a short paper of Edmund Beecher Wilson (1856–1939).

MENDEL'S PRINCIPLES OF HEREDITY AND THE MATURATION OF THE GERM-CELLS.

In view of the great interest that has been aroused of late by the revival and extension of Mendel's principles of inheritance it is remarkable that, as far as I am aware, no one has yet pointed out the clue to these principles, if it be not an explanation of them, that is given by the normal cytological phenomena of maturation; though Guyer and Juel have suggested a possible correlation between the variability of sterility of hybrids and abnormalities in the maturation-divisions, while Montgomery has recognized the essential fact in the normal cytological phenomena, though without bringing it into relation with the phenomena of heredity. Since two investigators, both students in this University, have been led in different ways to recognize this clue or explanation, I have, at their suggestion and with their approval, prepared this brief note in order to place their independent conclusions in proper relation to each other and call attention to the general interest of the subject.

Bateson, in his recent admirable little book on Mendel's principles, is led to express the surmise that the symmetrical result in the offspring of cross-bred forms 'must correspond with some symmetrical figure of distribution of gametes in the cell-divisions by which they are produced.' It is needless to remind cytologists that the study of the maturation-mitoses, especially in the case of arthopods, has revealed a mechanism by which such a symmetrical distribution may be effected; for the germ-cells in the great majority of cases arise in groups of fours, formed by two divisions, of which one is in many cases described as differing in character from the ordinary somatic mitoses in that it separates whole chromosomes by a transverse division ('reducing division' of Weismann). Wholly independently of Mendel's conclusions a considerable number of cytologists (vom Rath, Rückert, Häcker) early reached the conclusion that the chromatin-masses from which arise the 'Vierergruppen' (tetrad-chromosomes, or their equivalents) represent double or 'bivalent' chromosomes, each of which was conceived to arise by the union (synapsis), end to end, of two single chromosomes. An actual conjugation of chromosomes in synapsis was inferred by Rückert in some cases (e. g., in *Pristiurus*), and more recently described in a far more detailed way in *Peripatus* and certain insects by Montgomery (1901), who reached the remarkable conclusion that 'in the synapsis stage is effected a union of paternal with maternal chromosomes,

From *Science* N. S. *16*: 991–993. 1902.

so that each bivalent chromosome would consist of one univalent paternal chromosome and one univalent maternal chromosome.' The ensuing transverse or reducing division, therefore, leads to *the separation of paternal and maternal elements and their ultimate isolation in separate germ-cells.* This conclusion rested upon evidence too incomplete to warrant its acceptance without much more extended investigation—it was, indeed, more in the nature of a surmise than a well-grounded conclusion. During the past year Mr. W. S. Sutton, working in my laboratory, has obtained more definite evidence in favor of this result, which led him several months ago to the conclusion that it probably gives the explanation of the Mendelian principle. In the great 'lubber grasshopper' *Brachystola* the chromosomes of the spermatogonia were found to be grouped in eleven pairs of different sizes, which reappeared in essentially the same relation through at least eight successive generations of these cells. In synapsis the graded pairs are converted into similarly graded bivalent chromosomes that appear to arise by a conjugation, or union at one end, of the two members of each of the earlier pairs. Cogent reason is given by Sutton for the conclusion that the chromosome-pairs consist each of a paternal and a maternal member. It is known that in fertilization chromosomes are contributed in equal numbers by the two gametes ('Van Beneden's Law'). Boveri's recent remarkable experiments on sea-urchins have proved that a definite combination of chromosomes is necessary to complete development, and strongly suggests, if they do not prove, that the

individual chromosomes stand in definite relation to transmissible characters taken singly or in groups. Every nucleus, however, contains two such combinations; for the facts of parthenogenesis and merogony prove that either the paternal or the maternal group alone may suffice for complete development. It is a natural conclusion from these facts that the constant morphological differences of the chromosomes observed in the grasshopper are correlated with constant physiological differences. If such be the case it appears highly probable, though the argument can not here be presented in all its weight, that those of corresponding size, associated in pairs, are the paternal and maternal homologues (*sit venia verbo*)! Sutton has pointed out that if this be indeed the case, the union of these homologues in synapsis, and their subsequent separation, which this preliminary union involves, in the reducing (second maturation) division, leads to the members of each pair being isolated in separate germ-cells; and this gives a physical basis for the association of dominant and recessive characters in the cross-bred, and their subsequent isolation in separate germ-cells, exactly such as the Mendelian principle requires.

A similar conclusion was subsequently, but independently, reached by Mr. W. A. Cannon, of the Department of Botany, though by a different and less direct path of approach. A study of hybrid cotton-plants, which are fertile, showed the maturation-divisions to be entirely normal, in contradistinction to the sterile hybrids of *Syringa*, where Juel has shown that the maturation-divisions are abnormal in character. It thus appeared that a sifting apart of paternal and maternal elements, such as Mendel's law demonstrates to occur, cannot be explained on the hypothesis of irregularities in the maturation-divisions (as had been suggested by Guyer's earlier work on pigeon-hybrids). Cannon therefore concluded, on this *a priori* ground, that such a separation of paternal and maternal elements must occur in the normal maturation-divisions, not only in the cross-bred, but also in the normal forms, and that in the character of these divisions must be sought the basis of the law. It is interesting that such a conclusion should have been reached by a botanist, on account of the fact that most recent botanical workers in this field have reached the result that transverse or reducing divisions do not occur in the maturation of the germ-cells in higher plants. It has, however, become clear that only the most exhaustive study of the most favorable material, particularly in the earliest stages of the maturation-divisions, can positively decide this question, and the importance of the most accurate and detailed further study of the phenomena is now manifest. The results I have indicated are already in part in press and will in due time be fully discussed by their authors. Should the study of the maturation-divisions indeed reveal the basis of the Mendelian principle we shall have another and most striking example of the intimate connection between the study of cytology and the experimental study of evolution.

EDMUND B. WILSON.
Zoological Laboratory of
Columbia University,
December 11, 1902.

Walter S. Sutton's (1877–1916) major paper was published in 1903; it is discussed in Chapter 4 of *Heredity and Development*. He pointed out with great clarity how Mendelian inheritance found an exact parallel in the behavior of chromosomes in meiosis and fertilization. Furthermore, he suggested that Mendel's second law could not always be true: for example, if two pairs of genes are on homologous chromosomes, they could not show independent assortment. Thus, if 'genes are parts of chromosomes' is the working hypothesis, the geneticist must find cases where two pairs of homologous genes do not show independent assortment. Cytology and genetics were becoming mutually stimulating fields. The predictions of the cytologist could be tested by the geneticist; the predictions of the geneticist could be tested by the cytologist.

During the next decade enormous progress was made in relating cytology and genetics. In 1914, E. B. Wilson was invited to London to summarize the field for the members of the Royal Society. He did so in one of the famous Croonian Lectures. This is what he had to say.

CROONIAN LECTURE: *The Bearing of Cytological Research on Heredity.** By EDMUND B. WILSON, Da Costa Professor of Zoology, Columbia University, New York

(MS. received and Lecture delivered June 11, 1914.)

The privilege of speaking in this historic centre of learning was first accorded to me more than 30 years ago, through the extraordinary kindness of Prof. Huxley to a young and unknown student. I would like to think it more than a fancy that to the same source, possibly, I may trace the distinguished honour of having been invited, after the lapse of many years, to speak here once again on the subject of cytology in its bearing on heredity. Of all Huxley's wise and felicitous sayings none has more persistently lingered in my memory or appealed to my imagination than one which vividly pictured, 35 years ago, the basic phenomenon that the cytologist seeks to elucidate. Suggested, no doubt, by the researches of Hertwig, Strasburger, and Van Beneden, then but recently made known, this well-known passage is as follows:—

'It is conceivable, and indeed probable, that every part of the adult contains molecules derived from both the male and the female parent; and that, regarded as a mass of molecules, the entire organism may be compared to a web of which the warp is derived from the female and the woof from the male. And each of these may constitute an indi-

---

* It has been impracticable to reproduce here the original photographs and some of the other figures by which the lecture was illustrated.

---

viduality, in the same sense as the whole organism is an individual, although the matter of the organism has been constantly changing' (1878).

The advance of modern cytology has been in some important respects a development of the germ contained in these words. For the aim of cytology, in so far as it bears directly upon the problems of heredity, is to trace out in the individual life the history of maternal and paternal elements originally brought together in the fertilisation of the egg. And the drift of latter-day research, while it has not precisely confirmed Huxley's conception, has nevertheless been quite in harmony with the essential thought to which he gave such picturesque expression at a time when the labours of cytology were but just begun.

This thought has been most nearly realised through the study of the cell nucleus, and in particular of the bodies known as chromosomes. I ask attention especially to these bodies in connection with certain problems of genetics, not because the chromosomes are the only elements concerned in heredity, but because they offer the most available point of attack and have in fact yielded the most definite results. The limitations of time compel me to take a good deal for granted, and to pass over, for the most part, the historical and controversial aspects of the subject. I must be content, in the main, to state briefly what I believe to be established or indicated by the evidence. My task is much lightened by Prof. Farmer's earlier presentation of many important aspects of the subject in his Croonian Lecture of seven years ago. Permit me, nevertheless,

From *Proceedings of the Royal Society B 88*: 333–352, 1914. Reprinted by permission.

for the sake of present clearness, to indicate briefly some of the essential facts determined prior to the rediscovery of Mendel's law in 1900.

(1) The work of cytology in its period of foundation laid a broad and substantial basis for our more general conceptions of heredity and its physical substratum. It demonstrated the basic fact that heredity is a consequence of the genetic continuity of cells by division, and that the germ-cells are the vehicle of transmission from one generation to another. It accumulated strong evidence that the cell-nucleus plays an important *rôle* in heredity. It made known the significant fact that in all the ordinary forms of cell-division the nucleus does not divide *en masse* but first resolves itself into a definite number of chromosomes; that these bodies, originally formed as long threads, split lengthwise so as to effect a meristic division of the entire nuclear substance. It proved that fertilisation of the egg everywhere involves the union or close association of two nuclei, one of maternal and one of paternal origin. It established the fact, sometimes designated as 'Van Beneden's law' in honour of its discoverer, that these primary germ-nuclei give rise to similar groups of chromosomes, each containing half the number found in the body-cells. It demonstrated that when new germ-cells are formed each again receives only half the number characteristic of the body-cells. It steadily accumulated evidence, especially through the admirable studies of Boveri, that the chromosomes of successive generations of cells, though commonly lost to view in the resting nucleus, do not really lose their individuality, or that in some less obvious way they conform to the principle of genetic continuity. From these facts followed the far-reaching conclusion that the nuclei of the body-cells are diploid or duplex structures, descended equally from the original maternal and paternal chromosome-groups of the fertilised egg. Continually receiving confirmation by the labours of later years, this result gradually took a central place in cytology; and about it all more specific discoveries relating to the chromosomes naturally group themselves.

All this had been made known at a time when the experimental study of heredity was not yet sufficiently advanced for a full appreciation of its significance; but some very interesting theoretical suggestions had been offered by Roux, Weismann, de Vries, and other writers. While most of these hardly admitted of actual verification, two nevertheless proved to be of especial importance to later research. One was the pregnant suggestion of Roux (1883), that the formation of chromosomes from long threads brings about an alignment in linear series of different materials or 'qualities.' By longitudinal splitting of the threads all the 'qualities' are equally divided, or otherwise definitely distributed, between the daughter-nuclei. The other was Weismann's far-seeing prediction of the reduction division, that is to say, of a form of division involving the separation of undivided whole chromosomes instead of the division-products of single chromosomes. This fruitful suggestion (1887) pointed out a way that was destined to lead years afterwards to the probable explanation of Mendel's law of heredity.

(2) Such, in bird's-eye view, were the most essential conclusions of our science down to the close of the nineteenth century. A new era of discovery

now opened. As soon as the Mendelian phenomena were made known it became evident that in broad outline they form a counterpart to those which cytology had already made known in respect to the chromosomes. Characters and chromosomes alike are singly represented (haploid or simplex) in the gametes, doubly represented (diploid or duplex) in the zygote and its products. In the formation of new germ-cells both alike are once more reduced from the diploid to the haploid condition. A parallelism so striking inevitably suggested a direct connection between the two orders of phenomena. And the hope was thus raised that the mechanism of heredity might be susceptible of a far more searching analysis than had yet been thought of.

It is a rather striking coincidence that almost at the moment of the rediscovery of Mendel's law, and apparently quite independently of it, microscopical studies were establishing the cytological facts upon which its explanation probably rests. Guyer's studies on hybrid pigeons led him, in 1900, to suspect a disjunction of maternal and paternal chromatin-elements in the reduction division, a conclusion which he developed further in 1902. But the real basis for an explanation of Mendel's law was laid by two conclusions announced in 1901 by Boveri and by Montgomery, independently of each other and apparently without knowledge of the Mendelian phenomena. Familiar as these conclusions are, I will dwell upon them for a moment, since they are fundamental to all that follows.

Boveri's masterly experiments on dispermic sea-urchin eggs gave the first conclusive proof that the chromosomes directly affect the process of development, and that they are qualitatively different in respect to their individual influence. Eggs into which two spermatozoa are caused to enter develop into larvæ that are almost always pathological, deformed or monstrous. The first cleavage of such eggs is by a tripolar or quadripolar division, and the cytological examination proves that this involves initial and apparently irreversible aberrations in the distribution of the chromosomes to the embryonic cells. Boveri's analysis, carried out with characteristic sagacity and thoroughness, seems to leave no escape from the conclusion that the abnormal combinations of the chromosomes thus produced are the cause, and the only cause, of the abnormal forms of development. The chromosomes must therefore be qualitatively different. This conclusion has been confirmed and rendered more specific by many later researches. It was proved, for instance, that in certain animals one of the chromosomes, or a small corresponding group of chromosomes, stands in some special relation to the determination of sex and the heredity of sex-linked characters. The study of hybrid sea-urchin larvæ by Baltzer, Herbst, and others, gives strong reason to conclude that many of the aberrations which they show in respect to the combination of maternal and paternal characters result from corresponding aberrations in the distribution of maternal and paternal chromosomes. In the evening primroses, the researches of Lutz and of Gates have shown that the *gigas* type of mutant has arisen in association with a doubling of all the chromosomes; recently the same observers show that the *lata* type is characterised by, and probably has arisen through, the presence of a single extra chromosome. To the still more recent important results of Gregory on the Chinese primrose I will presently refer.

The conclusion of Montgomery was not less important, but failed at first to receive the consideration that it deserved. Among the suggestions that immediately followed upon Weismann's speculations concerning the reduction division, one of the most fruitful was that of Henking (1891), that the reduction of the number of chromosomes in the germ-cells is initiated by their conjugation two by two in pairs during synapsis, to be followed by their disjunction in the reduction division. Montgomery drew the bold conclusion that in this process each chromosome of paternal descent unites with a corresponding or homologous one of maternal descent; and he suggested that this process, though occurring at the very end of development, might be regarded as the final step in the fertilisation of the egg. This surprising conclusion was based on a comparative study of the size-relations of the chromosomes in the diploid and haploid nuclei. I well remember the scepticism with which I, like many others, first received it. The conjugation of chromosomes, to say nothing of paternal and maternal homologues, has been obstinately contested; it must be admitted that the proof is still far from complete for the chromosomes generally. Nevertheless, in spite of all scepticism, the drift of later research has been, I think, steadily in its favour. Both in plants and in animals the diploid nature of the chromosome groups in the somatic cells is often clearly visible to the eye, owing to conspicuous size-differences among the chromosomes. In such cases, as was first urged especially by Montgomery and by Sutton, the chromosomes may be sorted out into pairs according to their size. In a few cases, of which the Diptera offer the most striking examples, the sorting out is performed by nature, all the chromosomes being actually grouped side by side in pairs according to their size (fig. 1).* The conclusion here becomes highly probable that each pair includes a maternal and a paternal member, and that these are destined to conjugate in synapsis. In the case of the sex-chromosomes, to which I shall return, the probability becomes a certainty.

The proof of the reduction division likewise remains incomplete for the chromosomes generally, and is fully demonstrative only in case of certain special kinds of chromosomes, in particular the sex-chromosomes and the 'm-chromosomes' of the coreid Hemiptera. Strong confirmatory evidence of both conjugation and disjunction has, however, been afforded for the chromosomes generally by studies on the maturation-process in hybrids, especially in Drosera by Rosenberg, in Œnothera by Geerts, and in Lepidoptera by Federley and Doncaster.

The promulgation of the conclusions of Boveri and Montgomery opened the modern period of cytological inquiry and, as has been said,

---

* These facts were illustrated by photographs of the chromosomes in Drosophila and Musca, from preparations by C. W. Metz, who has for some time been engaged with this problem in my laboratory.

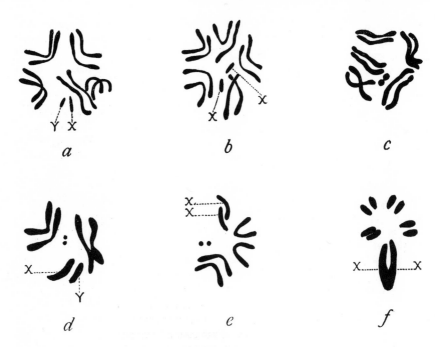

FIG. 1. Exact drawings of the diploid chromosome groups in various Diptera, showing the chromosomes grouped in pairs; *a, b, c, e,* from Stevens; *d, f,* from Metz.

a, Calliphora vomitaria, ♂ *; b,* the same, ♀; *c, Sarcophaga sarracina,* ♀; *d, Drosophila amœna,* ♂ *; e, D. ampelophila,* ♀[1]; *f, D. funebris,* ♀.

provided a substantial basis for the cytological explanation of Mendel's law. This explanation follows in the most simple and natural manner from the observed facts. It assumes primarily that the Mendelian phenomena result from the shuffling (to employ the phrase of Farmer) of chromosomes that are concerned in the determination of the so-called unit characters. More specifically, the main assumption is that Mendelian allelomorphs are borne by corresponding pairs of chromosomes, each consisting of a maternal and a paternal member. By the conjugation of the homologous members of these pairs two by two, to form bivalents or gemini, as assumed by Montgomery, the maternal and paternal homologues assume such a grouping that they may be disjoined in the succeeding reduction division (in general accordance with Weismann's early prediction); and from this follows the disjunction or segregation of the Mendelian allelomorphs which these chromosomes bear. The independent distribution or assortment of different units is explained by the assumption (in favour of which defi-

---

[1] [Now known as *Drosophila melanogaster.* (Ed.)]

nite evidence now exists) that the bivalents behave independently of one another.

The explanation as here outlined was first clearly and logically developed by Sutton in 1902–3, when a student in my laboratory. Naturally enough, however, several others came independently to more or less similar conclusions nearly at the same time—in particular Guyer, Correns, Boveri, Cannon, and de Vries. As will appear later, Sutton's elegant hypothesis was too simply framed to account for all the facts, and has had to undergo some modifications. In its main principle, however, it has received cumulative substantiation by later work in many directions. An important confirmation of the fundamental assumption is given by a discovery announced by R. P. Gregory before this Society only a few weeks ago. In certain plants of the Chinese primrose the usual number of chromosomes is doubled in both the gametes and the somatic cells. The genetic evidence obtained from such plants indicates that all the Mendelian units or factors thus far examined are correspondingly doubled. This result weighs strongly, I believe, in favour of the view that these factors are borne by the chromosomes, and may open the way to its crucial experimental test.

The full force of the hypothesis only becomes apparent when we come to closer quarters with the facts. I shall attempt to illustrate this by considering certain phenomena which now stand in the foreground of interest and bring home to us the intimacy of the relation that has been established between cytology and genetics.

(3) I first ask attention to certain facts relating to the cytological basis of sex, a subject with which my own researches have been especially engaged during the past ten years. To the cytologist the interest of the phenomena extends far beyond the special problem of sex. Nature has here performed a series of experiments which gives a crucial test of many of our earlier conclusions, provides a secure basis for further advances, and at the same time brings vividly before us the connection of the chromosomes with heredity. I will here touch only upon the main facts, especially in their bearing upon the phenomena of linkage, to which, I believe, they give the cytological key.

That the chromosomes are involved in the determination of sex was first suggested, in 1902, by McClung, who argued on *a priori* grounds that the so-called 'accessory chromosome,' which enters but half the spermatozoa, is a sex-determinant. A substantial basis for this conclusion was provided in 1905, when the late Dr. N. M. Stevens and myself, working independently on Coleoptera and Hemiptera, discovered that in some of these insects the sexes differ in the composition of the diploid chromosome-groups. In the simplest type, first worked out in the Hemiptera, the 'accessory'—or, as I have preferred to call it, the X-chromosome, or sex-chromosome—is unpaired in the male, but paired in the female. Since the sexes are identical in respect to the other chromosomes, the latter may be disregarded, the sexual formulas being written simply as XX for the female and X (or X0) for the male. All of the other chromosomes are paired; hence the male possesses an odd number of chromosomes, one less than that of the female. Thus is explained the fact, first discovered in

1891 by Henking in Pyrrhocoris, that in the reduction division of the male the X-chromosome passes undivided to one pole, so that two classes of spermatozoa are formed, one with X and one without. In the female, on the other hand, the two X-chromosomes conjugate to form a bivalent, as usual, and then disjoin in the reduction division, so that every egg receives one X. This fact, at first inferred from the other relations, was soon afterwards demonstrated by direct observation, first by Morrill in insects, on rather scanty evidence, later fully established by Boveri, Gulick, Mulsow, and Frolowa in nematodes. It thus became clear that fertilisation of the egg by the X class of spermatozoa will produce the female combination XX, by the no-X class the male combination

X (or X0) (fig. 2). From these relations, and those found in a second type, described below, Miss Stevens and I concluded that the determination of sex in these animals depends upon which class of spermatozoon enters the egg, and that the X-chromosome plays some special *rôle* in the process. In a general way this substantiated McClung's earlier suggestion, though he reversed the actual significance of the two classes of spermatozoa.

I discovered in my first researches on Hemiptera a second type, independently found and fully worked out in the Coleoptera by Miss Stevens, in which the X-chromosome of the male is accompanied by a mate of different type, often much smaller that X, which I called the Y-chromosome. This was the first discovery of

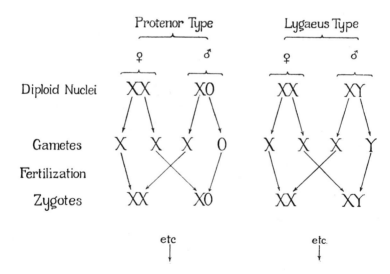

FIG. 2. Diagram of the relations of the sex-chromosomes to sex-production, showing the two main types represented by the Hemiptera *Protenor* (Y-chromosome absent) and *Lygæus* (Y-chromosome present). In either case random union of the maternal and paternal gametes reproduces the original forms (males and females) in equal numbers.

a heterogeneous chromosome-pair in any animal or plant. In this case X and Y conjugate and disjoin like any other chromosome pair—a fact here shown with incomparable clearness—so that half the spermatozoa receive X and half Y. The X-class are, as before, female-producing, while the Y-class are male-producing; and the sexual formulas become XX for the female and XY for the male (fig. 2). Owing to the small size of Y, these differences are in many species conspicuously visible in the diploid nuclei, despite the fact that the sexes here possess the same number of chromosomes. The two types are connected by a series of intermediate conditions, varying with the species, and X may consist of two or more separate chromosomes. The Y-chromosome, on the other hand, is single in all cases thus far accurately known.*

At the time these conclusions were announced it seemed unlikely that the fertilisation of the egg by the two classes of spermatozoa could ever actually be followed out. This has, nevertheless, been accomplished recently by Mulsow in the case of a nematode, Ancyrocanthus, where the chromosomes remain distinct in the mature spermatozoa and can readily be counted, even in the living object. Both classes were here traced into the egg, and the sexual differences were clearly shown in the germ-nuclei at the time of their union.

---

* In an extreme case, now under investigation by Mr. Goodrich in my laboratory, the X-element consists of not less than eight distinct chromosomes, opposed by a single Y. The females here show seven more chromosomes than the males (photographs).

I will not enter upon the many interesting modifications of detail which these phenomena exhibit. In principle, the facts are the same in many insects and nematodes, probably in the myriapods and arachnids, perhaps also in the mammals and in man, though the demonstration here still leaves much to be desired. An extremely interesting series of researches by Morgan, von Baehr, Schleip, Doncaster, and others have proved that the same principle applies also to the parthenogenetic forms, such as the aphids, bees, and ants. Hardly less interesting are the investigations, especially of Boveri, Schleip, Krüger, and Zarnik, which show that this principle may even be extended to certain types of hermaphrodites. The results of genetic experiments on Lepidoptera and on birds lead us to expect the existence in these forms of a different cytological type, in which the eggs, instead of the spermatozoa, are of two different classes; but the cytological facts have not yet become sufficiently clear to warrant any definite conclusion. In the case of birds, indeed, a conspicuous contradiction still appears between the cytological and the genetic results; but the cytological observations have not yet produced evidence that can compare in cogency with that available in case of the insects or the nematodes.

I turn to the broader significance of the cytological facts that have been made known in this field. They constitute a very definite advance upon Boveri's general demonstration of the qualitative differences of the chromosomes; for it is impossible to doubt that the X-chromosome stands in some special causal relation with sex-heredity. A powerful argument for this is

given by the facts of sex-linked heredity, which I shall presently consider.
The riddle which this form of linkage
presents is solved by a cytological
phenomenon to which I first drew attention in 1906. The Y-chromosome,
when present, is confined to the male
line, and hence always passes from
father to son. The X-chromosome, on
the other hand, always passes from
father to daughter (because sperms of
the X class produce females), while
the sons receive their single X-chromosome from the mother (because the
male-producing sperms are of the no-
X or Y class) (fig. 2). I will show a
little later that on this curious fact
probably depends the 'criss-cross'
type of sex-linked heredity in which
the sons are like their mothers, the
daughters like their fathers.

The cytological phenomena of sex-
production lend strong support to the
theory of the genetic continuity of
chromosomes. They give unquestion-
able proof, in case of a particular
chromosome pair (XY), of the conjugation and subsequent disjunction of
corresponding maternal and paternal
chromosomes. They thus substantiate
the conclusions of Henking and
Montgomery and confirm Weismann's
earlier conception of the reduction
division. They are in full accord with
genetic studies, which prove that one
sex is homozygous, the other
heterozygous, with respect to a sex-
determining factor. They give the first
direct evidence of a difference of
nuclear constitution between the
homozygous and the heterozygous
conditions, and of corresponding
gametic differences. And finally, in the
case of a particular chromosome pair,
they fully substantiate the general
cytological explanation that has been
offered of Mendel's law.

(4) The facts just considered now
lead us to some of the most intricate
and interesting of current problems.
No phenomena appealed more
strongly to the interest of earlier
naturalists than those of correlation.
A very interesting light is thrown
upon this problem by the phenomenon
now widely known as gametic coupling or linkage, and it is here, perhaps,
that we may best appreciate to what
an extent cytology and genetics reciprocally illuminate each other.

In the second of Sutton's original
papers (1903) he pointed out what
seemed to be an obstacle in the way
of his own hypothesis, of which much
has been made by later critics. The
number of chromosomes is probably
always much smaller than that of
Mendelian units in any given case;
hence each chromosome must bear
many such units. From this it follows
that if the composition of the chromosomes be fixed, or even fairly constant, the units should cohere in definite groups, equal in number to that
of the chromosomes; but the earlier
studies on heredity gave little definite
evidence that such was the fact.*
Sutton did not, I think, meet the
difficulty, which he himself had
pointed out. It was perhaps the same
difficulty that led Correns (1902) and
De Vries (1903), in their attempts to
explain Mendel's law, to treat the

---

* In a general way, of course, this fact
was known to earlier observers, e.g., 'We
appear, then, to be severally built up
out of a host of minute particles, of
whose nature we know nothing, any one
of which may be derived from any one
progenitor, but which are usually transmitted in aggregates, considerable groups
being derived from the same progenitor'
(Francis Galton, 'Natural Inheritance'
1889).

chromosomes as of quite secondary importance. Their explanations operated almost wholly with smaller elements, of which the chromosomes were supposed to consist. It is now clear, however, that only when the chromosomes are taken into account do all the facts fall into line. For the most recent studies in genetics have produced indubitable evidence that Mendelian units are often, in fact, more or less definitely linked together in groups, as they should be under the chromosome theory.

Linkage was first clearly recognised in sex-limited or sex-linked heredity, to which I have already referred. A form of linkage having no relation to sex was brought to light a little later by Correns and by Bateson and Punnett in certain plants, and is now known to be of rather wide occurrence. I will confine my attention mainly to the case in which both these forms of linkage are now most accurately known, that of the fruit-fly, *Drosophila* [*melanogaster*]. In this species a very extended experimental analysis of the genetic phenomena has been carried on during the past four years in the laboratory of Columbia University by my colleague, Prof. T. H. Morgan, and his pupils and co-workers, Sturtevant, Bridges, and many others, from the investigations of whom the following results are reported. Drosophila (to paraphrase the words of Lacaze Duthiers) seems made for the experimental study of genetics. It passes through a complete generation from egg to egg in about twelve days. A single female not infrequently produces upwards of a thousand eggs. These fortunate circumstances have made it possible, in the course of four years of continuous study, to accumulate a prodigious mass of data, far surpassing in extent any others thus far made known. During this period these flies have given rise to more than a hundred definite mutations which are inherited in accordance with Mendel's law. They are of many different kinds, affecting the colour, shape, and structure of the body and of the eyes, the structure of the wings, legs, antennæ, and so on. Up to the present time 72 of these characters have been more or less completely tested as to their behaviour when crossed with the normal or 'wild' form, and with one another. The all-important fact which these tests have established is that the characters fall into four definite linkage-groups, of which the first now includes 31 characters, the second 23, the third 17, the fourth, so far, but a single one. These numbers represent of course only a beginning. They steadily increase as observation continues.

As the elaborate experimental analysis has proceeded, carried on by a number of specially trained co-operating observers, it has been more and more conclusively demonstrated that the units of each group are more or less firmly linked together in heredity, while those belonging to different groups are quite independent. This at once suggests that the units of each group (or corresponding things on which they depend) are borne by a particular chromosome which constitutes their common vehicle of transmission, and that to this fact is due their cohesion or linkage in heredity. Conversely, the several groups are independent of one another, because of the independence of the chromosomes which bear them. This hypothesis would have been a plausible one even were the number of chromosomes in Drosophila unknown. In point of fact, however, the gametic number of chromosomes in this species (or of chromosome-pairs in the diploid groups) is actually the same as that of the linkage-groups, namely, four (fig. 1, *d*, *e*). It is at least an old coincidence that one of these chromosomes, like one of the linkage-groups, is extremely small. One is tempted to guess that this may explain why for a long time but three linkage-groups could be identified, and that the fourth thus far contains but a single character, recently discovered by Muller.

Thus far, admittedly, the hypothesis presents a somewhat speculative aspect; but fortunately there is a means of testing it specifically, for the cytological evidence demonstrates that one of the four chromosomes is definitely connected with the determination of sex. One of the four groups of units, therefore, should likewise exhibit some special relation to sex. And this is in accordance with the facts, for every one of the 31 characters of the first group exhibits sex-linked heredity, of the same type as that which appears in colour-blindness or hæmophilia in man. It was pointed out, in 1910–11, by Morgan, Gulick, and myself that the heredity of sex-linked characters of this type exactly follows the course of the X-chromosome; that is to say, that the history of such characters is precisely such as it should be if they were dependent upon factors borne by this chromosome. Like the latter, the sex-linked units are always simplex or haploid (hence heterozygous) in the male; and they zigzag between the sexes in exactly the same way. No other group shows this relation.

Without entering far into the detail, let me illustrate these phenomena by a single example, that of the so-called 'criss-cross' heredity. The normal Drosophila possesses red eyes; a common mutant has white eyes, recessive to red. If a pure-bred white-eyed female be paired with a normal red-eyed male, all the resulting sons are white-eyed, like their mother; all the daughters red-eyed, like their father. Exactly analogous results appear when, instead of white eyes, any other units of the first group, such as yellow body colour or miniature wings, are similarly tested. The results at once lose their apparently bizarre character if we assume that the production of each sex-linked character depends upon something (which we may call a 'factor' or a 'gen'[2]) that is borne by the X-chromosome. I have already emphasised the fact that the sons derive this chromosome from their mother. In the cross just considered the sons therefore inherit with this chromosome the white eyes of their mother; the daughters, on the other hand, are red-eyed like their normal father, because they receive from him in every case a normal X-chromosome bearing the factor for the dominant red colour (fig. 3, A).

This has been tested in many ways, with results always in accordance with the hypothesis. Very convincing evidence in its favour has recently been obtained by Bridges through the study of a particular race of Drosophila, which regularly shows about 10 per cent. of exceptions to the 'criss-cross' type of heredity. This holds true for all the sex-linked characters thus far tested. Bridges' analysis[3]—too intricate to be entered upon here—led to the

---

[2] Genes. [Ed.]

[3] Discussed in *Heredity and Development*, Chapter 5.

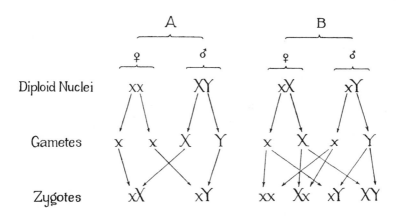

FIG. 3. Diagram of the relations of the sex-chromosomes to sex-linked heredity. Any normal (dominant) sex-linked character (*e.g.*, red eyes) is assumed to depend on the presence of a particular factor contained in the X-chromosome. Loss or modification of this factor produces a corresponding recessive (*e.g.*, white eyes), X now becoming *x*.

A. Criss-cross heredity, when the double recessive, white-eyed female (*xx*) is paired with the normal, red-eyed male (XY). The heterozygous daughters (*x*X) are red-eyed, with white-eye recessive; the sons (zygotes *x*Y) white-eyed.

B. History of the following generation, produced by crossing *x*X and *x*Y. The offspring of both sexes are now indifferently white-eyed (*xx*, *x*Y), or red-eyed (X*x*, XY).

conclusion that these exceptions are due to a failure of the two X-chromosomes to undergo disjunction in the reduction division of the female, so that the mature egg sometimes receives both X-chromosomes, sometimes neither. Eggs of the XX type might be expected to produce females even if fertilised by the no-X type of sperm, the XX combination (characteristic of the female) being supplied solely by the egg. Sex-linked characters shown by such females must be derived from the mother. On the other hand, eggs of the no-X class fertilised by the X class of sperm (normally female-producing) should produce males, and these should show only sex-linked characters derived from the father. This hypothesis was first tested, very ingeniously and thoroughly, by combining different sets of sex-linked characters derived respectively from the mother and the father. The results uniformly sustain the hypothesis. Very recently Bridges has tested his assumption cytologically. The expectation is that eggs of the XX type fertilised by sperm of the X class should produce females with three X-chromosomes, while if fertilised by sperms of the Y class the females should possess two X's and a Y. The cytological examination has demonstrated that certain females of this race actually possess three of these chromosomes.

Taken as a whole, the foregoing evidence gives almost crucial proof in favour of the conclusion that both the sex-determining factor and the sex-linked ones are borne by the X-chromosome. Sex-linked heredity of the type seen in birds or in the Lepidoptera requires an explanation somewhat different in detail but similar in principle (Spillman, Castle). In the facts of sex-linkage generally the chromosome hypothesis finds, I think, its strongest support, for the linkage of sex-linked factors with one another is of quite the same type as that which appears in other groups that are independent of sex, and the conclusion can hardly be avoided that in both cases linkage is due to the same cause.

We now take a final step in order to consider a seeming difficulty which introduces us to the most recent inquiries in this field. If our hypothesis is correct, how does it come to pass that linkage is not complete? How can we explain the variations in the so-called strength of linkage? Let me again illustrate by a single example taken from Drosophila, showing the heredity of two pairs of sex-linked characters of the first group. One pair comprises the normal grey body colour (G) and its recessive mutation yellow (Y), the other the normal red eye-colour (R) and its recessive mutation white (W). Let the pure-bred dominant female RG be crossed with the pure-bred recessive male WY and the hybrid offspring be inbred. Were linkage complete we should expect to find in the grandchildren the same combinations as those which entered the hybrid, and these alone—that is to say, RG or WY. In point of fact, this expectation is nearly always realised, but about one individual in eighty shows one or the other of the new combinations RY or WG. Genetically this means that R and G, or W and Y, are strongly linked, yet now and then may dissolve their union and recombine. Cytologically it means that the original X-chromosomes, bearing in one case RG, in the other WY, usually maintain their original constitution, yet occasionally may undergo an exchange of units so as to produce the new combinations observed. How is this possible?

In attempting to answer this, it is necessary to bear in mind that the recombinations with which we are dealing affect units that are usually linked together, and hence belong to the same group—in the example just given to the sex-linked or X-group. The recombination or exchange of units must accordingly take place between the corresponding or homologous chromosomes of a pair (here the X-pair). It therefore becomes probable that the exchange is effected during the intimate association of these chromosomes in conjugation or synapsis. It was long since suggested by Boveri that an exchange of particular elements might take place between conjugating chromosomes as between conjugating Protozoa, but this suggestion is too vague for our present purpose. The basis for a more specific explanation was offered by Janssens in 1909 in his theory of the 'chiasmatype,' more recently elaborated in a remarkable manner by Morgan and by Sturtevant. This explanation is as follows:—

It has long been known that subsequent to the process of conjugation —sometimes, as now seems probable, during this process—the two chromosomes of each pair, while still in the form of long threads, often twist about

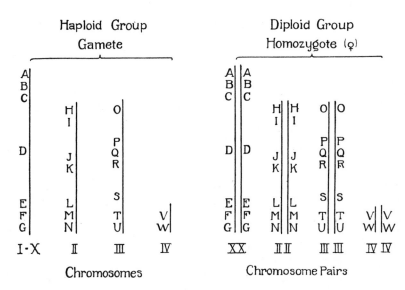

FIG. 4. Diagram of chromosomes and linkage-groups, based on the relations observed in *Drosophila ampelophila*. Heavy vertical lines represent chromosomes (or the chromatin-threads from which they arise), letters different factors or gens, assumed to be aligned in linear series in the threads. In the diploid groups corresponding chromosomes are paired side by side in the position assumed by them during conjugation or synapsis.

Each of the four series A–G, H–N, O–U, V–W, forms a definite linkage-group in which the factors tend to cohere, while independent of all other factors.

In the male diploid groups one X-chromosome is missing, its place being taken by a Y-chromosome (*Lygœus* type). The nature of the latter is still unknown.

each other, thus producing the so-called 'strepsinema' stage. Janssens concluded from a careful study of the facts in the Amphibia that the double spirals thus formed may in some cases come into contact and fuse at certain points where the threads cross, and then may be separated again by a straight longitudinal split through the points of fusion. Such a process would lead to an exchange of certain regions between the two threads. Janssens named this the 'chiasmatype,' and briefly called attention to the possible bearing of such a process on the Mendelian phenomena. Morgan afterwards very ingeniously developed this thought as follows: The 'determining factors' or 'gens' of unit-character are assumed to be aligned, as Roux long since suggested, in linear series in the chromatin-threads and in a definite order. In the process of conjugation corresponding or allelomorphic gens (large and small letters in the diagrams) are assumed to lie opposite one another in the two threads, as is shown in diagram in figs. 4 and 5. If in a certain proportion of cases the two threads (linear series of gens) twist together, unite, and separate in the manner described by Janssens, the

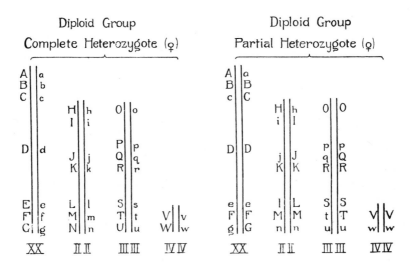

FIG. 5. Diagram illustrating heterozygous conditions. Homologous or allelomorphic factors (*e.g.,* C and *c*) occupy corresponding loci or levels in homologous chromosomes. The condition at the left is heterozygous for all factors; at the right, heterozygous for some (A*a*, *g*G, etc.), and homozygous for others (BB, *cc*, etc.).

result will be an exchange between them of certain gens, as shown in fig. 6. A simple and elegant solution of the problem of recombination or of 'crossing over' is thus given.

It should be pointed out that Janssens' actual observations on the chiasmatype still lack adequate confirmation, that the strepsinema has been observed in the longitudinally split single chromosomes of the somatic divisions, and that nearly all cytologists have hitherto believed the twisted threads to untwist again before actual separation takes place. I have not yet been able fully to satisfy myself concerning the facts; but there is no doubt that in the Amphibia many of the appearances seem to be in favour of Janssens' conclusions.

The most ingenious part of the explanation relates to the varying strength of linkage. It is obvious that if the twisting be not too close, the likelihood of a chiasma taking place in the interval between any two units (and hence of their separation or dissolution of linkage) increases with the distance between them in the thread. Conversely, the nearer together two units lie the greater the probability of their remaining in association; in other words, the greater the 'strength of linkage.' Hence the astounding possibility which this suggests of using the 'strength of linkage' as an index of the serial order of the units in the threads and the relative distances between them. This, it seems to me, is the most remarkable result to which these researches have led, for it opens the possibility of a detailed experimental analysis of the nuclear organisation—almost, we might say, of the topographical anatomy of the germ-plasm.

By the application of this method to an immense body of experimental data, Morgan and his co-workers, Sturtevant in particular, have actually plotted the location of most of the units in each chromosome, and constantly use the diagrams thus obtained as working models for further analysis. The order and relative distances of the units in each linear series, once determined, are found to be remarkably constant when tested by varied experiments designed to this end. The practical value of the hypothesis is attested by the fact that when the distance (strength of linkage) between any two units, A and B, is known, and also that between B and a third unit, C, the relation between A and C may be *predicted* with considerable accuracy. This fact gives reality to the assumption that each unit has a definite locus in the linear series (chromatin-thread), and that allelomorphic units occupy corresponding loci in homologous chromosomes. Further corroboration is found in the interesting phenomena presented by multiple

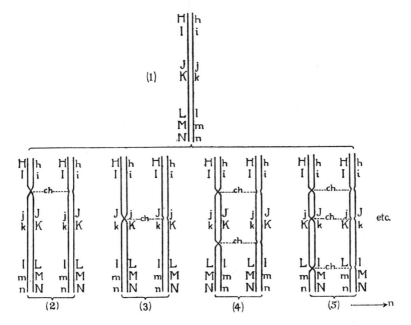

FIG. 6. Diagram of the exchange of factors (dissolution of linkage) as explained by the chiasmatype hypothesis. The second chromosome-pair of fig. 5 is here employed as an example. The original condition is assumed to be that shown above (1). The lower figures (2–5) show a few of the many possibilities of exchange or 'crossing over' between the two homologous linkage-groups or chromosomes, H–N and *h–n*. In each case the point of crossing, fusion, and subsequent splitting is shown at the left, with the result at the right. The position of the chiasma indicated in each case by *ch*. In (2) and (3) but one chiasma is present, in (4) two, and in (5) three. A very large number of recombinations is possible, even with so small a number of factors as is here represented.

allelomorphs, of which an example is given by the four eye-colours: white, eosin, cherry, and the normal or red colour, all of which belong to the first or sex-linked group in Drosophila. Any two of these are allelomorphic to each other, and the important fact is that all exhibit the same strength of linkage with all other sex-linked units. The inference is that each of the units in question must occupy the same locus in the sex-chromosome. Since no two can occupy the same locus at the same time, it follows that not more than two of them can co-exist in any particular female, and not more than one can be present in the male. And this corresponds with the facts as actually observed.

To those not actually engaged in such investigations this hypothesis will, perhaps, seem of highly speculative character. But is it more so than many working hypotheses of experimental physics or organic chemistry that have proved themselves fruitful in the past? I will not pretend to answer. There is no doubt that it provides us with a simple, easily intelligible and effective means of handling enormous masses of intricate data, of devising new experiments, of predicting results. Such an hypothesis, venturesome though it may seem, is something more than a speculation.

I have endeavoured to show how the chromosome-theory, first outlined in very general form, has been more and more specifically developed until it has become an important instrument for the detailed analysis of intricate genetic phenomena. I am well aware that some eminent students[4] of genetics are still reluctant to accept this

theory, at least in its more detailed applications. I am not disposed to reproach them for such scepticism. The cytologist suffers under the disadvantage of working in so unfamiliar a field that some of his conclusions, even among those most certainly established and most readily verifiable, are apt to give a certain impression of unreality, even to his fellow naturalists. It is undeniable, too, that in this subject, for better or for worse, hypothesis and speculation have continually run far in advance of observation and experiment. It is quite possible that some of my hearers may consider some of the views I have touched upon as a fresh illustration of this fact. If so, I beg them to bear in mind that no conclusion which I have considered has been reached as a merely logical or imaginative construction. I have endeavoured to limit myself to matters of observed fact, and to conclusions that are either demonstrated by facts or directly and naturally suggested by them.

To those who have had opportunity to come into intimate touch with both cytological and genetic research the conclusion has become irresistible that the chromosomes are the bearers of the 'factors' or 'gens,' with the investigation of which genetics is now so largely occupied. What are these gens? How do they operate? We do not know what they are. We assume only that a gen is *something* that is necessary to the development of a particular character. We do not know how they operate; for, despite all that experimental cytology and embryology have taught us concerning development, we are still without adequate understanding of its mechanism. We may nevertheless guess that gens play

---

6 Bateson was one of them. (ED)

their several *rôles* by virtue of their specific chemical nature, and that the study of chemical physiology as applied to development is destined to take an important part in the future investigation of this problem. In the meantime it would be well to drop the term 'determiner' or 'determining factor' from the vocabulary of both cytology and genetics. What we really mean to say is 'differential' or 'differential factor,' for it has become entirely clear that every so-called unit character is produced by the co-operation of a multitude of determining causes. Embryologists long since demonstrated by direct experiment that the cell-protoplasm as well as the nucleus is concerned in the determination of development. Our whole study of the cell leads us to the conclusion that it is an organic system, in the operation of which no single element can be wholly dissociated from the rest. When, therefore, we speak of nuclei or chromosomes as the 'bearers of heredity' we are employing a figure of speech. They are such just to the extent that they are necessary to development and heredity; but how far this conclusion carries we are as yet unable to say. Genetic experiment has already given some ground for the conclusion that definite types of hereditary distribution may be immediately dependent upon elements contained in the protoplasm. Recent advances in our knowledge of the 'chondriosomes' or 'plastosomes' provide this conclusion with at least a possible cytological basis.

Our conceptions of cell-organisation, like those of development and heredity, are still in the making. The time has not yet come when we can safely attempt to give them very definite outlines. It is our fortune to live in a day when the business of observation and experiment leaves little time or inclination for *a priori* speculations concerning the architecture of the germ-plasm or of the cell. Nevertheless it is impossible not to be struck with the fact that recent advances in cytology and genetics are in certain important respects in line with theoretical views put forward nearly 30 years ago by Roux, Weismann, and de Vries. These views were, it is true, almost purely imaginative or logical constructions. Some of them, especially as applied to the mechanism of embryological development, have been experimentally disproved, others are incapable of verification, and hence have fallen into disrepute. We have become chary of theories which assume all parts of the cell to be built up of ultimate, self-propagating, vital units, such as 'gemmules,' 'pangens,' or 'biophores.' The working hypothesis that has here been considered must not be identified with those far-reaching speculations; it is at once more limited in scope and more flexible in form. And yet, as fare as the cell-nucleus is concerned, those visions of a bygone speculative era are now beginning to seem more real than would have been thought possible by some of us ten or even five years ago. We read in the latest productions of cytology and genetics of the division and genetic continuity of factors or gens, of their linear alignment in the chromatin threads, of their conjugation and disjunction, of their linkage or independent distribution, in heredity. We find such conceptions no longer treated as belonging to an age of cytological romance, but employed every day in the most matter-of-fact way as practical instruments of laboratory experiment, analysis, and prediction. We are bound to no speculative systems or extravagances of an earlier day if we recognise in this, let the outcome be what it may, a triumph for the men who first endeavoured to bring cytology and the experimental study of heredity into organic relation.

Students today seem to have little difficulty in understanding how the behavior of chromosomes in meiosis and fertilization, so ably described by Wilson, can explain Mendelian heredity. This was certainly not the case in the years immediately after Sutton published his major paper. In fact, even two decades later there were still numerous and powerful doubters (see Darlington, 1960). Oskar Hertwig, who had been such an ardent proponent of the hypothesis that the nucleus was mainly responsible for inheritance, vigorously opposed the notion that there was a relation between chromosomes and the Mendelian genes. Bateson remained skeptical for more than a decade (see Punnett, 1950). An even more surprising case is that of E. B. Wilson, who, at first, had difficulty in understanding Sutton's hypothesis. Wilson was the master student of the cell and in 1896, and again in 1900, he had marshalled the evidence to show that the nucleus was the prime candidate for the controller of inheritance. One would surely think that in 1902 his mind would be prepared to understand his student, Sutton. This is Wilson's account of his conversion.

. . . I well remember when, in the early spring of 1902, Sutton first brought his main conclusions to my attention, saying that he believed he had really discovered 'why the yellow dog is yellow.' I also clearly recall that at that time I did not at once fully comprehend his conception or realize its entire weight.

We passed the following summer together in zoological study at the sea side, first at Beaufort, N. C., later at South Harpswell, Me., and it was only then, in the course of our many discussions, that I first saw the full sweep and the fundamental significance of his discovery. Today the cytological basis of Mendel's law, as worked out by him, forms the basis of our interpretation of many of the most intricate phenomena of heredity, including the splitting up and recombination of characters in successive generations of hybrids, the phenomena of correlation and linkage, of sex and sex-linked heredity and a vast series of kindred processes that were wholly mysterious before their solution was found through Mendel's law. Subsequent to the appearance of Sutton's papers, Boveri stated, 1904, that at the time they were published he had himself

From an appreciation that Wilson wrote following Sutton's death. It was published in a little volume that included a brief biography of Sutton, an account of his funeral service, and tributes and appreciations from his friends. The title page of the book reads: Walter Stanborough Sutton. April 5, 1887. November 10, 1916. Published by His Family. 1917.

already reached the same general result. This does not, however, in the smallest degree detract from Sutton's fine achievement, which will take its place in the history of biology as one of the most important advances of our time. He made an indelible mark on scientific progress, and his name is known wherever biology is studied. [pp. 69–70] . . .

During this summer Sutton had fully worked out his theory of the chromosomes in relation to Mendel's law and upon his return to New York he immediately set about the preparations for its publication. His first paper, as already stated, appeared late in 1902, the second early in the spring of the following year. These two brief papers were intended to be of a preliminary nature, a fuller presentation of his conclusions, together with a large number of beautiful drawings, already finished at that time, being reserved for a later work which he had expected to offer as a dissertation for the Ph.D. degree at Columbia. It was a source of profound regret to us that circumstances prevented the realization of that plan and brought his cytological investigations to a close. In spite of his brilliant talents as an investigator it would perhaps be more accurate to say because of them—the career of a teacher did not tempt him. Could he have been assured of a reasonable means of support from a life devoted to pure research, he would not, I believe, have hesitated. But he had his own way to make in the world and from the first had a strong inclination towards the study of medicine. The combination of circumstances proved irresistible; and after a year or two spent in business he returned to Columbia, entered the Medical School, and graduated with the highest honors two years later. Others can speak with a greater competence than I concerning his brilliant career as a physician and surgeon; but I can testify to the strong impression which, as a student of medicine, he made upon the faculty of the College of Physicians and Surgeons at Columbia. He was generally recognized as one of the ablest medical students of his time, and at the end of his medical course was awarded one of the most coveted hospital positions in the City of New York, where the final preparation for his professional practice was completed. [pp. 71–73] . . .

Once convinced, however, Wilson never faltered in his support of the hypothesis that genes are parts of chromosomes. Interestingly enough, one of Wilson's colleagues at Columbia University, Thomas Hunt Morgan, was very skeptical at first of the usefulness of this hypothesis but eventually Wilson convinced him. This is the same Morgan who, as we shall see in the next chapter, was to advance the science of genetics so greatly. His success was due, in no small measure, to the fact that he based his experiments on the hypothesis that genes are parts of chromosomes. It is important to remember, therefore, that he worked in the laboratory where this hypothesis was first proposed by Sutton and then documented so carefully by Wilson.

So the story turns to Morgan.

BIBLIOGRAPHY

Refer to the Preface for a list of the classical papers included in other anthologies.

BALTZER, FRITZ. 1964. 'Theodor Boveri.' *Science 144*: 809–815.

BATESON, W. 1913. *Mendel's Principles of Heredity*. Cambridge: At the University Press. Chapter 10.

CREW, F.A.E. 1965. *Sex-determination*. New York: Dover.

DARLINGTON, C. D. 1960. 'Chromosomes and the theory of heredity.' *Nature 187*: 892–895.

DARLINGTON, C. D. 1969. *Genetics and Man*. New York: Schocken Books. Chapter 6.

HUGHES, ARTHUR. 1959. *A History of Cytology*. New York: Abelard-Schuman.

MCKUSICK, V. A. 1960. 'Walter S. Sutton and the physical basis of Mendelism.' *Bulletin of the History of Medicine 34*: 487–497.

MORGAN, T. H. 1903. 'Recent theories in regard to the determination of sex.' *Popular Science Monthly*. December 1903. Pages 97–116.

MORGAN, THOMAS HUNT. 1913. *Heredity and Sex*. New York: Columbia University Press.

MORGAN, T. H. 1940. 'Bibliographic memoir of Edmund Beecher Wilson 1856–1939.' *Biographical Memoirs of the National Academy of Sciences 21*: 315–342.

MULLER, H. J. 1943. 'Edmund B. Wilson—an appreciation.' *American Naturalist 77*: 5–37, 142–172.

WILSON, EDMUND B. 1909. 'Recent researches on the determination and heredity of sex.' *Science N.S. 29*: 53–70.

WILSON, EDMUND B. 1912. 'Some aspects of cytology in relation to the study of genetics.' *American Naturalist 46*: 57–67.

WILSON, EDMUND B. 1928. *The Cell in Development and Heredity*. New York: Macmillan.

5 / *Morgan and Drosophila*

The genetics of transmission, that is, knowledge of how the genes pass from generation to generation, was completed by the Morgan school in a period that began in 1910 and ended in the 1920's. Some of the milestones of that scientific endeavor are recounted in Chapter 5 of *Heredity and Development*. Wilson described the beginnings in his Croonian Lecture of 1914 to the Royal Society (reprinted in Chapter 4). Eight years later, Morgan was invited to continue the story in another Croonian Lecture, which is reprinted below.

THOMAS HUNT MORGAN

Thomas Hunt Morgan (1866–1945) was a scientist of broad training and interest. Much of his early work was concerned with the experimental analysis of development. Even though he was a close friend and fellow professor at Columbia with E. B. Wilson, he was skeptical of much of the early work in genetics and its relation to cytology (recall his 1909 paper reproduced in Chapter 3; also refer to his 1910 paper listed in the bibliography at the end of this chapter).

In 1909 he began to experiment with *Drosophila melanogaster,* a remarkable organism that, in its evolution, seems to have anticipated most of the geneticist's requirements. The work was carried out in collaboration with a notable group of students and associates, the chief ones being Alfred H. Sturtevant (1891–1970), Calvin B. Bridges (1889–1938), and Hermann J. Muller (1890–1967). All began as students of Morgan at Co-

155

lumbia University and Sturtevant and Bridges spent their entire scientific careers with him, first at Columbia and after 1928 at the California Institute of Technology. So Morgan continues where Wilson had left off:

CROONIAN LECTURE:—
On the Mechanism of Heredity.
By T. H. MORGAN, For.Mem.R.S.,
Professor of Experimental Zoology in
Columbia University.
(Lecture delivered June 1, 1922.—
MS. received August 3, 1922.)

Several years ago I ventured to state that with the demonstration of the wide applicability of Mendel's two laws, and with the later discoveries of linkage and of crossing-over, the traditional problem of heredity had been solved. On account of this statement, I have been rebuked for arrogantly affirming that there was nothing more to be learned about heredity!

Perhaps I had given less offence if I had made clear that I realised, as fully as another, that these discoveries in heredity were of such a sort that a whole new world for investigation opened before us. My critics, however, attempted to put me in the wrong by pretending that I implied that there were no new worlds to conquer. It may, therefore, not be out of place to attempt to show how the solution of the traditional problem of heredity has led to further discoveries, and has given us a glimpse at least of still newer problems that may in time lead to even more far reaching consequences.

Mendel's principles of heredity may be said, I think, to be mechanistic in principle, by which I mean that the coming together and separating of spe-

cific elements are concepts characteristic of physical events.* It is true that Mendel did not state that the elements that he postulated as coming together and separating are material particles. For all that we know to the contrary, the good abbot may have had something more spiritual or mystical in mind. Nevertheless, whatever it is that meets and separates, whether spiritual or material, the *process* is in the nature of a physical event.

The new data relating to heredity might be treated as a series of purely statistical problems without regard to any special mechanism that has given the data. There would be admittedly a certain security in treating the problems in this way; but there are several

---

* It is interesting in reading Mendel's paper to note how, in almost every instance, when the opportunity arises for stating that the members of a pair of allelomorphs *separate* to pass to their respective germ cells—it is interesting to note that he does not say this explicitly, but gives rather the result of such a separation. His usual method is to state that the number of germ cells containing the one element is the same number as that containing the other. At the end of his essay, however, the following statement occurs: 'Since in the habit of the plant no changes are perceptible during the whole period of vegetation, we must further assume that it is only possible for the differentiating elements to liberate themselves from the enforced union when the fertilising cells are developed.

From the *Proceedings of the Royal Society. B. 94*: 162–197. 1922. Reprinted by permission.

reasons for hesitating to divorce the results from the animal and plant that supplies them. In the first place experience has shown that many suggestions of value have come from a knowledge of what happens in the egg and sperm. For example: occasionally, in the ripening of the germ-cells specific irregularities are known to recur. It is, therefore, instructive to find that certain rare events in Mendelian inheritance—events that are not immediately deducible from these principles as such—that these events are expected as necessary consequences of such irregularities; or, stated conversely, there are some types of genetic events that have been shown by actual cytological demonstration to be connected with exceptional behaviour of the chromosomes.

In the second place, need I urge that, as biologists, we are curious to

In the formation of these cells all existing elements participate in an entirely free and equal arrangement, by which it is only the differentiating ones which mutually separate themselves. In this way the production would be rendered possible of as many sorts of egg and pollen cells as there are combinations possible of the formative elements.' There is another procedure that Mendel follows that obscures the clarity of his view as to the nature of the process of 'segregation.' The dihybrid formula of an individual homozygous in one of the pairs of elements in question is written aAB instead of aABB as we write it to-day. The latter presents to the eye two pairs of elements, each of which segregate; the former formula (aAB), if taken literally, seems incomplete from the point of view of segregation. In practice, however, Mendel did not fail to introduce B into every gamete (aB, AB) of such an individual.

find out how Mendel's principles tie up with the rest of our knowledge concerning the changes in the germ-cells, the agents through which heredity takes place.

I will even venture to suggest that it is not a bad plan, for biologists at least, to maintain a guarded attitude towards abstractions that ignore the sources of the data from which the abstractions were made. Experience has shown, I think, that in an undeveloped subject such as ours progress has come less from unverified speculations about living things than by putting every new idea to the test of a critical experiment on the material itself before regarding a new idea as a serious contribution to science.

I.

During the thirty-five years that followed the publication of Mendel's paper (1865 to 1900) (while it still remained unnoticed) students of the germ-cells—cytologists—found that some extraordinary events take place at the time of maturation of the egg and the sperm. Weismann made use of some of the newly acquired facts in his well-known attempt to explain both development and heredity, but, except for the idea of the continuity of the germ-plasm, his attempt was little more than ingenious guessing and could scarcely have succeeded so long as the fundamental principles of heredity were unknown; and to Weismann, as well as to the rest of the biologists, Mendel's results were at that time a sealed book.

Mendel's paper was discovered in 1900, and even then for two years no one seems to have clearly realised that a mechanism had already been found that furnished an explanation of Men-

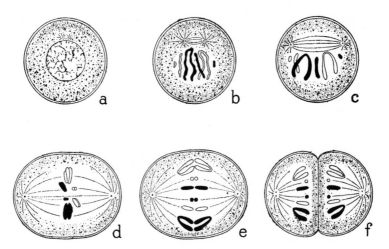

FIG. 1. Cell division.

del's laws. A young student, William Sutton, working in the Columbia Laboratory with Prof. E. B. Wilson, first stated clearly in 1902 the application of the then known facts of cytology to Mendel's laws.

I beg that you will allow me at this point to review rapidly certain rather well-known discoveries relating to the ripening of the germ-cells—discoveries familiar to biologists, but perhaps not so familiar to physicists and chemists.

In any cell about to divide (fig. 1), whether body-cell or germ-cell, the nuclear wall disappears, the chromosomes are resolved into threads, and a spindle appears in the protoplasm. The chromosomes move to the equator of the spindle. Here each splits lengthwise into daughter halves, one daughter chromosome moves to one pole of the spindle and the other daughter chromosome moves to the other pole. Two new resting nuclei begin to form and the protoplasm constricts into two parts. By this process, repeated over and over again, all the

cells of the developing animal or plant are produced.

In these familiar processes of cell-division, there are two events of paramount interest to those students of genetics who hold that the chromosomes are the carriers of the hereditary elements; first, that every cell in the body contains the sum total of all of the hereditary elements; and second, that at every division, both in the early development, when the organ-forming regions are being developed, as well as in the later development when the specific tissues have appeared, the chromosomes divide lengthwise into daughter halves that are always exactly equivalent. It is obvious, therefore, that any theory that relates to the chromosome mechanism must recognise that no appeal can be made to a sorting out of the hereditary elements in the fertilised egg during its development—that every part of the body at all times must contain the entire hereditary complex. This means, of course, that development and differentiation must take place in the presence of all the hereditary material at every point (nucleus) of the embryo. The necessity of such an interpretation of the facts of cytology, while generally admitted, is seldom carried to its logical conclusion, with the result that the problems of heredity have become lamentably confused with the problems of embryonic development.

For each species there is a definite number of chromosomes (fig. 2). In every cell the chromosomes are duplex, that is, they are in pairs. One member of each pair has come from the father and the other from the mother. There is also a considerable body of evidence showing that these chromosomes persist from one cell-generation to another, although in the resting phase they cannot, as a rule, be identified. While observation cannot be said to have established the continuity of the chromosomes, the evidence from genetics is overwhelmingly in favour of such an interpretation, but adds one important qualification, that, at one stage at least in the ripening of the germ-cells, an interchange may take place between the members of the same pair. I shall come back to this point later, for it is this interchange that has opened up to us far-reaching possibilities.

Two peculiar divisions take place in the germ-cells just before they be-

come transformed into ripe sperm or into ripe eggs, and leading up to these divisions a process occurs that is unique, something that never happens in any other cell, either in the body or in the earlier cells of the germ-track itself. The chromosomes conjugate in pairs—each paternal unites with the corresponding maternal chromosome. The number of chromosomes appears to be reduced to half. In reality the chromosomes of each pair have come to lie side by side throughout their entire length.* Their reduction in number is apparent, not real. When this conjugation is accomplished the nuclear wall disappears, a spindle develops, the chromosomes pass to its equator, and then each double chromosome separates into its component halves. This process, except for details, is the same in the sperm-cells and in the egg. Let us follow each in turn.

In the sperm-cell the conjugated chromosomes separate and move to one or to the other pole of the spindle, and the cell divides (fig. 3). Then, without a resting stage, another spindle develops, the chromosomes pass into it, and now each splits lengthwise as

---

* The possibility that end to end conjugation may take place in certain plants does not affect the question here at issue, except possibly the interpretation of crossing over.

FEMALE   MALE

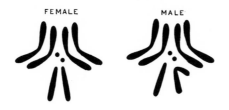

FIG. 2.

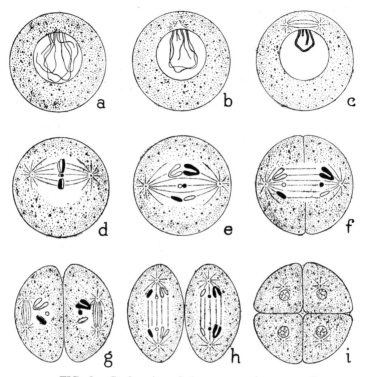

FIG. 3.   Conjugation of chromosomes in sperm cell.

in ordinary cell-division. The daughter halves move to opposite poles and the cell divides. After these two maturation divisions have been accomplished four sperm cells are produced, each with half of the total number of the original chromosomes.

When the *egg* matures (fig. 4) the conjugated chromosomes pass to the spindle, and the spindle then moves to the pole of the egg. At one division, the members of each pair separate and move towards the poles of the spindle. A protrusion of the protoplasm takes place near the pole into which the outer set of chromosomes passes. The protoplasm constricts and the first polar body is formed. A new spindle

develops about the chromosomes left in the egg, and each chromosome splits lengthwise. The daughter halves separate, half going into the second polar body, half remaining in the egg. The latter pass into a resting stage to produce the mature egg-nucleus. The first polar body also divides at the same time. Here also four cells are produced, but only one, the egg, is functional. It contains the half number of chromosomes.

For our present purposes the important event in the maturation of the germ-cells is the conjugation of the maternal and paternal chromosomes in pairs and their subsequent dispersal in such a way that each germ-cell gets one member of each pair.* By means of this mechanism Mendel's two laws can be stated in terms of chromosome behaviour. Two examples will quickly show this.

If a pomace fly (*Drosophila melan-* *ogaster*) of a vestigial winged race is crossed to a wild type (long winged) fly (fig. 5), the offspring (F_1) are long winged. If these are mated, three longs to one vestigial appear in the offspring (F_2).

If a factor for vestigial (v) is present in the vestigial race and is carried by each member of a given pair of chromosomes, then after maturation each sperm will have one of these chromosomes. In the long winged fly the corresponding factor (V) is carried by the same pair of chromosomes, and after maturation each egg will contain one of these chromosomes. Hence the hybrid offspring (F_1) will receive one long and one vestigial-bearing chromosome (vV), and, since the hybrids have long wings, we say that long dominates. When the hybrids mature and their germ-cells in turn are formed, each egg will contain either the long (V), or the vestigial (v) chromosome. Simi-

---

* In this description of the two 'maturation divisions' no account is taken of 'crossing-over,' *i.e.*, interchange between members of the same pair of chromosomes. The genetic evidence clearly establishes this relation, even though the cytologist has not yet been able to furnish convincing evidence of its occurrence. It is important to emphasise that the genetic evidence has shown how arbitrary was the distinction made by cytologists between a reduction and an equation division (the first and second maturation divisions), for after interchange it is not the two original chromosomes that separate at reduction, but part of each. In other words, while for the tetrad there is a reductional and an equational division, the process as now understood relates to genes, not to whole chromosomes.

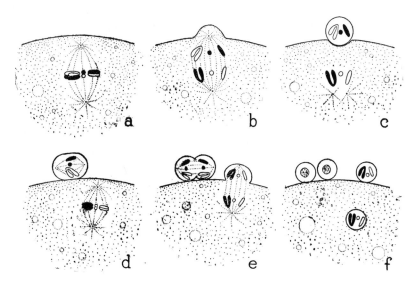

FIG. 4. Maturation of egg.

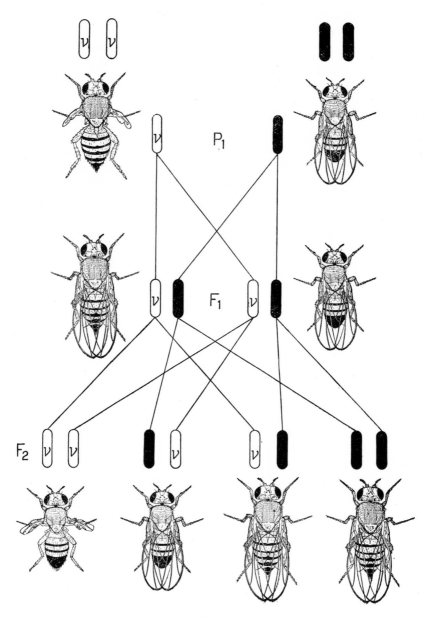

FIG. 5.

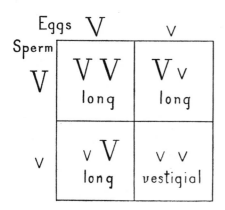

FIG. 6.

larly each sperm will contain either a long (V), or a vestigial (v) chromosome (fig. 6). Chance meeting of any egg and any sperm will give the F₂ results, viz.: three long to one vestigial. The genetic and cytological results agree. The behaviour of the chromosomes gives an explanation of Mendel's law of segregation.

When two pairs of factors are present, each carried in a different pair of chromosomes, their inheritance is as follows (fig. 7):—Suppose that a fly that is ebony as to body-colour (e) and has vestigial wings (v) is used

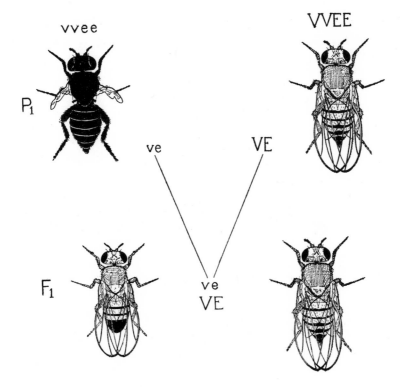

FIG. 7.

as one parent, and a fly that is grey (E) and has long wings (V) is used as the other parent. Here the two factor-pairs are grey (E) *versus* ebony (*e*), and long (V) *versus* vestigial (*v*). The offspring (E*e*V*v*) are grey long. If the factor-pairs (E*e* and V*v*) are carried in different pairs of chromosomes, as shown in fig. 7, and if each pair is sorted out to the germ-cells of the hybrid independently of the way in which the other pair is sorted out, four kinds of germ-cells result. Chance union of any egg and any sperm (fig. 8) will give the sixteen kinds of individuals which, reduced to classes, are in the ratio of 9:3:3:1. Here, again, the chromosome behavior is the same as the genetic, provided the pairs of chromosomes are sorted out independently.

Fortunately we have evidence showing that such independent assortment of chromosomes takes place. Miss Carothers has studied in several grasshoppers the behaviour of chromosomes at the reduction division (fig. 9). Slight differences in the mode of attachment of the chromosomes to the spindle fibres (differences that are constant) enabled her to show that the sorting out of the members of the pairs is, in reality, independent.*

---

* In this figure (fig. 9) each of the four horizontal lines (1*b*, *c*, *d*, *e*) shows the twelve *double* chromosomes (tetrads) present in each spermatocyte, *i.e.*, the twenty-four diploid chromosomes that have conjugated giving these twelve tetrads. The conjugants (paternal and maternal chromosomes) are about to separate in the first maturation division. One of these (No. 4) is the X-chromosome, and has no mate in the male. It is placed in such a position in the drawing that the later migration to the pole

Eggs →	VE	Ve	vE	ve
Sperm				
VE	VE VE	Ve VE	vE VE	ve VE
Ve	VE Ve	Ve Ve	vE Ve	ve Ve
vE	VE vE	Ve vE	vE vE	ve vE
ve	VE ve	Ve ve	vE ve	ve ve

FIG. 8.

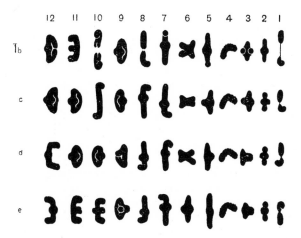

FIG. 9.

II.

Even if it is admitted from such evidence that the chromosome mechanism suffices to explain Mendelian segregation and assortment, as Sutton pointed out, it still remained to be shown that we are not dealing with analogy or coincidence; but that the chromosomes are specifically related to genetic events. Fortunately we have to-day such evidence. Let me give a few illustrations.

There is a race of Drosophila called 'Diminished-bristles' (fig. 10), that was discovered by Bridges, in which

all the 'Diminished' individuals carry *only one* small, or IVth, chromosome. Half of the ripe sperm-cells carry this IVth chromosome, and half do not carry it. Now if 'Diminished' is crossed to a fly from the 'eyeless' stock (that carries a recessive gene in the IVth chromosome) there should be two kinds of offspring corresponding to the two kinds of sperm. Such in fact is the case, and the offspring that are 'Diminished' have been shown to have only the IVth chromosome. Moreover, since the single IVth chromosome that they carry is de-

---

would be upwards, *i.e.*, towards the top of the page. The other chromosomes are oriented with respect to this one. Three of these pairs (viz., 1, 7, and 8) are each composed of two unlike chromosomes, in the sense that one member of each pair is straight (terminal attachment of spindle fibre), the other, bent (sub-terminal attachment). These differences are found in every pair of these particular chromosomes in this individual.

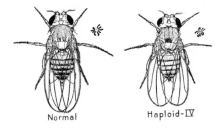

FIG. 10.

rived from the recessive stock, we can understand why these first generation Diminished flies are eyeless; in other words, when Diminished flies are out-crossed to any IVth chromosome recessive stock, the recessive character appears as though it were a dominant.

There is another stock of Drosophila, also discovered by Bridges, in which certain individuals contain *three* IVth chromosomes. These flies may be distinguished from wild type by their smaller eyes, darker body colour, narrower wings, etc. (fig. 11). Owing to the presence of three IVth chromosomes, half of the matured germ-cells will contain two IVth chromosomes, the other half only one IVth chromosome (fig. 12). If such a triploid-IVth-chromosome individual is mated to eyeless, half of the offspring will be triple IVths. If two such triples are mated to each other they do not give a three to one Mendelian result, but give a ratio of about 26 to 1, which is the expectation for such a chromosome situation.

In back-crosses such F_1 flies give a 5:1 ratio (fig. 13), instead of a Mendelian 1:1 ratio.

FIG. 11.

Here we have an unusual genetic and cytological behaviour that checks up at every point; I say at every point, because there are several other possible tests, most of which have been made, and the genetics conform to the expectation based on the known distribution of the IVth chromosomes.

Other kinds of irregularities also occur at times in the division of the germ-cells. One kind is called non-disjunction of the Ist chromosome, because at one maturation division the two X-chromosomes do not always *disjoin*, but may both go out of the egg into the polar body or both may remain behind.

If both should remain in the egg and this egg is fertilised by a Y-bearing sperm (fig. 14), an XXY individ-

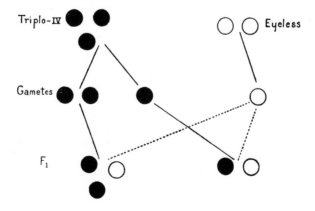

FIG. 12.

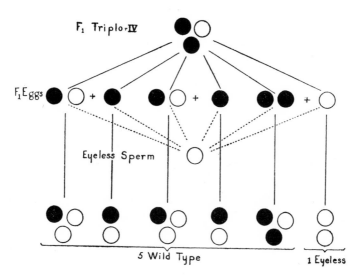

FIG. 13.

ual will result, having these three sex-chromosomes in all the cells of the body, including, of course, the eggs before maturation. Such XXY females will then exhibit secondary non-disjunction, for when an egg reaches the maturation stage a sort of triune relation will exist in the three sex chromosomes, and since only two can conjugate, the third must do a *pas seul*.

Fig. 15 shows the four possible kinds of eggs that result from this non-disjunction. If the four kinds of eggs are fertilised by X-bearing sperms, four kinds of individuals are expected. In order to reveal the nature of these individuals, it is advantageous to use an XXY female in which both X's carry a recessive character, such as white eye-colour, and to cross her to a normal red-eyed male. This combination gives four classes of offspring. One kind of female (1) is XXY, and should show non-disjunction when bred. This has been proven. The next

female (2) is normal. The third (3) has three X's, and usually dies, but rarely she comes through, is always sterile and *has* three X's. The fourth individual is a normal red-eyed male. He behaves as such.

If, on the other hand (fig. 16), the four kinds of eggs are fertilised by the Y-sperm of the same male, four kinds of individuals are expected. The first (5) is white-eyed male with one X and two Y's. If he is tested he is found to produce some XXY daughters that are non-disjunctional. The second (6) is a normal white-eyed male. The third (7) is a white-eyed female that is non-disjunctional. When tested she is found to give such results and has been shown to have two X's and a Y-chromosome. The fourth (8) never appears, because obviously it has no X-chromosome at all.

My next illustration concerns what we call the double yellow females, recently discovered by Lilian V.

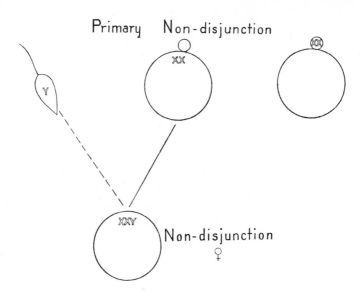

FIG. 14.

Morgan. This stock is descended from a fly that showed a reversal of the usual results in sex-linked inheritance. The double yellow females are now known to contain two X-chromosomes joined to each other, end to end. Each carries the factor for yellow. When fertilized by a grey male, such females, as a rule, produce yellow daughters and grey sons. Figure 17 shows why

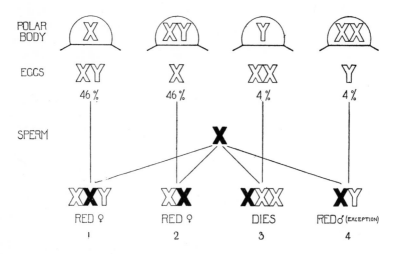

FIG. 15.

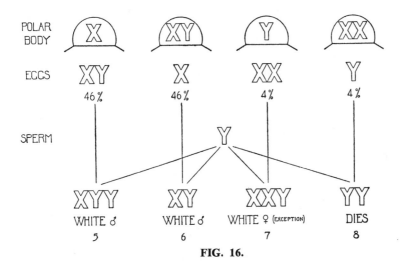

FIG. 16.

this takes place. After the elimination of the polar bodies there are two kinds of eggs. One has the double X the other a Y. Fertilized by a normal male there are four kinds of offspring expected. The first (XXX) has a double yellow-bearing-X and a grey-bearing-X. She generally dies, but occasionally comes through. When she does she is grey, because one grey gene dominates two yellow ones. She is sterile, and cytologically she has been shown to have a double-X (yellow) and a single-X chromosome.

The second individual is a double yellow female that repeats the story. The third is an XY male like its father. The fourth, YY, dies.

Thus half the offspring die, and the normal 1:1 sex ratio of females to males is maintained. Aside from the great theoretical interest of this case, it has a very practical side also; for by using the double yellow females we can carry on certain stocks of Ist chromosome characters, which have infertile females. The males of such

stock, bred to double yellow females, reappear in the offspring, as well as do double yellow females. The stock is self-regulatory, since the rare XXX females are infertile and require no further attention.*

III.

There is a corollary to the view, that the genetic factors are carried by specific chromosomes, that has far-reaching consequences, which from the first were foreseen.

If there are many factors in the

_____

* One reservation must be made. Occasionally the two X's break apart in an egg before the polar bodies are produced. Then one goes out, and such a ripe egg behaves like an egg with a single yellow bearing X. If, for instance, it should be fertilized by a Y sperm, it will give a yellow male, as in ordinary sex-linked transmission. If it should be fertilized by an X sperm, carrying wild type genes, it will give a normal wild type (XX) female, that breeds as a heterozogote for yellow.

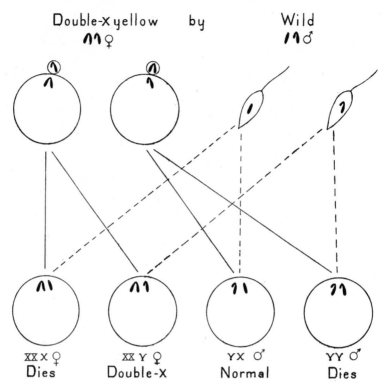

FIG. 17.

same chromosome, we should expect these characters to be inherited in groups. Such groups are known.

In *Drosophila melanogaster* there are only four pairs of chromosomes (fig. 2). In the female there are two X-chromosomes, two large pairs of autosomes and the minute fourth pair. In the male there is an X and its mate called Y, and the same three pairs of autosomes.

There are about 400 races of *D. melanogaster* whose characters have been sufficiently studied to show how they are inherited. They are inherited in four groups. This means that if two or more members of the same

group go in together, *i.e.*, if they lie in the same chromosome, they tend to be inherited together through successive generations.

Fig. 18, shows some of the characters of the first group (I) carried by the X-chromosome. Some of these characters are eye-colours, others are wing, or leg, or bristle characters, etc. In group II we find other eye characters (fig. 19), wing, leg, and bristle characters.

In group III again, we find still other modifications of the same parts of the body (fig. 20). In group IV only three characters have so far been identified (fig. 21). The experimental

FIG. 18.

evidence shows, as I have pointed out, that these last characters are carried by the small chromosome.

It is interesting to note that the number of characters in each group is about proportional to the known sizes of the chromosomes. It is true there are somewhat more in group I than in

FIG. 19.

FIG. 20.

the other two large groups in proportion to its length, but we understand why this is so, since a new mutant recessive gene appearing in the X-chromosome is more likely to be found than a recessive mutant in any other chromosome.

If we turn to other species of Drosophila we find, as far as the evidence goes at present, that there is in

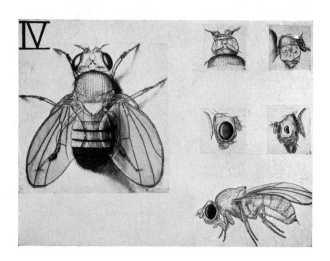

FIG. 21.

them also a correspondence between the number of chromosomes and the number of linkage groups.

The species that is most like *D. melanogaster* is *D. simulans*. So similar are the two that they have until recently been supposed to be the same species. Now we recognise many minute differences between them, and know that they give sterile offspring when crossed. There are four pairs of chromosomes in simulans (fig. 22), identical in shape and size with the chromosomes of melanogaster. Sturte-

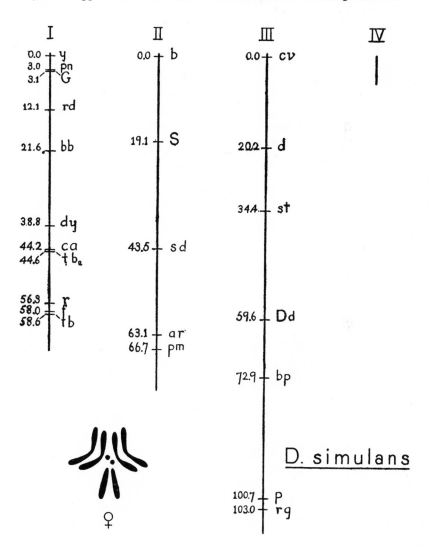

FIG. 22.

vant has found, to date, three groups of linked characters in this species.

In *D. willistoni* there are, according to Metz, three groups of linked genes and three pairs of chromosomes (fig. 23).

In *D. virilis* there are, according to Metz, five known groups of linked genes and six pairs of chromosomes (fig. 24).

In *D. obscura* there are, according to Lancefield, five groups of independent genes and five pairs of chromosomes (fig. 25). The first group is almost twice as long as the first group in melanogaster (the yellow, notch, white loci in the middle instead of at the end), and the X-chromosome is a bent chromosome nearly twice as long as the X in the other species.

In other animals and plants evidence is slowly coming in showing that the linkage groups correspond with the number of the chromosomes. In the edible pea, for instance, there is evidence, according to White, that there are seven independent factors and seven chromosomes.

In a wild California plant, Clarkia, Burlingame has found two linkage groups, and there are only two pairs of chromosomes.

In cases where there are numerous chromosomes it is difficult to find enough independent factors to test out the relation of linkage groups to

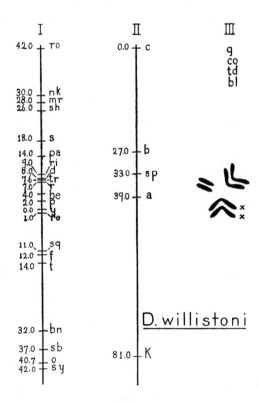

FIG. 23.

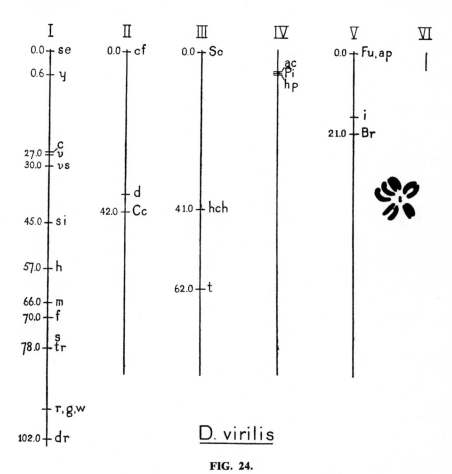

I   II   III   IV   V   VI

D. virilis

FIG. 24.

chromosomes. We shall have to wait for further evidence until more extensive work has been done on such forms. Nevertheless it may, I think, be claimed without exaggeration that the facts so far obtained are consistent with the view that the linkage groups correspond in number to the number of chromosomes.

IV.

The evidence so far considered tells us little or nothing as to how the genes are situated in the chromosomes. Here a new phenomenon enables us to carry our analysis further. The essential evidence can be presented most easily by a few illustrative cases.

As we have seen, there is a mutant race of Drosophila with vestigial wings (v). The gene for vestigial is in the IInd group. There is another mutant race called black (b), whose gene is also in the IInd linkage group.

We can easily make a race that is both black and vestigial. If we cross (fig. 26) a black vestigial (bv) to a wild-type fly (BV), we get wild-type

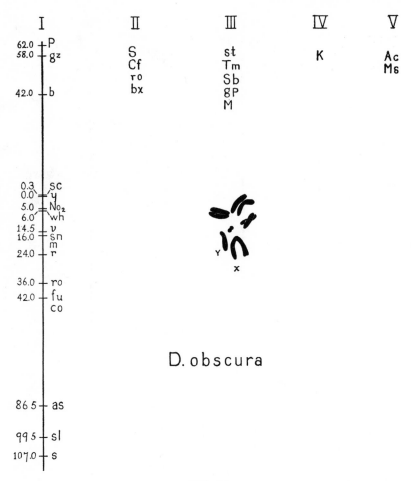

FIG. 25.

hybrid offspring (B*b*V*v*). If now we take such a hybrid female and backcross her to a black vestigial from stock, we get, not two classes of offspring, as expected, if black and vestigial were completely linked; but, on the contrary, we get four classes—two large classes, representing linkage (*bv* and BV) and two cross-over classes (*b*V and B*v*). In other words, while there is still a tendency for the linkage to hold, there is also evidence that it breaks in a certain number of the cases. We may say that black colour has crossed over to long wing in 8½ per cent. of cases, and reciprocally wild-type colour has crossed over to vestigial in 8½ per cent. of cases. Taken together we say that there is 17 per cent. of crossing over.

My second illustration of crossing over relates to the first or X-chromosome. If a fly with white eyes and yellow wings (both characters carried in the X) is crossed to wild type (fig. 27), all the daughters are wild type, and all the sons are yellow white. If we cross the hybrid female to her white-eyed yellow brother (or else to a white-eyed, yellow stock male), we

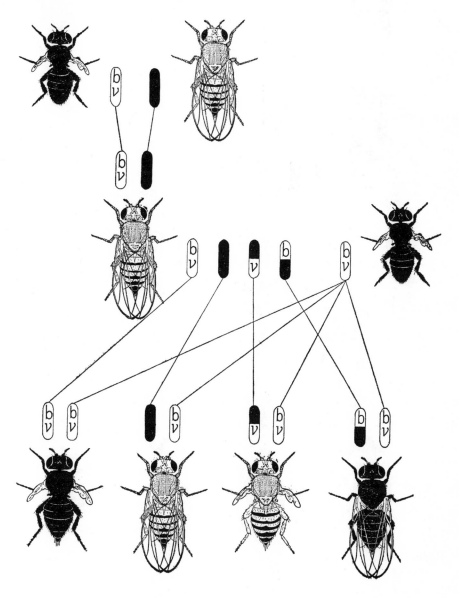

FIG. 26.

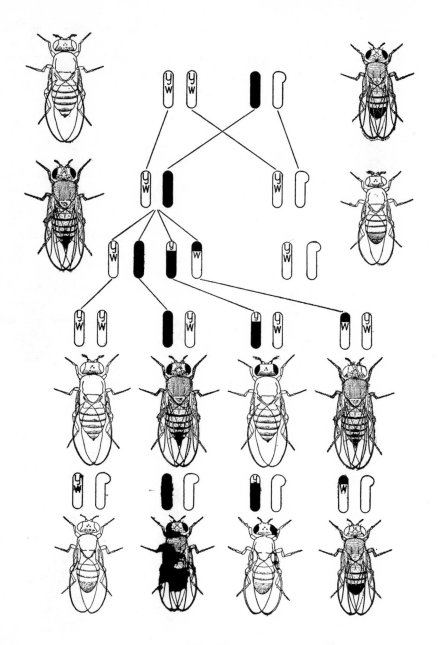

FIG. 27.

find that 99 per cent. of the offspring show linkage. In addition there is 1 per cent. of crossing over.

Hundreds of similar cases could be cited. We find every degree of crossing over, from 50 per cent. (which approaches free assortment of characters) to cases much less than 1 per cent., but there is a constant percentage of crossing over for each two linked genes under the same environmental conditions.

We find such interchanges are always within the same linkage groups. Evidently we must revise our first conclusion that the chromosomes remain intact. We must revise this first conclusion, in so far that members of the same pair sometimes interchange pieces. More important is the fact that the percentage of interchange is constant for each particular combination. It is from such data that we have been able to form a theory of the location of the genes in the chromosomes.

This question of crossing over is so important for what is to follow that the evidence relating to it calls for most careful consideration. I am especially anxious to make clear how far the evidence that is necessary for the construction of the chromosome maps is independent of later attempts to discover the method by which crossing over takes place. To do this, it will be necessary to recall certain general features of the situation:

(1) The chromosome theory, namely, that the hereditary units are carried by the chromosomes, means to-day much more than that the chromatin is the material basis of heredity. The theory carries with it the idea of the individuality and continuity of the chromosomes. The individuality was held by cytologists largely as a matter of faith until genetics showed that the chromosomes have specific relations to the body characters. But here again the extraordinary exchange of equivalent parts between the maternal and paternal members of a pair demonstrates what was to Boveri only a casual speculation or a possibility with almost no evidence in its favour and no proof at all to support it. As to the continuity of the chromosomes from one cell division to the next—a relation that has proven an almost insuperable difficulty to cytology—it matters not at all to genetics whether in the resting stage the hereditary elements separate (break apart or dissolve) so long as they come together again at the next cell division in the same chromosome and in the same order that they had previously occupied. That the 'genes' should actually be dissolved and come together with the extraordinary regularity indicated by all the work on crossing-over is almost beyond belief, yet I repeat it is not essential to the continuity idea to suppose that they remain united.

(2) It seems to me that there is no escape from the conclusion that interchange of equivalent blocks of genes (*i.e.*, pieces of chromosomes) takes place in a perfectly orderly manner for anyone who accepts the view that the 'chromosomes' are the bearers of the hereditary units. It shows a failure to grasp the situation, to be willing to accept the chromosome view and assume a critical attitude towards the *evidence* for crossing over—since the evidence for the one is of the same nature and as cogent as it is for the other. By 'marking' specific chromosomes (both members of the same

pair) it has been possible to demonstrate that when crossing over takes place each chromosome breaks apart at the same level and an interchange takes place. For example, if the loci of a chromosome are indicated by the letters A, BC, D, E, F, and its mate by *a, bc, d, e, f,* then whenever the first series breaks between D and E (let us say) the other one breaks between *d* and *c*. As a result of such a breaking followed by interchange, the two series that result are A, BC, D, *e, f,* or *a, bc, d,* E, F.

(3) Such being the fact, it necessarily follows that the nearer together two units lie the less likely is a break to occur between them; and conversely the farther apart two units lie the more likely is a break between them. It is not necessary to make any further assumption (except that where crossing over occurs at one level it is less likely that another crossover will occur simultaneously near the first) in order to deduce from the situation the arrangement of the genes on the chromosomes.

I should like to emphasise, as strongly as I can, that the chromosome maps, representing the arrangement of the genes, are derived directly from the genetic evidence relating to crossing over. It is true that we have gone further and have tried to find out how the genetic evidence ties up with cytological processes known or supposed to take place in the germ-cells, but the localisation of the genes has been determined independently of these attempts to discover the mechanism of crossing over. The latter might be entirely erroneous without at all affecting the validity of the methods by which the genes have been located.

It may be best therefore at this point to say something about these maps (fig. 28) before passing to a consideration of possible interpretations of the mechanism of crossing over.

The maps* enable anyone to predict to within a small degree of error how a *new* character that appears in Drosophila will be inherited with respect to the other characters already known. This means that all we have to do is to determine, first, the linkage group of the new character, then the crossing over within this group between the new character and any two other members of its series. After this, by means of very simple calculations that take at most a few minutes, we can predict how the new character will be inherited with respect to every other known character.†

---

* In the figure of the maps, giving the principal loci of the four chromosomes of *Drosophila melanogaster,* the more common loci are named, but the others are also indicated by cross lines. The spacing gives the relative position of the loci. 'Distance' is used in a figurative sense for crossover value. The distance apart of the loci on the map ('map distance') is only a relative matter, and rests on the assumption that crossing over in one part of the 'chromosome' is as frequent as in all other parts—an assumption that we have found reason to think is not entirely accurate.

† Actual length of the section between the loci is only one of the factors determining the amount of crossing over between the loci, and, consequently, the map distance. Both environmental and genetic factors are known to influence the frequency of crossing over within a given length. Therefore, the length of a section of the chromosome represented by a unit distance (1 per cent.) may be different regions of the chromosome. A parallel to the maps is found in a rail-

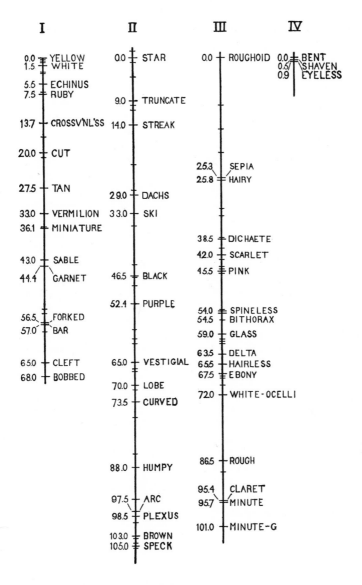

FIG. 28.

road timetable, where the number of minutes between stations is given. From such a table one can judge accurately the sequence of the stations and roughly the actual number of miles between

them. Knowledge of the normal speed of the train and the condition of the road bed and of the grades would make it possible to judge more accurately the number of miles between the stations from the number of minutes between the stations.

This ability to predict would in itself justify the making of such maps. But the maps have for us a somewhat wider interest, because they furnish evidence as to how the hereditary factors lie in the chromosomes. A single illustration will, I hope, make this clear (fig. 29). Two points, yellow and white, in the first chromosome are represented as 1·2 units apart. Now if a new type, bifid wing, should turn up and its crossover value with white should be found to be 3·5, experience shows that it is expected to give with yellow either the sum of 1·2 and 3·5 (=4·7) or the difference between 3·5 and 1·2 (=2·3). In other words, if it lies south of white it should give 4·7 per cent. crossovers with yellow; if north of white it should give 2·3 per cent. crossovers. The same principle is known to hold for four, five, six or any number of points in the same genetic series.

This relation of three or more points to each other is a relation of linear order, and cannot be represented in space in any other way than by a series of points arranged in a line like beads on a string.*

---

* It is perhaps hardly necessary to explain that all the results with Drosophila that are given in the text are not the result of one individual's work—certainly not my own. The work has been done by a small band of collaborators, in part under the auspices of the Carnegie Institution of Washington, and in part as members of the Zoological Laboratory of Columbia University. Sturtevant, Bridges, Muller, Metz, Weinstein, have held the advanced line, supported by twenty or more other investigators and students, each of whom has made one or more worth-while contributions to the Drosophila work.

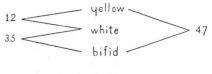

FIG. 29.

V.

Once in each generation in the whole course of the germ-track cycle of cell divisions, the maternal and the paternal members of each pair of chromosomes come together as thin threads and appear to fuse into a single thread. Thus, as has often been pointed out, the conjugation of the chromosomes—the meeting of the maternal and paternal genes—takes place before and not immediately after fertilization.

In searching for a time in the history of the germ-cells when interchanges (crossing-over) might take place, it is to this stage (conjugation) that one's attention would naturally turn. It is very fortunate, therefore, that it has been found possible to obtain experimental evidence in Drosophila that crossing-over does occur at about the time of conjugation of the chromosomes. I refer to Plough's temperature experiments. Plough found that a change in the temperature causes a change in the amount of crossing-over. By subjecting flies whose eggs were known to be at certain stages of maturation to differences in temperature, he showed definitely that crossing-over occurs at this late stage in the history of the egg-cell.

If now we turn to the evidence furnished by cytology no one would fail to have his attention arrested by certain descriptions of an event that occurs when the chromosomes conjugate. At this time, according to

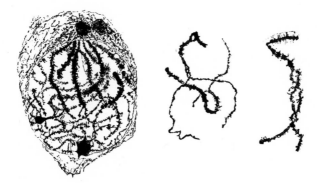

FIG. 30.

Janssens and other observers, the thin threads that are united at one end and extend out into the nucleus sometimes appear to overlap or even to twist around each other (fig. 30). Now if they come together while still in this condition it follows that a maternal thread will lie on one side of its paternal partner for a part of its course, and on the other side for the rest of its course. This gives a mechanical model that fulfils all the re-

quirements of crossing over, *provided* the parts of the thread that come to lie on the same side unite with each other to form a continuous chromosome (fig. 31). In the present stage of cytological research it is extremely improbable that this event— the breaking of the old threads at the crossing over level and the reunion of the ends on the same side—could ever be actually followed, since it would require continuous observation on liv-

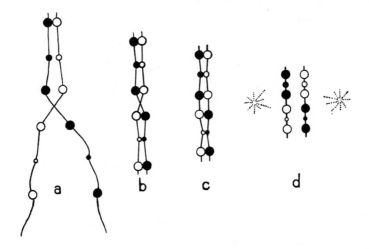

FIG. 31.

ing material.* At best one might hope to find evidence that crossed threads are sometimes actually present when the conjugation is taking place, and that at the time of separation of the condensed thread (reduction) the chromosome threads are no longer twisted about each other.

There is, it is true, a certain amount of evidence of this kind. For example, Janssens has published certain figures of the separating thread at the reduction division that seemed to him to show that they represented an earlier condition of overlap of the thread with partial fusion at the crossing of the inner threads.† It has been pointed

---

* Recently Seiler has found in one of the moths, Solenobia, two pairs of chromosomes that may join temporarily in the male to form only one pair at the maturation stages. If the union is at random the outcome would be the same as for free assortment; but if certain unions are more likely than others, then the results will appear to give the same result as crossing-over in Drosophila. If the union and subsequent separation of the two pairs of chromosomes in question is always at a given point, the phenomena will be quite different from what is observed in Drosophila. Other hypotheses to account for crossing-over are not in harmony with the facts to be explained. For a detailed criticism of this see 'The Mechanism of Mendelian Heredity,' Morgan, Sturtevant, Muller, Bridges.

† Crossing-over is here supposed to take place between the two inner threads of the four-strand stage. Janssens has pointed out the necessity of two divisions to bring about the reduction of the diploid number of chromosomes to half, in order that there shall be a single line of genes in each chromosome (*i.e.*, a chromosome not heterozygous). Aside

out, however, that another interpretation of the crossed threads is possible that does not involve crossing-over. Nevertheless, it still seems to me that Janssens' interpretation, while not conclusive, is the more probable interpretation of the later condition that he finds. In addition to Janssens' earlier work a recent account by Gelei (fig. 32) seems to show very clearly the crossed threads at the time of the conjugation of the thin threads.

Many cases are known in which the chromosomes, *after* they have conjugated and have begun to condense, are twisted about each other (fig. 33); but it is generally supposed that this latter twisting is a secondary process and need not *necessarily* cor-

---

from the appearance of a teleological view implied perhaps in this explanation, it is not obvious why two divisions should be necessary. Only when four strands are present at the time of crossing-over (when two of the strands are interchanged) is Janssens' 'necessity' for two divisions called for, but then it would be equally 'needed' even if no crossing-over occurred.

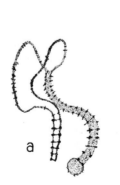

FIG. 32.

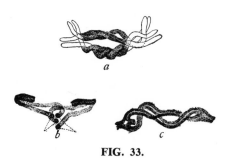

FIG. 33.

respond to any earlier twisting of the chromosomes. Such evidence by itself cannot be used either for or against the twisting hypothesis unless it can be shown, as Janssens tried to show, that certain kinds of crossed threads, observed in these later stages, are sometimes traceable to an earlier interchange. Fortunately there is a little evidence that also at the thin thread stage the chromosomes are sometimes twisted around each other. Janssens thinks that his observations show this, but it has been found very difficult to make certain of this relation.

It is unfortunate that it is the figures resembling those of the *later* coiled threads that are beginning to be copied into text books as the stages supposed to furnish the evidence for the twisting of the chromosomes at the time of crossing-over. This is unfortunate because these figures do not really conform to the evidence of crossing over furnished by Drosophila. In the first place single crossing-over has been shown to occur for the X-chromosome of Drosophila in only 43 per cent. of the possible cases, and double-crossing over in about 12 per cent., and triple-crossing over (an extremely rare event) in about 2 per cent. Yet if these late twisted chromosomes had any relation to crossing-over, it would

be expected to occur many times in the length of each chromosome pair. In the second place, it appears that all, or nearly all, of these late coils are straightened out as the two chromosomes condense, preparatory to entering the spindle.

While it is evident, therefore, that genetics has far outstripped cytology in regard to the evidence of interchange between members of the same pair of chromosomes, yet cytologists have described a series of events taking place at the time when geneticists expect to get evidence of crossing-over. This creates a very favourable situation so far as genetics is concerned, but, for the present, geneticists may have to wait until the cytologists can make further advance in the study of these stages.

There is one fact about genetic crossing-over that is of prime importance for any interpretation of the process, namely, the fact that whole series of identical (or allelomorphic) genes are exchanged whenever crossing-over occurs. In other words, great pieces of the chromosomes are involved and always identical pieces. Moreover, measurements of these 'blocks of genes' show that they have a certain modal length, there are very few very short blocks, more larger ones, and very few very large ones, etc. This discovery fits in excellently with the view that crossing-over is brought about by twisting of the chromosomes about each other, for if the chromosomes twist about each other in loops, then, owing to the rigidity of the chromosomes, very short loops will be less likely to occur than somewhat longer ones. Regions on each side of a crossover are, therefore, expected to be protected from

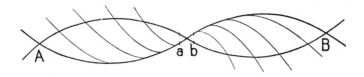

FIG. 34.

another crossing-over. We call this interference, because a crossover that might otherwise occur is interfered with. Now Muller put this matter to a test by taking into account all those cases in which crossing-over occurs at two points at the same time. For example, if the first crossing-over takes place at the left of the chromosome (fig. 34), the chance that a second crossing-over should take place at some other point will be the greater in proportion as the other point is distant from the first. This we find is actually realised.

Furthermore, on the twisting hypothesis interference decreases until it vanishes at a certain distance. Beyond that distance, crossing-over should again be interfered with, because the more frequent length of loop has been surpassed* (fig. 35). Weinstein put this possibility to a crucial test. He

---

* Figure 35 is a diagram to show the frequency of interference. The basal line stands for the chromosome; the arcs (arising at one point to the left) are intended to show where a second cross-

found that there is a decrease in the amount of crossing-over as anticipated.

VI.

The evidence from crossing over has led to the conclusion that the hereditary elements, the genes, are arranged in linear order in the chromosomes. If we think of these elements as material particles they must be supposed to have the power of self-division, and to remain unchanged through long periods. More than this we need not postulate. How they affect the cells in which they lie we

---

over may fall. The heavy curved line indicates the frequency of a second crossover, or, in other words, the frequency of blocks of different lengths. The modal length is indicated by the distance to the base of the heavy, vertical line. It is evident that, from any given point (crossover level) taken as a starting point, there is first a region in which no crossing over occurs; then the chance of one occurring increases up to the modal length, after which the chance of a second crossover decreases again.

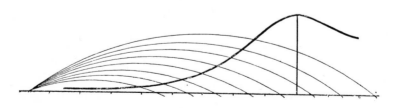

FIG. 35.

do not know. Nor do we know whether they are functioning all the time, or only under certain specified conditions. However they do their work, we must regard the organism in large part as the outcome of the sum total of their activities.

It is, however, important to emphasise that these ultimate units are not necessarily to be thought of simply as the representatives of each part of the organism, for every part of the organism must result the activity of a large number of the elementary units. But, obviously, if something should come in that brought about a change in any one of these units, the end product made by them all might be affected, and the effect might hit some one particular part of the body harder than other parts. It is this indirect effect that we see when a mutation occurs and we refer it back to the ultimate source from which it arose, that is, to a modification of some part of the germ-material.

I mention this in particular, because I find that many people get the impression that we suppose that each part of the adult organism must have a single representative in the germ-track, and sometimes the impression is produced, unfortunately, that we imagine the chromosomes as representing a sort of miniature of the animal. One of my friends has laboured for years under the impression that our idea of the germ-plasm is that of a mysterious insect in the nucleus, a sort of insectulum, and, under the circumstances, I cannot blame him for thinking that we are on the way to the madhouse.

It is hazardous to make any statement as to the size of these ultimate units, because, in the first place, we do not know how long the chromo-

somes are at the time of crossing-over, and also because we do not know how near the nearest genes may lie to each other. We have, however, speculated a little about them; and, with the understanding that it is largely speculation, I may venture to state what the outcome has been.

Several methods have been tried. One of these may be illustrated by the following example:—If there were a known number of balls in a bag, say 5000, and one was taken out and put back, and this was done over and over again for 1000 times, it would sometimes happen that the same ball was drawn again or several times over. It is obvious that the greater the number of balls the less likely is the same one to be drawn more than once. Conversely, if we know the number of times balls have been drawn (twice, three times, etc.), we can find out how many balls there must be in the bag. Now in Drosophila most of the mutations have appeared only once, a few twice, fewer still three times, etc. From these data we have calculated that there are, roughly, more than 2000 genes in Drosophila. Since each individual has two sets of genes, there will be 4000 in all.*

If we take the number of loci at 2000 (which is a low estimate), and calculate the volume of the four chromosomes (whose total length in pre-

---

* This method of calculating the number of genes was first suggested by Muller for the X-chromosome, where data were then most easily obtainable. Here the calculation is based on data for mutation in all of the chromosomes. Multiple allelomorphs are counted as so many identical (and repeated) mutations of the same locus; but the reversion in Bar has been omitted as exceptional. No doubt many repeated mutations have

served material is taken at 7.5 microns in length and 0.2 in breadth), the total volume will be 0.236 cubic microns. If we divide this volume by 2000 and estimate the diameter of the resulting gene, it turns out to be 60/1000 of a micron in diameter.

In this calculation the measurements of the chromosomes were made when they are condensed in the equatorial plate stage. Now, since it is more probable that crossing-over takes place when each chromosome is fully extended (thin-thread stage of conjugation), it seemed better to make another estimate on this basis. If we assume that the second chromosome of Drosophila expands as a loop, its length must be about twice the diameter of the nucleus. Let us allow 5.5 microns for the diameter of the nucleus of this stage, and then double it for the length of the thread. There are 108 genetic units in this second chromosome. If we allow 1/5 of a unit as near the minimum cross-over value, there will be room for 540 genes in this chromosome,* which makes each gene 20/1000 of a micron in diameter.

In still another way an attempt was made to get some idea of the size of the gene. Since the head of the spermatozoon is commonly supposed to contain only chromatin material, it would be possible to determine from the volume of the material in the sperm-head how large each gene might be (the maximum limit) by dividing this volume by the postulated number of genes.† From this volume the diameter of the gene could be determined. The result of such a calculation gives the diameter of the gene 77/1000.

The three estimates give 77, 60, and 20 thousandths of a micron respectively for the size of the gene. Different as they are, it is still surprising that the range is not wider when the many possible sources of error are considered. They may at least have an interest as the first crude attempts to get some idea of the size of the material elements postulated by genetics.

It is not without interest to compare these estimates with the estimated sizes of organic molecules. The molecule of hæmoglobin has been given at

---

been overlooked, and the error of random sampling is large, which reduces the reliability of the figures. Moreover, all lethal mutations have been left out of account, because one cannot be sure of their recurrence.

* This calculation assumes that the shortest distance between the genes, treated as points, is 1/5 of a unit, but we have data for smaller distances even than this; in other words, 540 intervals is a relatively low estimate of the number of genes. It also assumes that the genes are evenly spaced, and there is some evidence that this is not the case.

† This calculation assumes that the sperm head is composed of chromatin alone, for which there is no real evidence. If other material is present, then the calculated size of the gene is too large, i.e., its true value will be closer to the results of the other calculations. On the other hand, if the sperm head is composed entirely of chromatin, then the value obtained for the size of the gene 77/1000 is a maximum, since the number of genes used in the calculation is a minimum. It should be stated that the measurements of the sperm head were made from preserved material (sections of the testes, mounted in balsam), and no allowance was made for shrinkage. The calculation should be made again on living sperm, if possible. The sperm head measured about 7.0 microns in length, and 0.3 microns in breadth.

2½/1000 of a micron, and that of casein is almost the same. The size of the gene on the basis of these tentative estimates seems to be larger, but not much larger than that of some protein molecules.

VII.

We have covered a good deal of ground, and I realise and regret that the mass of experimental data on which most of the conclusions rest has necessarily been left in the background. The data are, however, published and accessible. Except for the questionable attempts to estimate the size of the gene, the main conclusions concerning the mechanism of heredity and the orderly arrangement of specific genes in the chromosomes rest on quantitative data and on analytical deductions that are tested by further experiments wherever possible. The theory that the chromosomes carry the hereditary factors, and that these factors lie in linear order in the chromosomes, enables us to predict, with a high degree of certainty, how any new character will be inherited with respect to all the other 300 known characters of Drosophila. If the accuracy of prediction is a test of the usefulness of a theory, then we may claim some justification for our view. More than this, I think, it is not necessary to claim for a scientific theory.

The evidence has given us a glimpse at least of processes that are so orderly and so simple as to suggest that they are not far removed from physical changes; and the order of magnitude of the materials is so small as to suggest that its component parts may come within the range of molecular phenomena. If so, we may be well on the road to the promised land where biological results may be treated as physical and chemical events.

Twice each decade the geneticists of the world assemble to appraise their field. The Sixth International Congress of Genetics was held in 1932 at Ithaca, New York. Morgan was the president and his address was in many ways the valedictory for himself as a geneticist and for the field of classical genetics as well. Several years before he had returned to experimental embryology: his type of genetics was essentially complete. Details were to be added but there were to be no great additions to general theory. Thereafter, genetics was concerned with two other problems: how genes act and the molecular nature of the genetic material.

Here is Morgan's retrospective evaluation, together with some predictions for the future.

THE RISE OF GENETICS
By Professor T. H. MORGAN
CALIFORNIA INSTITUTE OF TECHNOLOGY

The new developments in science that occur from time to time can generally be traced either to the invention of a new method or to the discovery of a new fact that has far-reaching consequences, or to the elaboration of a new theoretical principle that suggests new lines of investigation. In the

From *Science 76*: 261–267; 285–288. 1932. Reprinted by permission.

latter case, it is the prerogative of science, in comparison with the speculative procedure of philosophy and metaphysics, to cherish those theories that can be given an experimental verification and to disregard the rest, not because they are wrong, but because they are useless.

In the case of genetics the situation was in some respects different from any of these procedures; for it began with the discovery of a discovery that had been made 35 years before. We can date the beginning of genetics, then, from the resurrection of Mendel's paper in 1900. Its rehabilitation was not, however, due to a literary find, but to a need resulting from similar experiments by de Vries, Correns and Tschermak that unveiled a series of phenomena identical with the facts of Mendel's earlier work.

The significant fact is that when the time was ripe to appreciate its fundamental significance, Mendel's forgotten paper was discovered with the result that the activities of hundreds of biologists, as the program of this present Congress bears witness, had the direction of their scientific careers entirely redirected, or begun along new lines. The discoveries that rapidly followed, showing that the same laws applied widely to the other plants and to animals also, brought about realization that a great step forward in biology had been made.

But before we consider the rise of genetics after the year 1900, it is proper on this occasion to pay tribute to the earlier work in hybridizing that furnished the background of procedure to which Mendel himself probably owed a considerable debt. Let us pause for a moment and recall a bit of history, for it would be unfair to forget or to underrate everything prior to the first year of the present century.

If to-day we express surprise that Mendel's paper remained unnoticed for 35 years, let us recall that this is a not unfamiliar experience in biological science. Between the experimental proof of sex in plants by Camerarius (1694) and the prize essay of Linnaeus on the sex of plants (1760) sixty-six years elapsed.

At about this time the scientific study of hybridizing may also be said to have been begun by Linnaeus and his students, and especially by Kohlreuter in several memorable papers (1760–66).

Then, thirty-three years elapsed before Sprengel's (1793) observations on the natural crossing of plants by insects, which made clear that cross-fertilization is of wide-spread occurrence in flowering plants.

More interesting, perhaps, to modern geneticists are the pioneer experiments on peas that, in a very real sense, were the precursors of Mendel's work. It is not as generally known as it should be that some of the facts on which Mendel's results with garden peas rested had been recorded by several earlier experimenters. In 1823 Thomas Knight, 42 years before Mendel, described a cross between a pea with a gray seed-coat and one with a white that gave seeds which were uniformly gray-coated. These seeds when grown produced in the next year both gray and white seeds. John Goss had in 1822 also reported experiments with garden peas and found the first generation of offspring had seeds like the paternal race. From these in the next generation he obtained peas of two kinds, one like those of the original grandpaternal race, the others like those of the grandmaternal. Separating these he found that the blue peas produced in F_3 only blues, and the white peas both blues and whites. Here is an example of what to-day we call dominance and recessiveness, as well as segregation in F_2. In the same year (1822) Alexander Seton reported similar results. Nearly fifty years later Thomas Laxton (1868–72), working with peas, recorded numerous facts similar to those first spoken of, and in addition he mentioned cases in which two pairs of contrasted characters were present. Assortment between the pairs was found—which result is familiar to students to-day and which Mendel's work established.

Amongst the earlier hybridists the name of Naudin (1861–64) is most often referred to as a forerunner of Mendel, and it is sometimes stated that he anticipated Mendel's discoveries. His principal prize paper appeared in 1863, two years prior to Mendel's paper before the Brunn Society, and was followed by two others in 1864 and 1865. Naudin laid emphasis on the identity of individuals of the first generation hybrids, including reciprocal crosses. He insisted on the intermediate character of the F_1 hybrid, with the important reservation that the intermediate forms do not stand always equally distant from the two parents. We now know that, taken character by character, sometimes an intermediate condition, sometimes complete dominance, may be found. But whichever condition holds for a particular character, the phenomenon of segregation in the germ-cells of the F_1 hybrid remains unaffected.

Naudin stated explicitly that in the second and later generations there is a mixture of forms, including some which are like the original parents and other that approach these in various degrees. Then follows his most important deduction, namely, that the second generation results find their explanation in the disjunction of the two specific essences derived from the parents in the ovules and in the pollen of the hybrid. Here we have a highly significant contribution, for, not only did Naudin see clearly that the results are explicable on the principle of disjunction (or, as we say now, segregation), but that this, taking place both in the egg and in the pollen, gives the kinds of characters that appear. So important historically is this fact that there should be included his specific statement showing that he had a perfectly clear idea as to how disjunction accounts for the diversity in the second generation. If, he says, a pollen grain bearing the characters of the male parent meets an egg of the same kind, a plant that is a reversion to the paternal species will result; similarly for the maternal species. But if a pollen grain of one kind meets an egg of the other kind, a true cross-fertilization takes place like that of the first generation and an intermediate form will result. It will be agreed, I think, on all hands that this was a brilliant interpretation of results based on first-hand experience. It falls short of Mendel's work in two or three important aspects: (1) The failure to put the hypothesis to a test by back-crossing; (2) the failure to see what the numerical results should be on the basis of disjunction of the elements in the hybrid. His use of the words 'disordered variation' in the F_2 and later generations brings out the essential difference between Naudin and Mendel. It is the orderly result of disjunction or segregation that is the important feature of Mendel's work; and finally, the clearness with which Mendel stated and proved the interrelation between character-pairs in inheritance, when more than one pair is involved, places his work distinctly above everything that had gone before. Nevertheless, the genial abbot's work was not entirely heaven-born, but had a background of one hundred years of substantial progress that made it possible for his genius to develop to its full measure.

If, in this brief review, I have neglected to bring in the names of a number of well-known selectionists whose work has been in the main in the field of agriculture, it is not because I do not realize the importance of their work or the great difficulties they overcame, but because, for the moment, we are interested especially in the development of our theoretical knowledge of genetics.

So far I have spoken only of plants. What part, may be asked, has the study of animals played in the pre-Mendelian history of genetics, i.e., down to 1865?

The question of sex in plants that took botanists a hundred years to decipher was not so difficult for zoologists. If we may accept the traditional story, it was not unknown in the Garden of Eden. Aristotle had a good deal to say about it. The credit of finding a sex-determining mechanism can properly be claimed by zoologists, but this happened only in the opening years of the present century.

Hybridizing was also familiar to zoologists, but in pre-Mendelian times occupied only a relatively small part of their interest. What was known has been recorded by Darwin in his 'Animals and Plants under Domestication.' This scattered and loose information was incorporated after 1859 in the discussions of the theory of evolution.

The chief contribution of zoologists to present-day genetics was along different lines. In the latter half of the last century there was great activity in the field of cellular morphology. The important facts concerning chromosome-division and the extraordinary changes that take place at the time of maturation of the germ-cells and at fertilization were first made out by zoologists. The names of Kölliker, Flemming, Fol, Van Beneden, Hertwig and Boveri are landmarks in the history of cytology. Correspondingly for plants the names of Hofmeister, Strasburger, Du Bary and Guinard run a parallel course.

Weismann's theoretical contributions have also played an important historical rôle. 'The Continuity of the Germ-plasm' served to counteract the all-too-prevalent influence of Lamarck and his successors, whose views if correct would undermine all that Mendel's principles have taught us. Weismann's speculations on the origin of new variations by recombination of elements in the chromosomes, while not to-day acceptable as stated by him, nevertheless focused attention on an important subject. His discussion of the interpretation of the maturation divisions played, I believe, a leading rôle in directing attention to a subject that was destined very soon to have great importance for genetics.

Thus at the end of the last century some extraordinary advances had been made in unraveling the changes that take place in the maturation of the germ-cells. These advances led to the recognition of a mechanism that was to place the theoretical elements of Mendel's hypothesis on a firm foundation of fact. But this, however, was not apparent until 1903.

GENETICS AT THE BEGINNING OF THE CENTURY

We come now to the fateful year 1900, when three lines of fundamental significance for genetics were ready to be brought together. I refer, of course, to the mutation theory of de Vries, to the rediscovery of Mendel's paper, and to the application of the discoveries in cytology to the new theories.

The intimate connection between the mutation theory, as first propounded, and the origin of the characters that follow Mendel's laws was not immediately evident, since de Vries laid emphasis on the many character changes that result from each progressive mutational step. In fact, at about this time de Vries recognized three types of mutational changes: Progressive changes—changes that introduce something new, leading to the sudden appearance of a new elementary species; retrogressive changes, the result of something lost or becoming latent; and degressive changes, in which old characters are revived.

This nomenclature, in so far as it is purely descriptive and based on characters rather than changes in the germ-plasm, covers broadly many of the facts with which we are familiar to-day. But in the light of the work of the last 30 years, especially when applied to genes, this description can no longer be accepted as fundamental; for now we have information that gives a more consistent picture of the changes produced by the genes. For example, the evidence from hybridizing elementary species, on which de Vries based in part his distinctions, has to-day a different interpretation. We no longer hold that a progressive change introduces an entirely new, unpaired element into the germ-track, for the unpaired chromosome in cases of heteroploidy can surely not be regarded as the usual step for progressive evolution. Again, the permanence of certain hybrid combinations, whenever such exceptional cases arise, are not now regarded as due to the introduction from each parent of a new unpaired element, but can be interpreted in different ways in different cases.

It was the emphasis that de Vries laid on mutational changes in the germinal material as sharply discontinuous, irrespective of the effect on the character, that has had important and far-reaching consequences for genetic work and theory.

The groundwork for discontinuous phenotypic variation had in 1894 been laid by Bateson's contribution on discontinuous variation. While we recognize that some of the examples Bateson collected are not inherited but phenotypic (which confused the picture), nevertheless his insistence on the importance of discontinuity prepared the way for the acceptance of the more fundamental distinction that de Vries had made.

But I wish to emphasize that the revolution in our ideas that took place at this time was not so much due to the insistence on discontinuity of somatic structures, but discontinuity in the hereditary elements. An example will serve to illustrate the difference. When a gene changes, its effects on new characters, taken individually, are generally very different. Some of them may be sharply marked off from the original character. The character showing the greatest effect is the one generally picked out for genetic work. But at the same time there are changes in other organs that are less conspicuous—some of the characters are so little affected or so variable that, taken by themselves, they would give a picture of continuity rather than of discontinuity. They would often pass unnoticed were not attention drawn to them by the discovery of the major change.

For the theory of evolution some of these inconspicuous changes may be more significant than the more obvious discontinuous change. In fact, if evolutionary advances are more often through invisible physiological mutational changes rather than morphological ones, we can better understand the paradoxical situation in which taxonomists find themselves, to wit, that the sharp structural differences, that are used for diagnostic separation of species, relate to characters that seem often to be unimportant for the well-being of the individual. The new point of view is a complete inversion of much of the thinking in which the evolutionary theory indulged in the past.

As I have said, the rapid expansion of genetics after 1900 has been intimately connected with the applications of the chromosome theory to the experimental work in genetics. The integrity of the chromosomes and their continuity from one cell-generation to the next, the constancy in number of the chromosomes in each species and the absence of mixing of the materials of the conjugating chromosomes at the time of meiosis

have furnished the basis on which genetics rests.

I think we can not overemphasize the significance of this relation between the theoretical side of genetics and the factual side as observed in the known behavior of the material basis of heredity. To put the matter bluntly, the recognition that there is a mechanism to which genetic theory must conform, if it is to be productive, serves to keep us on the right track and acts as a check to irresponsible speculation, however attractive it may seem in print.

Some one may reply that it is not always an advantage to keep one's nose to the grindstone. Granted! but realizing how often ingenious speculation in the complex biological world has led nowhere and how often the real advances in biology as well as in chemistry, physics and astronomy have kept within the bounds of mechanistic interpretation, we geneticists should rejoice, even with our noses on the grindstone (which means both eyes on the objectives), that we have at command an additional means of testing whatever original ideas pop into our heads.

EXPANSION SINCE 1900

I come now to the expansion of the Mendelian theory that has taken place in the last 30 years. If I refrain from giving the names of the numerous contributors to this advance, it is because many of the discoverers are before me in person; or, if not, will get reports of the congress. Future congresses will probably be better able to evaluate individually the merits of those who have made the significant contributions in this generation.

It must have been evident to many geneticists after 1903 that if the chromosomes are the bearers of Mendel's elements, there would be only as many independent characters as there are chromosomes, provided the then current idea of the integrity of the chromosomes were true. This would place limitations on Mendel's second law—the law of independent assortment. In fact, the genetic evidence can now be said to have firmly established that there are more characters independently inherited, within stated limits, than there are chromosomes.

But linkage also turned out to have its limitations, and these very limitations made it possible to determine the localization of the genes in the chromosomes. I refer, of course, to crossing over. Since localization of the genes is to-day the basis of much of the quantitative work in genetics, I may be allowed to elaborate it somewhat.

The outstanding genetic fact is that these interchanges take place only between homologous chromosomes—i.e., between members of the same pair.

The second important genetic fact is that when the interchange takes place, large blocks of the chromosomes are exchanged. This can be proven only in cases where more than two loci are involved, and best when a considerable number of well-spaced genes have been located. Until recently the evidence that large blocks of genes are involved in crossing over was known only genetically. No certain cytological proof was known. To-day, however, the proof has been found. No doubt this cytological evidence will be presented and discussed at this congress.

It has also been determined on genetic evidence that more than one interchange may take place between a pair of chromosomes, which can be checked only in cases where there are enough intermediate loci between two pairs to serve as markers.

A moment ago I said that crossing over has furnished the basis for the theory of localization. May I give an illustration, in the hope of removing a criticism of the localization technique that is based, I believe, on a misunderstanding? It has been said, for example, that the changes made from time to time in the genetic map of the Drosophila chromosomes discredit the method by which the localization is determined. It might as well be said that the method by which the atomic weights in chemistry were gradually improved discredited the procedure of the chemist.

Two illustrations will serve our present purpose. Let us suppose a new mutant character is found and its chromosome group—i.e., its linkage group—determined by familiar methods. We may proceed, then, to find its relation to two known loci in that chromosome. If these are far apart, the cross-over data will give only its approximate position. Having found this, it may turn out that the locus lies near another gene in that region, but whether above or below may be uncertain. We next proceed to find its more exact location with respect to this third gene, using either of the other two genes as a second point. In this way the new gene is more accurately placed with respect to the third locus. Further work will then be necessary if there are other genes in this region.

The second illustration concerns a

distinction between crossing-over data given in the actual experiment and its conversion into map distance. For very small values, say 5 points, the two are the same because double crossing over is not present. But in longer distances the crossing-over data may depart widely from the map figures because double crossing over makes the figures too low. In Drosophila, for example, the sex chromosome is 70 units of map distance, but the crossing-over data are found to give not over 50 units. In this case the map distance has been built up piece by piece through the summation of crossing-over data of loci so near each other that double crossing over is eliminated.

In other animals and plants, where few loci have as yet been found, the cross-over data are generally put down as map distance. This may be far from the real map distance, and since the actual amount of double crossing over in such less-worked-out forms is unknown, and since crossing over is different in different species, the loci must be regarded as only provisional.

There is another factor to be taken into account. The theory of localization was based in a general way on the assumption that crossing over in one region of a chromosome has the same frequency as in other regions. The Drosophila workers have long known that this is not exact, and in fact they had invented methods to show that crossing over is different in different regions.

The crowding of the genes in some regions of the genetic map and their scarcity in other regions has been shown to be due to the different frequencies of crossing over per unit of absolute distance in the cytological chromosome. This seems to be a fundamental relation for all chromosomes. In the X-chromosome of Drosophila, which appears to be a special case, most of the genes are crowded at the two ends of the chromosome with a middle region of undetermined length having few or no genes, in the sense that the Y-chromosome is empty of genes. These facts do not invalidate the purpose for which the maps were invented, since the relative position of the genes remains the same. It is their position relative to each other, allowing very precise prediction of the topographical relation of a new gene to all other known genes, as determined by corrected cross-over data, that is important.

This brings us to one of the most recent fields of modern genetics—the study of the redistribution of the linkage group by translocation. Treatment with x-rays has been found to be a prolific source for material of this kind, but it should not be forgotten that translocation had been discovered and utilized for genetic interpretation several years before x-rays were used. Even to-day, with much evidence before us, the way in which x-rays bring about this result puzzles us. In a crude way we might picture the electron shooting holes in the chromosomes, thus breaking them apart. But when the relative sizes of the electron and the chromosome are considered, it is difficult to see how such a disruption would result from a single shot.

Even more surprising is the fact that the broken end of a piece may reunite with the end of some other chromosome and, acquiring thereby an attachment fiber, form a new linkage group. Of course it does not follow that such a reunion occurs whenever a chromosome is broken. It is only those cases where reunion does occur that are recovered and studied by geneticists. When no such union is brought about the piece, lacking an attachment point, will be lost, and the zygote containing it will probably die.

As I have said, the astonishing fact remains that the broken end becomes at times attached to the end of another chromosome. Without the objective evidence of this union that we have to-day, it might have been supposed that the broken-off fragment would rather have made, or retained, a side-to-side union with a corresponding part of its homologous chromosome. However, since the conditions of the cell that permit conjugation of like chromosomes occurs only once in the life-cycle, such a union is not, then, to be expected if the breaking has occurred after that event. If it had occurred earlier in the germ-track the piece would no doubt have been lost before meiosis came on. Here, then, we have a field inviting speculation. Let us hope that it will not long remain there, but that evidence concerning these puzzling relations may soon be forthcoming.

In this connection I need hardly recall to mind that, on the current theory of crossing over, the linear order of the genes is broken at the same level in two of the strands, and a new lengthwise reunion of the broken ends takes place. Whether this breaking and reunion is a comparable process to that seen in translocation we do not know, and it would be unprofitable at present even to make guesses.

POLYPLOIDY

In even a passing review of present-day genetics, the numerous problems connected with the increase in number of the chromosomes, or polyploidy in technical language, can not be ignored. But how can one hope even to summarize the work that is pouring in with the arrival of every new number of the genetics journals? The importance of polyploidy for the evolution doctrine is perhaps clear, but needs cautious handling in the light of the past history of phylogenetic interpretation of the facts of comparative anatomy. I hope that that history, at least, will not be repeated when the story of genetics comes to be written, for, in the light of recent work on the exchange of limbs between non-homologous chromosomes, and on translocations, the comparison of chromosome numbers without this knowledge may be very misleading. The determination of the linkages of the genes is the only safe basis for such comparisons.

At present I can do no more than briefly indicate some of the obvious and salient points. In many families of plants, and also in a more limited number of animals, chromosome groups are present that are multiples of a basal number, usually of the haploid number of the lowest member of the group. These are frequently double or triple, or quadruple groups of a basal number, generally assumed to be the haploid number. A good many of our cultivated plants are also known to show multiples of a real or postulated basal number of chromosomes. It is natural to assume that, in many cases, this has come about by the actual doubling of the whole chromosome group rather than by breaking of the chromosomes, that would also lead to doubling their number. It is more consistent to assume that doubling is the method by which the number of chromosomes is increased, because of the evidence from the sizes of the chromosomes, from their method of conjugation and from the relation of chromosomes to the attachment-fiber.

There are several known ways in which we can bring about a doubling of the number of chromosomes in a cell. The usual way is to suppress the cytoplasmic division of the cell at the time when the chromosomes divide. When this is done the chromosomes do not reunite, but the descendants of that cell will forever possess twice the original number of chromosomes. Theoretically, the process might go on forever, unless there are upper limits of a physiological nature preventing an indefinite increase. Doubling of diploids gives tetraploids. These crossed to diploids give triploids. Double tetraploids (or octoploids) crossed to tetraploids give hexaploids, and so on.

This work furnishes an opportunity for the solution of certain genetic problems of theoretical interest, for, without this knowledge, some of the known genetic ratios would have been difficult to interpret. With this knowledge they are found to conform to recognizable general principles.

It is perhaps ungracious to point out that the mere study of chromosome numbers in different species may in itself become mere hackwork. It looks as though it may become as popular for academic work as section-cutting of embryos was at an earlier period. It is more generous, perhaps, to regard the work on chromosome counts as pioneering and therefore preliminary work in the search for new materials, some of which will certainly be of value for deeper-lying genetic problems. This is especially evident in the study of hybrids whose parents, whether cultivated or wild types, have different numbers of chromosomes. The erratic behavior of chromosomes, often seen in the maturation of the germ-cells of such hybrids, clearly explains the exceptional and often abnormal results that follow. Without this information we might be tempted to indulge in much profitless and arbitrary speculation.

Not only are we familiar with cases where a multiplication of the same group of chromosomes is brought about within the species, but there are a few cases where an increase has been brought about by crossing distinct species with different numbers of chromosomes, and chromosomes that do not mate at meioses. These situations are full of interest for students of genetics, presenting a wide range of new possibilities.

Of great importance for the genetic interpretation of polyploidy in terms of chromosomes is the identification of chromosomes that carry specific genes. Only a few years ago this was known in only one animal, but the number of cases is steadily increasing. Until information of this kind becomes more general there will be, as at present, a good deal of guessing as to the relation of chromosome groups having different numbers of chromosomes.

INFLUENCE OF THE GENES ON THE CYTOPLASM

If another branch of zoology that was actively cultivated at the end of the last century had realized its ambitions, it might have been possible today to bridge the gap between gene and character, but despite its high-sounding name of *Entwicklungsmechanik* nothing that was really quantitative or mechanistic was forthcoming. Instead, philosophical platitudes were invoked rather than experimentally determined factors. Then, too, experimental embryology ran for a while after false gods that landed it finally in a maze of metaphysical subtleties. It is unfortunate, therefore, that from this source we can not add, to the three contributory lines of research which led to the rise of genetics, a fourth and greatly needed contribution to bridge an unfortunate gap. I say this with much regret, for, during that time and even now, I have not lost interest in this fascinating field of embryological experimentation. It is true that a great deal of factual evidence came to light, and it is true that many misleading ideas were set aside, but the upshot was negative so far as the formulation of any of the factors of development, whether mechanistic or otherwise, are concerned. This may be because the work was pioneer and largely qualitative. Perhaps my disappointment at the outcome of the work has led me to an overstatement of its failures. Something did emerge that the future may show to be of fundamental importance for genetics. I mean the experimental demonstration that the immediate factors in the differentiation of the embryo are, at the time of their activity, already in the cytoplasm of the cell. Second only in interest was the discovery that, within certain limitations, the already determined specificity may be reversed, or rather, shall I say, the initial steps already taken are reversible by factors extraneous to the individual cells.

These statements call for further elaboration, because they are unconsciously in the background of much of our thinking about genetic problems, and should if possible be more sharply formulated.

That the form of cleavage of the egg is determined by the kind of chromosomes it contained before the egg reached maturity has been sufficiently proven; and since the foundations of all later differentiation are laid down at this time, the demonstration is of first-rate importance for genetics, because it shows that we are not obliged to suppose the genes or chromosomes are functioning at the moment of the visible appearance of characters.

This is demonstrated by introducing into the egg foreign sperm of a species having another type of development. Although the chromosomes from the sperm are present from the first cleavage onward, they produce at first no effect on the cleavage; only after a time do they succeed in bringing about changes in the embryo. This evidence is, as I have said, important for our genetic analysis, for it serves as a warning that the time relations between gene and cytoplasm may have a relation different from that of an immediate dynamic change in the cytoplasm. The preparation for the effect may have taken place long before the actual event.

The second inference is no less significant. I need not labor the point at this late date that the characters of the individual are the product both of its genetic make-up and its environment. The earlier, premature idea, that for each character there is a specific gene—the so-called unit-character—was never a cardinal doctrine of genetics, although some of the earlier popularizers of the new theory were certainly guilty of giving this impression. The opposite extreme statement, namely, that every character is the product of all the genes, may also have its limitations, but is undoubtedly more nearly in accord with our conception of the relation of genes and characters. A more accurate statement would be that the gene acts as a differential, turning the balance in a given direction, affecting certain characters more conspicuously than others. But let us not forget that the environment may also act as a differential, intensifying or diminishing, as the case may be, the action of the genes.

The best illustration of this double relation is seen in the determination of sex. When an unpaired chromosome is present, in one or in the other sex, its genes determine, as a rule, whether a male or a female develops from each egg. Under environmental conditions which, as we say, are normal, the differential acts almost perfectly; but under other unusual conditions and in a few special cases its power may be partially overcome and even a reversal may take place. These unusual environmental conditions may be external agents, such as temperature or . light. They may also be internal factors, such as hormones. Even 'age' itself may bring about a reversal of sex in certain types. These statements are commonplaces to-day. The only

differences of opinion concern the emphasis that one theorist places on the environment, and another on the genic composition.

In passing, a word may be said about the genes as sex factors or differentials. All through the 32 years of the present century there have been attempts to isolate (in a genetic sense) the sex-determining factors. At first, when the chromosome mechanism was discovered, the idea prevailed that one X, let us say, made a male, and two X's a female. The sex-chromosome itself was then taken as the differential. Very soon after this the idea that the sex chromosome was the carrier of a gene for sex suggested itself, and a search was started to locate such a gene or genes in this chromosome. More recent work on translocations has shown the probable futility of such an interpretation. The tendency at present is rather to look upon all the genes, or at least many of them, as sex-determining in exactly the same sense, as all or many of the genes have an effect on the development of each character. It may well be, however, that certain genes in the sex-chromosome (as in other chromosomes) are more influential than others in turning the balance one way or the other, but even so, it does not at the present moment—in the light of recent evidence—seem probable that a single gene for sex-determination is to be found in the X-chromosome any more than, in the contrary sense, there is a single gene for sex in any special autosome. Here again, some one or a few genes may be more influential than others, but this is also true to varying degrees of the gene for any other character. The theory of balance between the intracellular products of the genes is the most direct contribution to physiology that modern genetics has made. It is an idea familiar to classical physiology as applied to organ systems, but a distinctly new contribution to cellular physiology. It may be a long time before these intracellular genic substances are isolated and purified (since there may be many steps between the actual primary substances and the end-product of such substances in the cell-plasm); nevertheless as a point of view the presence of genic materials rather than a dynamic action of the nucleus is supported by some analytical evidence. Already there is afoot in several quarters, and by methods partly genetic, partly physiological, partly embryological, partly physical and chemical, a decided effort to approach this problem.

If we could obtain these substances in pure condition we might then be in a position to speak more confidently of a quantitative study of gene-activities in the sense that chemistry is quantitative. Meanwhile there are other more practical methods by which we may construct provisional hypotheses as to the nature of the intracellular substances that are the products of the genes, namely, through a study of triploids, trisomic types, fragments of chromosomes and by analysis of crosses between different species. These openings do not, of course, exclude the possibility of the discovery of entirely new methods of approach.

Let us not forget that the idea of balance, as seen in the character, is really an old and familiar one to geneticists. For example, the intermediate character of the F_1 hybrid was generally interpreted as due to a conflict between the old and the new gene. Again, the familiar statement that characters are often affected by modifying gene-action that enhances or diminishes the effect of the primary gene, is another example of the intracellular balance of the activity of the genes.

What has been said so far relates to the action of the gene on the cytoplasm of its own cell—its intra-cellular action. Those of us working with insects or plants are apt to think of genetic problems in this way, and are inclined to consider mainly the effects that do not reach beyond the cells in which they are produced. But in other groups, especially birds and mammals, the effects of the genes are not always so limited. We are on the threshold of work concerned with the isolation of the so-called sex-hormones, the end-products of the thyroid gland, the pituitary, the thymus, and the substances isolated from the suprarenal bodies. All these substances produce their effects outside of the cells that manufacture them. In themselves they are far removed from the primary action of the genes.

In this connection certain work carried out by experimental embryologists should not be overlooked, beginning with the early experiments of Lewis in 1904 and culminating in the more recent work of Spemann. Here it appears as the result of grafting experiments that certain organs of the body develop in response to the vicinity of other organs, as when, for example, the lens of the eye of the frog is shown to be a response to the presence beneath the skin of the optic lobe. Similar and more far-reaching effects have been recently found for other organs of the embryo. The simplest interpretation, perhaps, is the setting free of a hormone by an embryonic organ or group of cells that calls forth a response in neighboring regions.

This and other evidence goes to show that gene-activity may produce results outside of the cells in which the first steps are initiated. The problem at present is one of immediate importance in the study of gynandromorphs, mosaics and intersexes.

EVOLUTION

Sooner or later every geneticist is asked what bearing this work has on the theory of evolution. In the early years of the century when genetics was new, some of us tried to sidestep the question, partly on the grounds that genetics was not ready to discuss the bearing of the new work on evolution, but mainly because it seemed unfortunate to compromise the precise results of the new procedure with the evolution doctrine which, because it dealt with a historical problem, was largely speculative. After 32 years of activity, caution may still be the wiser course to pursue; yet, on the other hand, we are now prepared, I think, to make a more definite commitment. It is, of course, obvious that only those characteristics that are inherited can take part in the process of evolution. The only characters that we know to be inherited are those that arise first as mutants, *i.e.*, discontinuously, or, as we say, by a change in a gene. Here genetics has made a very important contribution to evolution, especially when it is recalled that it has brought to the subject an exact scientific method of procedure. If we compare our present status in this respect with the discussions of the old school of evolutionists concerning variability, there can be no question but that genetics has contributed valuable information.

In the second place, the objection has been not infrequently made that geneticists are dealing only with aberrant or abnormal characters—hence their results, however accurate, can have nothing to do with the kind of progressive changes that have made evolution of new types possible. Such objections have come largely from those who ignore what geneticists have done and are doing. The same objections have also come from those whose minds are closed to new evidence, or who can not distinguish between the value of tested and verifiable theories and vague views or juvenile impressions with a teleological background or bias.

Without elaborating, I wish to point out briefly that there is to-day abundant evidence showing that the differences, distinguishing the characteristics of one wild-type or variety from others, follow the same laws of heredity as do the so-called aberrant types studied by geneticists.

Even this evidence may not satisfy the members of the old school because, they may still say, all these characters that follow Mendel's laws, even those found in wild species, are still not the kind that have contributed to evolution. They may claim that these characters are in a class by themselves, and not amenable to Mendelian laws. If they take this attitude, we can only reply that here we part company, since *ex cathedra* statements are not arguments, and an appeal to mysticism is outside of science.

There remains still the question of

the causal origin of mutations. Here also some progress has been made, but the subject is admittedly by no means on the same footing as is our knowledge of the laws of inheritance. It behooves us, then, to be careful, for our progress in this respect has been slow and to some extent erratic. I mean by this that we have not yet found a method of producing specific results—*i.e.*, a method by which particular genes can be changed in a particular way.

Even here, however, something has been done. In the work with x-rays and heat the same mutants appear that are already known, and that have come up without treatment. In addition, new mutants appear, as they do also without treatment. If it can be shown on a large scale that the same ratio for known mutations holds for x-ray and for spontaneous mutations, we may have found an opening for the further study of the causes of certain types of mutation.

I have been challenged recently to state on this occasion what seemed to be the most important problems for genetics in the immediate future. I have decided to try, although I realize only too well that my own selection may only serve to show to future generations how blind we are (or I have been, at least) to the significant events of our own time.

First, then, the physical and physiological processes involved in the growth of genes and their duplication (or as we say their 'division') are obviously phenomena on which the whole process of reproduction rests. The ability of the new genes to retain the property of duplication is the background of all genetic theory. Whether the solution will come from

a frontal attack by cytologists, geneticists and chemists, or by flank movements, is difficult to predict, although I think the latter more promising.

Second: An interpretation in physical terms of the changes that take place during and after the conjugation of the chromosomes. This includes several separate but interdependent phenomena—the elongation of the threads, their union in pairs, crossing over, and the separation of the four strands. Here is a problem on the biological level, as we say, whose solution may be anticipated only by a combined attack of genticists and cytologists.

Third: The relation of genes to characters. This is the explicit realization of the implicit power of the genes, and includes the physiological action of the gene on the rest of the cell. This is the gap in our knowledge to which I have referred already at some length.

Fourth: The nature of the mutation process—perhaps I may say the chemico-physical changes involved when a gene changes to a new one. Emergent evolution, if you like, but as a scientific problem, not one of metaphysics.

Fifth: The application of genetics to horticulture and to animal husbandry, especially in two essential respects; more intensive work on the physiological, rather than the morphological, aspects of inheritance; and the incorporation of genes from wild varieties and species into strains of domesticated types.

Should you ask me how these discoveries are to be made, I would become vague and resort to generalities. I would then say: By industry, trusting to luck for new openings. By the intelligent use of working hypotheses (by intelligence I mean a readiness to reject any such hypotheses unless critical evidence can be found for their support). By a search for favorable material, which is often more important than plodding along the well-trodden path, hoping that something a little different may be found. And lastly, by not holding genetics congresses too often.

The next genetics congress was held in Scotland on the eve of World War II, which was a turning point for the world and for genetics—each started on a new path.

BIBLIOGRAPHY

Refer to the Preface for a list of classical papers included in other anthologies. Many of the references given for Chapter 1 are useful for topics covered in this section. Some of the references in *Heredity and Development* have not been repeated here.

ALLEN, GARLAND E. 1969. 'T. H. Morgan and the emergence of a new American biology.' *Quarterly Review of Biology 44*: 168–188.

AUERBACH, C. 1957. 'The chemical production of mutations.' *Science 158*: 1141–1147.

BABCOCK, E. B. 1949. 'The development of fundamental concepts in the science of genetics.' *Portugaliae Acta Biologica*. Série A, Volume R. B. Goldschmidt. Pages 1–46.

BATESON, W. 1916. 'The mechanism of Mendelian heredity (a review).' *Science 44*: 536–543. A review of Morgan et al., 1915.

BATESON, W. 1926. 'Segregation: being the Joseph Leidy memorial lecture of the University of Pennsylvania, 1922.' *Journal of Genetics 16*: 201–235.

BLAKESLEE, A. F. 1936. 'Twenty-five years of genetics.' *Brooklyn Botanical Garden Memoirs 4*: 29–40.

BRIDGES, C. B. 1925. 'Sex in relation to chromosomes and genes.' *American Naturalist 59*: 127–137.

BRIDGES, C. B. 1935. 'Salivary chromosome maps.' *Journal of Heredity 26*: 60–63; also *29*: 11–13; *30*: 475–477.

BRIDGES, CALVIN B., and KATHERINE S. BREHME. 1944. *The Mutants of Drosophila melanogaster*. Washington: Carnegie Institution of Washington. Publication 552. A catalogue.

CARLSON, ELOF AXEL. 1966. *The Gene: A Critical History*. Philadelphia: W. B. Saunders. Note especially Chapter 11.

COOK, R. C. 1937. 'Chronology of genetics.' *U. S. Department of Agriculture Yearbook 1937*: 1457–1477.

CREIGHTON, H. B., and B. MCCLINTOCK. 1931. 'A correlation of cytological and genetical crossing-over in *Zea mays*.' *Proceedings of the National Academy of Sciences 17*: 485–497.

DOBZHANSKY, TH. 1929. 'Genetical and cytological proof of translocations involving the third and fourth chromosomes of *Drosophila melanogaster*.' *Biol. Zentralbl. 49*: 408–419.

DUNN, L. C. 1969. 'Genetics in historical perspective.' In *Genetic Organization*. Volume 1. Edited by Ernst W. Caspari and Arnold W. Ravin. New York: Academic Press. Pages 1–90.

GLASS, H. B. 1963. 'The establishment of modern genetical theory as an example of the interaction of different models, techniques and inferences.' In *Scientific Change*. Edited by A. C. Crombie. New York: Basic Books.

HALDANE, J. B. S. 1938. 'Forty years of genetics.' In *Background to Modern Science*. Edited by Joseph Needham. Cambridge: At the University Press.

HAYES, H. K., and C. R. BURNHAM. 1959. 'Suggested literature for students in plant breeding and genetics.' *American Naturalist 93*: 17–25.

JANSSENS, F. A. 1909. 'La théorie de la chiasmatypie.' *La Cellule 25*: 389–411.

KOMAI, TAKU. 1967. 'T. H. Morgan's times: A Japanese scientist reminisces.' *Journal of Heredity 58*: 247–250.

LANSTEINER, KARL. 1901. 'Uber Agglutinationserscheinungen normalen menschlichen Blutes.' *Wiener Klinische Wochenschrift 14*: 1132–1134. The ABO blood groups.

MORGAN, T. H. 1910. 'Chromosomes and Heredity.' *American Naturalist 44*: 449–496. A pre-Drosophila essay in which he expressed skepticism about the relation of chromosomes to inheritance.

MORGAN, T. H. 1911. 'An attempt to analyze the constitution of the chromosomes on the basis of sex-limited inheritance in Drosophila.' *Journal of Experimental Zoology 11*: 365–413.

MORGAN, T. H., and CLARA J. LYNCH. 1912. 'The linkage of two factors in Drosophila that are not sex-linked.' *Biological Bulletin 23*: 174–182.

MORGAN, T. H. 1913. *Heredity and Sex*. New York: Columbia University Press.

MORGAN, T. H. 1914. 'The mechanism of heredity as indicated by the inheritance of linked characters.' *Popular Science Monthly*. January 1914. p. 5–16.

MORGAN, T. H., A. H. STURTEVANT, H. J. MULLER, and C. B. BRIDGES. 1915. *The Mechanism of Mendelian Heredity.* New York: Henry Holt. Revised Edition, 1923.

MORGAN, T. H. 1915. 'The constitution of the hereditary material.' *Proceedings of the American Philosophical Society 54*: 143–153.

MORGAN, T. H. 1917. 'The theory of the gene.' *American Naturalist 51*: 513–544.

MORGAN, T. H. 1919. *The Physical Basis of Heredity*. Philadelphia: Lippincott.

MORGAN, T. H., and C. B. BRIDGES. 1919. *The Origin of Gynandromorphs*. Carnegie Institution of Washington. Publication 278: 1–22.

MORGAN, T. H., C. B. BRIDGES, and A. H. STURTEVANT. 1925. 'The genetics of Drosophila.' *Bibliographia Genetica 2*: 1–262.

MORGAN, T. H. 1926. *The Theory of the Gene*. New Haven: Yale University Press. Reprinted 1964, Hafner, New York.

MORGAN, T. H. 1935. 'The relation of genetics to physiology and medicine.' Nobel Lecture, Presented in Stockholm on June 4, 1934. *Scientific Monthly 41*: 5–18.

MORGAN, T. H. 1939. 'Personal recollections of Calvin B. Bridges.' *Journal of Heredity 30*: 354–358.

MORGAN, T. H. 1941. 'Biographical memoir of Calvin Blackman Bridges 1889–1938.' *Biographical Memoirs of the National Academy of Sciences 22*: 31–48.

MULLER, H. J. 1916. 'The mechanism of crossing over.' *American Naturalist 50*: 193–221, 284–305, 350–366, 421–434.

MULLER, H. J. 1922. 'Variation due to change in the individual gene.' *American Naturalist 56*: 32–50.

MULLER, H. J. 1947. 'The production of mutations.' *Journal of Heredity 38*: 259–270. Nobel Prize lecture.

PAINTER, T. S. 1933. 'A new method for the study of chromosome rearrangements and the plotting of chromosome maps.' *Science 78*: 585–586.

STERN, CURT. 1970. 'The continuity of genetics.' *Daedalus.* Fall 1970. Pages 882–908.

STURTEVANT, A. H. 1925. 'The effects of unequal crossing over at the bar locus in Drosophila.' *Genetics 10*: 117–147.

STURTEVANT, A H. 1959. 'Thomas Hunt Morgan 1866–1945.' *Bibliographical Memoirs of the National Academy of Sciences 33*: 283–325.

WILSON, E. B. 1928. *The Cell in Development and Heredity.* New York: Macmillan. Chapters 10 and 12.

WILSON, E. B., and T. H. MORGAN. 1920. 'Chiasmatype and crossing over.' *American Naturalist 54*: 193–219.

6 / Genetics—Old and New

The term 'gene' can be defined broadly to signify a unit of the substance of inheritance. So defined, it has had a long history that includes Darwin's gemmules, Nägeli's idioplasm, and Mendel's factors. Genetics was made a science without its practitioners having any clear understanding of the ultimate nature of the gene. The geneticists of each generation were interested in two main questions about the gene: What is it? What does it do? Before 1940 little was known about genes other than the rules that describe their transmission from one generation to another and that they can mutate from one relatively stable state to another. There were no hard data on the size, shape, or chemical structure of the gene.

Considering the enormous importance of genes for life, it is not surprising that there was much speculation about genes and numerous attempts were made to define their nature. The geneticist who concerned himself most with this question was one of Morgan's students, Hermann J. Muller (1890–1967). From the early days of Drosophila genetics to the end of his life, he published a series of remarkable papers on the nature of the gene. One of the most famous is his Pilgrim Trust Lecture of 1945 to the Royal Society. That was an especially interesting year in the history of genetics. Classical genetics was essentially complete (*Heredity and Development,* Chapter 6) and radically new approaches to genetics were under way: Beadle and Tatum were developing the 'one gene–one enzyme' hypothesis (*H & D,* Chapter 6), and Avery and others were suggesting that the substance of the gene is DNA (*H & D,* Chapter 7). Muller was

207

able therefore, not only to sum up the old but to give his own insights into what was to come.

HERMANN J. MULLER

The Croonian Lecture of Wilson (Chapter 4) and that of Morgan (Chapter 5) plus the Pilgrim Trust Lecture of Muller, printed below, give the substance of classical genetics.

PILGRIM TRUST LECTURE

THE GENE

By H. J. MULLER, *Department of Zoology, Indiana University*
(*Delivered 1 November 1945—
Received 25 February 1946*)

1. GENERAL EVIDENCE OF A GENETIC
MATERIAL

The gene has sometimes been described as a purely idealistic concept, divorced from real things, and again it has been denounced as wishful thinking on the part of those too mechanically minded. And some critics go so far as to assert that there is not even such a thing as genetic material at all, as distinct from other constituents of living matter.

However, a defensible case for the existence of separable genetic material might have been made out on very general considerations alone. Despite the bafflingly complex and seemingly erratic character of biological things, and their change at the touch of analysis—characteristics which have made them such a happy hunting ground for obscurantists—it is obvious that the whole congeries of variable processes of each kind of organism tends to go in a succession of great cycles, or generations, or even alternations of generations, with the generations in turn made up of smaller or cell cycles, and that at the end of every greatest cycle something very like the starting-point is reached again. Now the finding of the starting-point in a complex course, were it observed in any other field, would be taken to imply the existence of some guide or guides, some elements that are themselves relatively invariable and that serve as a frame of reference in relation to which the passing phases of other features are adjusted. Such constants in other fields are, for instance, the fixed mass-energy relations in the motions of a pendulum, or in the recurrent track of a comet, the steady radio beam or the map in the journey of an airplane to its destination and back, and the stable nuclear structure amidst the changing electron patterns of an atom, as it finds its state of least energy once more. So, too, in the organism it would be inferred that there exists a relatively stable controlling structure, to which the rest is attached, and about which it in a sense revolves.

In the organism, however, there is

From *Proceedings of the Royal Society* B *134*: 1–37, 1947. Reprinted by permission.

the cardinal difference from other cyclic objects that the return to the starting-point finds all structures doubled in a cell cycle, or still further multiplied as a result of the succession of cell doublings comprised in a complete generation, or alternation of generations. And this in turn requires that the material furnishing the frame of reference, whatever it is, itself underwent such doubling or reproduction, and that this too must have taken place under its own guidance. It is for this reason that it may be called the *genetic* material. And it is important to note that, unlike the controlling factors in the other cases, it is not merely statically stable, but dynamically so, in that it usually succeeds in so affecting the heterogeneous other material with which it is supplied as to impress upon it precisely its own image.

That this ability to duplicate itself is based in unique properties of some relatively stable genetic material, rather than in the multitude of diverse substances and processes that engage in the cycles, may, curiously enough, be inferred more especially from the behaviour of the real exceptions to the principle of inner stability. These all-important exceptions are the comparatively rare cases in which, even in a 'pure line', sudden permanent deviations of type, or 'mutations', take place. Although of the most varied kinds, as judged by their unlike effects on the organism, it is characteristic of the great majority of these changes that in succeeding multiplications they become regularly incorporated, that is, they now take their place as part of the again stable, self-multiplying pattern. It is scarcely conceivable that, if the reproduction of every part of the organism were due primarily to the marvellous concatenation of a host of individual processes of the cycle, these could have been so arranged that, when disturbed in any one of innumerable ways, they would still be able to work effectively in reproduction, yet in a manner so *correspondingly* adjusted as now to effect a repetition of just the given alteration. Rather must it be inferred that the essential process of reproduction consists in the autosynthesis of a controlling genetic material, and that this occurs through some sort of laying down of the raw material after the model of the genetic material already present, no matter what—within certain very wide limits —the pattern of that genetic material happens to be to begin with. The building up of the non-genetic parts of the system would then take place, conversely, by a series of essentially heterosynthetic processes, that were ultimately controlled by the genetic material. It may be granted that this autosynthesis of variations of the genetic material would in itself be difficult enough to understand—it is, in fact, to be considered as the basic problem of living matter. Yet it does not lead to such an incomprehensible enigma as would the so-called 'organism-as-a-whole' view of the phenomenon which has, implicitly or explicitly, been adopted by some physiologists and others who would deny the existence of a special genetic material.

If the above argument is followed to its logical conclusion, all other material in the organism is made subsidiary to the genetic material, and the origin of life is identified with the origin of this material by chance chemical combination (Minchin 1916; Troland 1914, 1916, 1917; Muller 1921a, 1926; Alexander & Bridges

1929). Owing to its unique ability to duplicate its variations, this material will thereafter, in the course of protracted periods of time, be subject to the Darwinian principle of natural selection—that is, to the differential multiplication and accumulation, in series of particular variants—those that happened to have properties conducive to their own survival and further multiplication. This will carry the multiplying material step by step into further and further differentiations and complications.

Among the more successful forms will be those in which the genetic material has become organized into aggregates that produce, heterosynthetically, a system of companion substances—'protoplasm'—of such a nature as to aid in multiplication of the genetic aggregate itself. With increasing serviceability of this protoplasm, in some lines of descent, an ever-increasing range of substances will be made available for the raw material of the system, through the development of processes of conversion and utilization, as food, of material chemically further and further removed from itself,* as well as

---

* Horowitz (1945) has recently given convincing chemical illustrations of how the extension of synthetic activity 'backward' for utilization of ever-simpler substances, must have come about by mutations. This idea itself is not a new one but has long formed a necessary part of the theory that the gene constitutes the basis of life. So, too, has the idea, ably expounded in detail in recent years by Oparin (1938), that there must have been an extended accumulation of ever more complex organic combinations, permitted by the absence of living organisms that would break them down, before

through the development, in some lines, of powers of capturing its food and avoidance of becoming used as food itself or otherwise destroyed. Thus it would happen that even the simplest body deserving of being called a 'cell' has so outdistanced in complexity the most intricate product of inanimate nature as to make the 'living' world appear as a distinct category of nature. And it would explain the amazing fact that in any object of this living world the whole great system of materials and processes, unlike those in any natural inanimate system, is organized *adaptively*, that is, in such wise that all processes are focused, as if by forethought, towards just one end (a sometimes distant one), namely, the multiplication of the system itself—an end that in its turn constitutes another beginning, in the endless succession of cycles.

2. THE LINEAR DIFFERENTIATION OF THE CHROMOSOMAL NUCLEOPROTEIN INTO SELF-REPRODUCING PARTS

Of course the establishment of the modern genetic theory of the basis of life did not historically depend, in the main, upon such very general considerations concerning cyclic behaviour and the mutability of the cycles. As is now so well known, an enormous mass of observational and experimental evidence combined in establishing the fact that the material of heredity is to be found, in the main at least, in the chromosomes. It may

---

genetic material could accidentally arise from them and be suitably provided with the components needed for its own reproduction.

be recalled that the chromosomes are not really the lumpish bodies which they superficially appear to be at the time of cell division, but are extremely delicate, relatively very long threads which are not readily discernible in their ordinary more extended condition, but which take on the more conspicuous lumpish aspect by becoming compactly coiled into a helical spiral, or spiral of spirals. This constitutes a form of packing for their transportation at cell division. Arranged in fixed linear order in each long thread is a multitude of distinctive parts, each with its characteristic chemical effect in the cell. In preparation for each cell doubling, there is an exact duplication of each distinguishable part of each chromosome, so that two identical daughter threads are formed, not really by splitting as the process is often miscalled, but by the synthesis of new material.

It may also be recalled that, at the time of the preparation of the mature germ cells, or meiosis, the chromosome threads, then in relatively extended form, undergo the remarkable phenomenon of synapsis, that is, a side by side juxtaposition, by twos, of like parts of corresponding threads, and that this involves, as its most important consequence, crossing-over. In crossing-over there is a breaking of pairs of synapsed threads at one or more points, identical in the two members, just at or just after the time when each had doubled again, as if (as Darlington (1935) has suggested) the strain of reassociation into twos of the suddenly fourfold groups were at some points too much for two of them. Following or attendant upon the breakage is a reattachment between the broken ends of pieces not origin-

ally together, resulting in an exchange of corresponding segments. And finally (if other significant features are omitted) the segregation of the parts into different germ cells is carried out in such wise as to give rise to the known principles of Mendelian segregation and recombination, and of crossing-over as genetically observed.

The methods need not be recalled by which, through breeding tests, the relative positions of many of the parts in the chromosome threads, first, of the fruit fly *Drosophila melanogaster*, and later of various other species, have been mapped, or how direct verification of this mapping was obtained, through cytological observation of cases in which, through abnormalities of inheritance, given changes in chromosome number or structure were predicted, as well as through some cases in which the sequence of the testing was the reverse of this. Such work, in *Drosophila*, has been greatly aided by the discovery some twelve years ago, by Painter (1933) and by Heitz & Bauer (1933), that the relatively enormous cable-like bodies in the cells of the salivary glands of fly larvae are really chromosome formations. As Koltzoff (1934) first suggested, each cable is derived from one original pair of ordinary extended ('resting stage') chromosome threads that have undergone repeated duplication, but remain lying in parallel. Thus the appearance is given of a relatively extended chromosome that is greatly magnified in its transverse direction; so much so as to be able to show, along its length, characteristic variations in 'density' of nucleic acid (really caused, if Ris's interpretation is correct (Ris & Crouse 1945), by differences in degree of spiralization);

thereby, the different parts of the chromosomes can be recognized.

That the material of the chromosomes is of a specific type, chemically, being composed in predominant measure of nucleoprotein, a compound of protein with nucleic acid, was shown in analyses of sperm chromosomes by Miescher (1897) before the turn of the century, though only recently has it become reasonably certain—through the analogous finding in viruses—that it is really this major component rather than some elusive accompaniment of it which constitutes the genetic material itself. As protein is so potent, labile, and versatile chemically and has such possibilities for different permutations and combinations of its amino-acids and, as is now known, for the intricate folding of the chains and for their interconnexion into larger complexes, by all of which processes its properties can become radically altered, and as, on the other hand, the basic structure, at least, of nucleic acid is everywhere much the same, it has usually been assumed that the differences between genes reside in the protein component of the chromosomes. However, it should be noted that the nucleic acid also exists in highly polymerized form, and, as will be seen later, this reservation may be a very significant one.

The main point of all the above familiar work, for our present purposes, is that it has given incontrovertible evidence of the existence of definite genetic material, of a particulate nature, which certainly has, in each part, the property of self-determination in its own duplication. For, if any given part has somehow been removed, the same deficiency will be evident in succeeding generations. If,

on the contrary, an extra part has been added, this too will be found to have become reproduced later. And if, instead of mere removal or addition, there was any cytologically visible rearrangement of chromosome parts, or even a change too small for this and only detectable through its effects on the organism and placeable on the chromosome diagram purely by breeding tests, this alteration, too, will be found to have undergone multiplication. This, then, shows that it is the chromosome material itself that is already present, and not other material, which determines the type of chromosome material that is again to be produced, in duplication, and that is what is meant by saying that this material is 'self-duplicating', 'self-reproducing,' 'auto-synthetic,' or 'genetic' material. Most other materials in the cell are certainly not genetic in this sense. It is clear, secondly, that this self-duplication is not, primarily, a resultant of the action of even the genetic material as a whole, but of each part of it separately. That is, the material is potentially particulate, and each separable part, which determines the duplication of just its own material, may be called *a gene*.

Before considering further the properties of individual genes, as judged by results of their passage through generations, it should be emphasized that, though particulate in their self-reproduction, their products in the cell interact in the most complicated ways, both with one another and with the products of environmental conditions, in determining the characters of organisms, contrary to what many early Mendelians had assumed. Their integration, however, is essentially one of gene *effects* only, since in the immediate process of their autosynthesis they remain substantially independent. Very complicated integration occurs to be sure in the preparation of material for this final gene synthesis, but these processes, too, are properly to be considered as more remote gene effects, of a heterosynthetic nature.

3. THE INDEPENDENCE AND STABILITY OF THE GENES IN THEIR SELF-REPRODUCTION

Now consider some of the results of studies on the stability and mutability of genes. It has long been evident that in its mutation, just as in its self-duplication, the genetic material acts particulately. Early observation showed that the conspicuous, visible mutations, at least, usually affect only one kind of gene at a time. And after a given mutation has arisen, there seems to be no tendency for the cause of it to go on producing other obvious mutations,* and even the gene that has just become changed does not, ordinarily, show any heightened tendency to change again; that is, it appears to reproduce its new self in about as stable a manner as its predecessor had reproduced the original type. Any individual type of conspicuous mutation arose with such extraordinarily low frequency that, with the older genetic techniques at least, there was no thought of obtaining any

---

* Since this was written, exceptions to this principle have been found by Auerbach (personal communication) in cases of mutations produced by mustard gas (see pp. 13, 14); in these cases, given genes, and their descendant genes, are found to be unstable for some time after treatment.

quantitative estimate of it. The incidence of the mutation was quite unpredictable, and, even when all readily observable mutations were considered at once, they still appeared to arise quite randomly, and no relation was observable between the environmental or physiological condition of an organism and the kind of effect produced by the gene mutations that might arise in it.

It was a long time, however, before these results could be shown to be applicable to gene mutations in general. For there are many genes whose mutations may produce minor effects, of degree, confusable with the effects of changes in environment and in other genes, and such mutations cannot so easily be studied. Moreover, it was important to know whether the principles in question hold for these genes not merely when they are in 'pure lines,' as in Johannsen's work, but also under those conditions of cross-breeding which give rise to wide variations of an apparently continuous nature.

For tests of these less clear-cut cases, the development of elaborate techniques of breeding was necessary, in which certain known genes with conspicuous visible effects were utilized as so-called 'identifying factors' or 'markers,' to enable the distribution of the 'minor' genes (i.e. genes with minor differences in effect), present in the same chromosome with them, to be ascertained among a group of the descendants produced in cross-breeding. Thus descendants known to have received any given combination of the original minor as well as major genes could be distinguished, and subjected to quantitative tests for the determination of whether the minor genes had undergone change in the meantime. In work on the fruit fly, *Drosophila*, especially, there were a number of cases that had been put aside as 'skeletons in the closet,' in which it looked as though the genes themselves were undergoing continuous quantitative variation. But application of these methods showed in every case that this appearance had been deceptive. It was due, first, to temporary effects of environmental differences on the visible characters, not on the genes themselves, effects which were not inherited; secondly, to the fact that many minor genes were simultaneously undergoing segregation and forming new combinations; thirdly, to the fact that the so-called 'pure' or homozygous combinations formed were in these cases often unable to live or to breed effectively, so that most of the individuals that did breed continued to give rise to diverse types. At the same time the individual genes themselves, despite this superficial appearance of continuous variation, gave evidence of being highly stable, not subject to contamination by opposite types even on long-continued crossing with them, and subject to rare definite mutations only. It appears that so-called 'multiple factor' inheritance of this kind is typical for the great majority of characters in natural cross-breeding populations. For these are heterogeneous with regard to many minor as well as major mutations that have occurred within them and have accumulated through hundreds of generations. But this is no wise indicates a real instability of their individual genes.

It should be mentioned in this connexion that the same quantitative tests were adapted to disclosing inheritable variation of any kind whatever in the given characters, even if it were not due to mutation or recombination of chromosomal genes at all. Since the characters studied were unusually sensitive indicators of changes of all kinds, they were especially suitable for such a study. However, no such other kind of variation was to be found in them. Thus it became evident that other forms of genetic differences within the species must, if they exist, be exceedingly rare at best in this *Drosophila* material. Since then, essentially similar results have been obtained in other species of animals and plants, with certain important qualifications to be mentioned later. A series of analyses of the differences *between* species also have been carried out, by various investigators. These indicate, in a number of ways, that the vast majority at least (subject to the same qualifications) of the differences between species are of this same stable, particulate chromosomal nature. All this confirms us in the conclusion that most of evolution must have been built up of mutations of the individual chromosomal genes.

So rare are mutations of any given type under natural breeding of any kind that, before enough mutations could be collected under controlled, 'normal' conditions to put their study on a quantitative basis comparable to that of the older fields of segregation and crossing-over, a new set of genetic methods, again employing the principle of conspicuous 'marker' genes for given chromosomes as a whole, had to be developed. At the same time, special technical devices had to be introduced, that allowed the scale of the experiments to be increased by several orders of magnitude. Such work has now been carried on for nearly three decades. Its results are on the whole consistent, and have agreed in demonstrating a number of principles concerning gene mutation and gene stability. Only some of these can be considered here.

One of the first questions of interest concerns itself with the degree of stability of the gene. Here one is hampered by the fact that no one has discovered how many mutations are below threshold in their effects, or so drastic as to kill an individual before it can be observed, or in directions other than those subject to observation. There are, however, ways of making allowance for these, as by noting the effect on the data of extending the range of observation, and by taking especial note of genes in which large changes are seen to occur oftener than small ones that would also be readily detectable. Making such allowances, it appears probable that in *Drosophila* the over-all frequency of mutations, under ordinary conditions, is such that something between one in ten and one in thirty germ cells carries in some one of its many genes a new mutation that arose in a cell of the parent. This figure does not make mutation in general appear so rare. But in terms of the individual gene, and the single cell cycle, it means that there is an average chance of less than one in a million of any given gene undergoing a mutation in any given cell cycle.

Stating the same result from another point of view, it may be said that the gene makes no mistake whatever in building its daughter gene, the daughter gene in building its own daughter, etc., throughout something like a million copyings carried on in series, so accurate is the duplication process. This makes the half-life of the individual self-duplicating gene in *Drosophila*, traced through one descendant gene taken at random at each doubling, until a 50% chance for its mutation has accumulated, over a thousand years and possibly longer than ten thousand. For different mutations, and different genes, there is, to be sure, a great range. And yet, for some widely different types of higher organisms, the present meagre evidence indicates that the given average figure in terms of cell cycles holds very roughly true, in general. On the other hand, where the cell cycles are of much longer duration, as in a man as compared with a fly, preliminary figures, such as those of Haldane (1935) and others, indicate a correspondingly greater stability per unit of time, with a half-life, for the few genes studied, that approaches the order of a million years.

4. THE BLINDNESS AND MOLAR INDETERMINACY OF GENE CHANGES

If the primary cause of the adaptiveness of life processes lies in the natural selection of random mutations, then the individual mutations would not be expected to be adaptive, except as a result of a rare accident; in fact, the more elaborate a working organization already is, the more unusual would be a chance alteration that happened to improve it further. And in accordance with this it is found that the vast majority of the mutations that occur naturally are actually deleterious, and the larger the change involved the more harmful does it tend to be (Muller 1918, 1921*b*). This again confirms the inference that natural selection has been the indispensable guiding agency in biological evolution, and indicates that most of the evolution has come about through the prolonged selective accumulation of a multitude of individually small mutations, chosen to form a system of gene products ever better integrated, internally, for taking advantage of conditions that would originally have been inhospitable. Thus, as Huxley (1942) has pointed out, the range of conditions open to the collection of species as a whole has widened, and also the range open to certain particular kinds of species, that may legitimately be called the progressive ones.

That the determining causes of the ordinary individual mutations lie in the realm of essentially uncontrollable submicroscopic events, rather than in the gross conditions thought of by earlier generations of biologists as the causes of inherited variations, is more particularly indicated by the revealing fact, gained as a result of the application of specialized methods, that when a particular gene undergoes a mutation, not only do the genes of other kinds in the same cell remain unchanged, but even its partner gene of identical type, originally derived from the opposite parent but now usually lying but a fraction of a micron away from it, also stays undisturbed. This proves that the disturbing process has an ultramicroscopic degree of localization, just as in the case of changes of individual molecules by thermal agitation. Some evidence that the individual events of thermal agitation are themselves responsible for mutation was gained from the finding, since confirmed by others, that, within the range of temperatures normal to the

organism, a rise of 10°C causes a several-fold rise in the general gene-mutation frequency (Muller & Altenburg 1919; Muller 1928). In fact, as Delbrück (1935; see also Timoféeff-Ressovsky, Zimmer & Delbrück 1935) has since pointed out, the rise is of just about the amount to be expected on the basis of the absolute frequency at a given temperature, if the cause were thermal agitation.

But if any activations obtainable by thermal means at ordinary room temperatures can succeed in producing mutations, then surely the activations or ionizations of much higher energy content, produced by bombardment with X- or γ-rays or by particles of similar energy, and perhaps even by ultra-violet, ought to produce correspondingly greater results. All this has proved to be decidedly the case. Thus, it can be calculated from the results that the application of a half hour's treatment of 5000r. units of X-rays to *Drosophila* spermatozoa at say 17°C causes a rise in the mutation frequency of about 50,000-fold, over what it would have been in the same period without the treatment. This effect of the X- or γ-rays is produced by the activations or ionizations, or tiny ultimate clusters of ionizations, *acting individually*, since the frequency of the gene mutations induced is simply and directly proportional to the frequency of the induced ionizations, regardless, within very wide limits, of the size of the dose, the hardness of the rays, and their manner of distribution in time. As these individual 'hits' must certainly be random, it is important to observe that the gene mutations produced by them in the flies are as a group indistinguishable from those occurring 'spontaneously,' and have a sensibly similar distribution of qualitatively different effects. This similarity then confirms our inference as to the essential randomness of the natural mutations also. However, it can be calculated that these could not, except for a few, have been produced by the sparse radiation present in nature but must rather have been produced by some form or forms of thermal agitation.

5. MUTATIONAL EVIDENCE ON GENE STRUCTURE AND SIZE

A major purpose of our first radiation experiment had been to obtain evidence concerning the degree of inner particulateness of the gene, that is, to what extent it is compounded, like most macroscopic bodies, of several or many substantially identical, interchangeable components, such as molecules, all of them self-reproducing, as had often been assumed. It may in the first place be pointed out that such an assumption of compoundness seems opposed by evolutionary considerations. For if the gene were originally compound, in this sense, then by the separate mutation of any one of the parts, which should occur as a result of the individual impacts (unless one supposes a very specialized mechanism that always spreads such effects completely), a non-homogeneity would be introduced into the compound gene. If this non-homogeneity were to persist, then by successive mutations of one after another of the parts in different directions, accompanied by the selective survival of the more highly adaptive gene systems having greater complication, the genes of to-day would have become internally heterogeneous in regard to prac-

tically all their components and would no longer be merely compound, their compoundness having been replaced by complexity. Analogous to this is the evolutionary process whereby, within the chromosome as a whole, the genes themselves, originally alike, have differentiated from one another, as considerable evidence has shown.

But such non-homogeneity within the individual gene could thus survive repeated gene duplication only if the inner parts of the gene, like the genes as a whole within the chromosome, are in entirely fixed positions with regard to one another, and if even in duplication they may not ordinarily substitute for one another (Muller 1926, 1927). If, on the other hand, following the mutation of a given component, one daughter gene may sometimes receive the two representatives of this component while the other does not receive either of them, but receives normal components instead, then, by a continuation of this process of inexact apportionment, some descendant cells would finally come to receive nothing but the mutated type of component within the gene in question, and others nothing but the original type. In this case then each gene would become again truly compound. Because it is a matter of chance to which pole of a dividing cell a given daughter gene is pulled, there would come about in consequence an irregular distribution of the cells of all-mutant type among those of mixed and of all-original type, as in an old fashioned crazy-quilt. Further division must therefore result in whole batches of mutant cells interspersed among batches of non-mutant ones. Special attention was therefore paid to the distribution of mutant tissue with respect to normal tissue in cases in which mutations had been freshly induced, by irradiation, in a cell ancestral to the tissues studied. However, no crazy-quilt effects attributable to such an interchangeability of gene parts was found, among many cases in which it should have been found had it occurred.* It was therefore to be concluded that the parts of the gene, whatever they are, do have fixed positions and consequently are differentiated from one another, forming together one organized system. Either they constitute, together, just one molecule or megamolecule or, even if some of the parts are held together only by residual or van der Waals's forces, these parts are not merely repetitions of one another as in macroscopic bodies. Instead, the internal structure of the gene itself must be non-repetitive, or, to use a term recently applied by Schrödinger (1945) to the succession of genes in the whole chromosome, aperiodic.

The fact that both intra- and intergenic relations turn out to be aperiodic makes more acute the question, raised

---

* However, since the above was written, Auerbach has reported (personal communication) effects which might be interpreted in this way, obtained by the chemical treatment mentioned in the next section. (These have also been referred to in the previous footnote.) It will be important to attempt to devise criteria here which might distinguish between the putative shuffling about of gene parts above considered and the origination of a mutant gene which was unstable for some chemical reason. The latter seems more probable in view of the X-ray results. One possible criterion would be the obtaining of evidence for or against the affected gene passing through different grades of instability as the cells containing it proliferate.

by other considerations as well, as to how the limits of an individual gene may be defined. While the ultimate answer must be a chemical one, beyond present knowledge, it might meanwhile be held, theoretically, that one gene should designate an amount of genetic material so small that it cannot be further divided without loss of the property of self-duplication by at least one of the fragments. Substantially this definition, without use of the word 'gene,' was in fact given by Wilson in 1896, but there is no empirical test to establish the limits of the gene on these grounds and it is even possible that, so defined, the genes might be found to be overlapping.

There are at present, however, a number of empirical means for defining small bits of genetic material that remain capable of self-duplication as such, when detached from their original genetic neighbours and attached to others. In certain studies carried out for this purpose, there was evidence of a limit of size when X-rays were the agents used to cause the breakage (Muller & Prokofyeva 1934; Muller 1935a). In these studies, some dozen breaks in a minutely delimited region of a chromosome were found to fall into just four definite positions, so far as these positions might be judged by the admittedly imperfect criterion of expecting effects to be produced on the characters of the organism by removal of the genetic material lying between two different positions of breakage. By measuring the proportion of the dark region of the salivary chromosome corresponding to the total length of the three breakage segments thus constituted, and then estimating the volume that this portion would occupy in the small mitotic chromosome, it was found that the individual segment, or 'gene,' i.e. that part included between any two most nearly adjacent positions of breakage, would occupy a space smaller than one-twentieth of a micron in diameter, if it were considered as compressed into a cubical form. How much smaller than this it may really be there is no means of knowing by this method. For, in the first place, there might have been interstitial genes that escaped us because their removal caused no lethal or other detected effect. Secondly, one does not know how much of the cytologically visible chromosome is occupied by the gene material itself, even in a mitotic chromosome. And, thirdly, the ultimate possible division of the genetic material by this method may not have been attained.

The above qualifications show, however, that the given value is certainly a maximum one for the size of a gene as thus defined. And this maximum value proves to be that of an object too small to be visible by ordinary microscopic methods. Moreover, even if a gene of this size were padded in its transverse directions with adventitious material, its length in the salivary chromosome turns out to be too small for it to be optically resolvable there. This maximal gene might be termed an X-ray breakage gene, without commitment as to whether or not its limits would eventually be found to coincide with those determined by some future more refined method. It happens, however, that this estimate agrees in a very rough way with others of the maximum possible gene size, based on calculations dealing primarily with gene number.

It is evident that if one had a minimum value for the number of genes and divided it into the total possible volume which they could occupy (represented by the mitotic chromosome volume), a maximum value for gene size would be obtained; and vice versa, the above estimate of maximum size can be converted into a minimum for gene number. Minimum number values were first derived from data on the frequencies of crossing-over between given genes (Muller 1916), in comparison with over-all crossing-over frequencies, and later they were derived, in several different ways, from data on the frequencies of mutation of given genes (Muller & Altenburg 1919; Muller 1926), in comparison with over-all gene-mutation frequencies. These various methods (including the above method based on breakage) have agreed sufficiently to show that, in the fruit fly at least, the number of genes is well over a thousand, perhaps as high as ten thousand. And there is always the possibility that, by some special means, the division · might be pushed still higher. However, even taking the highest probable number of genes, the complexity and the degree of refinement in the determination of the organisms dependent upon them is so extreme as to require of the individual genes present in them to-day a very high degree of inner intricacy indeed, perhaps even greater than that which would be expected of the ordinary protein molecule.

6. CHEMICAL AND PHYSIOLOGICAL INFLUENCES ON THE MUTATION PROCESS

It has sometimes been supposed that gene size can be taken as equal to the average number of atoms in the volume that includes one 'hit' (one ultimate ion cluster) when a given mutation is produced, but this is probably an oversimplification. For one thing, the effective hit probably need not always strike the gene itself, for in the case of chromosome breaks at least it has been shown that two nearby breaks are sometimes caused simultaneously by a hit too small to have impinged upon both spots. In such a case, localized chemical changes must have been induced that spread a short distance beyond their point of origin and so, indirectly, caused the genetic change.* Thus, it is not only the amount of energy but its quality and form that count. This being the case, it is also unlikely that each hit on any part of the potentially 'sensitive volume' at any time would always be effective in causing a mutation.

That this second stricture is correct is shown by the fact that certain states of the cells or chromosomes are much more vulnerable than others to the mutational effects of radiation. For these reasons, it is not surprising that species appear to differ significantly in the vulnerability of their genes to radiation mutations (though this is also open to other interpretations), and that, even within the same species, genetically different stocks sometimes differ greatly in their so-called 'spontaneous' mutation frequency—a situation which by the way shows that the mutation frequency is itself subject to regulation through natural selection.

------

* The criticism that the effects might have resulted from different hits in the course of one electron path appears invalidated by quantitative considerations and by the fact that with γ-rays, which produce hits much farther apart, the results seem substantially similar.

In view of all this, it is also to be expected that the so-called 'spontaneous' mutations would tend to occur more often in some physiological conditions, regions of the body, or stages of development, than others, and this might be important in its bearing on evolutionary potentialities. Till recently no certain and substantiated evidence of such effects on normal genes had been obtained, and even the highly pathological conditions that attend nearly fatal doses of various chemicals have not seemed conducive to mutation. It is true that some slightly positive results have in the past been reported from chemical treatments, but because of technical doubts these have not been widely accepted. In addition, rather mild effects have also been produced by 'temperature shocks' in some apparently reliable experiments, yet in others even this agent has seemed ineffective. In the absence of more decisively positive and consistent effects, then, some authors have tended to treat mutations as the result of an inexorable statistical process, dependent purely on thermal agitation *per se*, together with such radiation as may be present, and therefore bound to accumulate at a practically constant time-rate at any given temperature, under ordinary conditions of natural radiation.

During the past two years, however, we have been conducting large-scale experiments on 'spontaneous' mutations in *Drosophila* in which the frequencies from young individuals and those aged in various ways have been compared, and these show definitely, in harmony with certain earlier results of Russian and Indian workers (Sacharov 1939; Olenov 1941; Zuitin & Pavlovetz 1938; Singh 1940), that there are decided differences in the mutation frequencies of different stages. These may in fact explain the apparent effects of some of the tests of chemical and heat-shock treatments previously referred to, if the time and conditions of breeding were not well controlled in these cases. A more precise analysis of the physiological factors at work in our ageing experiments is much to be desired, but meanwhile a sort of net result already reached is the conclusion that, in *Drosophila* at least, the germ cells of older adults are no more, and in some cases not as much, contaminated by mutations as those of very young ones. This may involve a mechanism which has helped to make it possible for species like man, whose individuals live to be so much older than those of lower organisms, to escape the genetically damaging effects of too high a mutation frequency. And further, the practical breeding problem arises, whether reproduction by older or younger parents is genetically more desirable. If the situation in man is like that in flies, then, in view of the relaxation of selection obtaining in civilized man, the answer would be that it is genetically preferable to have reproduction by the older people, for this would actually delay the accumulation of mutations in the population.

Attacking the problems of chemical effects on mutation from another direction have been the very remarkable studies of Auerbach & Robson (1944) carried on in Edinburgh during the war years, beginning in 1941. In these it has been possible to show for the first time that there are whole groups of known substances, the application of which results in diverse gene mutations, as well as chromo-

some breaks, with frequencies comparable with those produced by considerable doses of high-energy radiation.* Among the other very interesting features of their results are evidences of delayed action in the production of some of the mutations. I have no right, however, to speak for these investigators in more detail, as most of their publication has been delayed by war conditions. In a paper just sent to press by Hadorn, another substance, phenol, is reported to be moderately effective in *Drosophila*, the technique of dipping the larval gonad in a solution of it being employed. And last year Emerson (1944) announced the obtaining of varied mutations in the fungus *Neurospora* by treatment with rabbit serum. The rabbits used had been immunized against the fungus, although it is not yet known whether this had anything to do with the effect. These three series of experiments constitute the first decided break in the impasse that had developed in studies directed towards the chemistry of the mutation process. The leads provided by them appear far reaching in their promise.

7. EVOLUTIONARY INDETERMINACY?

Despite the success of these attacks it should be emphasized that in none of the cases has there been evidence of just specific mutations being induced. The chemical processes have affected the frequency of gene mutation in general, but each individual

---

* Since the lecture was given, it has become permissible to reveal that mustard gas was the substance primarily dealt with in this work (see Gilman & Philips 1946).

mutation remains a chance and un-controllable event, from the macro-scopic standpoint, and is no doubt the result of a quantum exchange caused by the impact of a suitable form of thermal agitation or radiation, as the case may be. For such macroscopic-ally indeterminate events to become expressed in those changes in the char-acteristics of the whole individual which can be seen directly there has to be an amplification, on each occa-sion, of the order of 10 to the 21st power times, when considering the size of the gene in relation of that of, say, a whole mammal. This amplifica-tion is, of course, attained by virtue of the gene's power of duplication dur-ing the growth of the individual, com-bined with its ability to affect catalytically the body of protoplasm far larger than itself which surrounds it in each cell. In reality, the amplifi-cation may be almost unlimitedly greater than this, since, if the muta-tion has been one of the beneficial ones, the organism containing it will tend to multiply many-fold during an unlimited succession of generations, and so will serve as the basis for further mutations. Thus the quanta of physics become the quanta of evolu-tion (Muller 1935b), and the ultra-microscopic events, with all the possibilities born of their statistical ran-domness and even of their ulterior physical indeterminacy, become trans-lated into macroscopic ones with a magnification vastly surpassing that of such an instrument as the Geiger counter, and almost approaching, if not possibly some day exceeding, that of the primordial cosmogenic quanta proposed by Haldane.

I do not wish to deny that in spe-cial cases mutations may occurr pre-dictably, but all the above evidence agrees in indicating that this ordinarily be so rare as to have played compara-tively little part in evolution. It would be of interest to show that, neverthe-less, because of the orderliness of se-lection working on these random mutations, combined with the statis-tically great amount of mutational ma-terial which it has to work on, the direction of evolution is in large meas-ure stable and determinate, up to a given point, It seems likely, however, that at certain special crises and turn-ing points, where unusual changes or combinations of changes become im-portant, evolution itself may sometimes become indeterminate, at least in a molar sense.

8. NON-CHROMOSOMAL GENES

Evidence concerning the nature of the gene is to be gained not only from the genes in the chromosomes. Despite what has been said concerning the preponderantly chromosomal nature of species differences, it has been known since the work of Correns over forty years ago that plastids of plant cells are self-duplicating bodies in the same sense as chromosomes. They differ very significantly from the chromo-somes in their transmission, however, in that: (a) they do not have their two daughter plastids regularly allotted to different daughter cells at cell divi-sion, (b) they are included to only a very limited degree or, in some species, not at all, in the male germ cells, and (c) they do not undergo synapsis and orderly segregation in germ-cell formation. But, like the chromosomal genes, they are capable of undergoing rare, sudden, definite mutations, of diverse kinds, that are transmitted to the daughter plastid in duplication. Under special cellular conditions, the mutations of some plastids at any rate become more frequent, and so become themselves subject to some extent to gene control. Owing to their inexact distribution in cell division the plastids undergo a chance process of sorting out, after any mutation, until all the plastids in a cell are again alike. Because of this, together with the limited opportunity for formation of whole-plastid recombinations in sexual reproduction, the lack of crossing-over, and the relatively small quantity of genetic material each single plastid contains, the amount of evolution due to plastid mutations must be very small as compared with that based on chromosomes, and this is confirmed by analysis of species differences in plants. Animals get along without them. They are most plausibly conceived as having had a common ancestry with the chromosomal genes, dating back to the period before the latter had become organized into typical nuclear chromosomes.

Only recently have most geneticists awakened to the fact that there is still other genetic material, self-reproducing and probably mutable, in the cytoplasm of some plant cells at least, in the form of much smaller, perhaps ultramicroscopic particles, called by Darlington (1944) 'plasmagenes.' Again there seems to be no mechanism for their exact apportionment or effective recombination. This limitation, as with the 'plastogenes,' would provide explanation enough of the great rarity of cases of inherited dif-

ferences due to this cause, as compared with cases of chromosomally inherited differences. It is not known to what extent the multiplication of different materials of this kind is subject to a unitary regulation, whether the amounts of each, or of all, present tend to reach stable equilibria, or whether they compete with one another in their multiplication, as some viruses do (Delbrück 1942), and so become mutally exclusive. At any rate, some of them are more vulnerable and subject to loss as a result of special conditions than are the chromosomal genes.

Cases of cytoplasmic inheritance in yeast found by Winge & Laustsen (1940) were postulated by them to have their seat in mitochondria, particles which are not supposed to be derived, in any direct manner at least, from the chromosomes, but are by some investigators classed with plastids. Yet in Lindegren's (1945) and Spiegelman's (1945) recent important work on a cytoplasmic enzyme of yeast that seemed to play a part in its own production, clear evidence was obtained that the 'cytogene,' as they called it, is in the first place derived from, or at least dependent for its origination on, a particular chromosomal gene. Being liable to loss when its substrate is absent, it is necessary to have it renewable, in this way, from a more dependable genic source, even though it may be self-reproducing. It would seem highly significant for gene theory that the material in this case is itself an enzyme, and hence too is presumably of protein nature. However, certain links of the evidence for proving that it is directly self-reproducing in the same sense as a gene, rather than of indirect aid for synthetic processes in general through its function in carbohydrate utilization, have yet to be made clear.

Until Sonneborn's (1943*a*, *b*, *c*, 1945*a*, *b*) remarkable findings in certain varieties of the peculiar one-celled · animal *Paramecium*, no instances of self-reproducing substances in the cytoplasm of *animals* have been known, apart from some which, by reason of infectivity or curability or visible structure, have rightly or wrongly been attributed to minute parasites or symbionts. The so-called 'kappa substances' in the *Paramecium* cytoplasm appear to be less fully self-reproducing than chromosomal genes, in that they require for their continued synthesis the simultaneous presence not only of themselves but also of specific chromosomal genes. Sonneborn believes that they are in fact parts or derivatives of these chromosomal genes, which have become detached for special reasons—reasons connected with the prior detachment and differentiation of a peculiar form of nucleus, the macronucleus, from a germinal or micronucleus. This work, like that on yeast, opens a new chapter in genetics and appears to afford a new angle of attack on gene problems in general. In common with the yeast work, it is at present in such a formative period that no attempt can be made to do it justice here. It should, however, be pointed out that, according to the theory proposed, the phenomena which may here be studied in the cytoplasm are largely reflexions of those which, in ordinary organisms, and even in other existing varieties of *Paramecia*, more primitive in this respect, are proper to the genes in the

chromosomes. But, in the chromosomes, they are not isolable, and therefore escape detection as such.*

There remains room for the speculation, put forward by several authors, that in the cytoplasm of the body cells even of higher animals there may be plasmagenes, or 'kappa substances,' causing differences between tissues, and perhaps capable of mutating unfavourably so as to give rise to virus-like bodies, such as those of some cancers. Darlington (1944) has postulated that such genes and viruses might arise by the metamorphisis of ordinary cell proteins, and their combination with the cytoplasmic nucleic acid. There are additional possibilities, not necessarily exclusive of this or of one another, which require investigation. For example, Altenburg (1946; personal communication early in 1945) has suggested that viruses which originally entered as invaders may in the course of evolution have become symbiotic or even necessary 'viroids' that thereby take their place as part of the normal cytoplasmic equipment, subject, however, to mutations that may give abnormal effects.

But despite all present uncertainties regarding these crucial questions, the mass of evidence concurs in pointing

___

* As proof of this article is returned to the press, Sonneborn has, in an address in the *Cold Spring Harbor Symposium on Heredity and Variation in Microorganisms* (12 July 1946), undertaken a revision of the interpretation of these phenomena. He now regards the cytoplasmic substances in question as plasmagenes, not necessarily of nuclear origin though, like other known plasmagenes, capable of survival only in a given protoplasmic environment, in part conditioned by the nuclear genes.

to the exceeding rarity, at best, of effective genes in the cytoplasm of the *germ cells* of multicellular animals. Certainly this is not because genes in the cytoplasm are impossible, but rather, as Darlington has pointed out, because they lack those potentialities for persistence and for evolution, dependent on fixed linear arrangement, that the genes which are tied together into chromosomes have developed. For this fixed linear arrangement has been a prerequisite in allowing, for the chromosomal genes, controlled duplication and apportionment, instead of competition or sorting out, for their progressive increase in number of types and differentiation from one another (see below), and for the effective means they have of forming new combinations.

9. INCREASE IN THE NUMBER OF KINDS OF CHROMOSOMAL GENES

So far as the chromosomal genes themselves are concerned, there is every reason to infer that, being the product of an extended evolution, they cannot arise *de novo* by any sudden change from extra-chromosomal substances. True, it is known that the number of genes in the chromosomal germplasm can increase in evolution, but in these cases there is evidence that each chromosomal gene arises from a pre-existing one. Occasionally this happens by a mere doubling of a whole chromosome or set of chromosomes, followed by their failure of separation into two daughter cells, but in cases that become permanently established it happens more often by the misplacement of a small piece of a chromosome, which becomes broken out and then inserted somewhere else

into the line of a set of chromosomes that already contains the duplicate of this piece in its older position as well. Such an increase in gene number does not for ever remain merely a change in the quantity or proportionate number of these genes. For in the course of geological time different mutations are bound to accumulate in the inserted group of genes from those that become established in the group that occupies the original position. Thus the germplasm becomes not merely more compound but more complex and, other things being equal, the possibilities of organizational complexity for the body in general should rise also.*

10. ON THE NATURE OF THE SELF-DUPLICATION PROCESS

All the above-mentioned studies and numerous others, many of them of a cytological nature, have helped to furnish a biological setting that should be of use in the coming *chemical* attack on the nature of the gene, on the mechanism of its self-duplication, its mutation, its behaviour in meiosis, and its action on the organism. On the chemical side, it has become clear that the gene is really of nucleoprotein nature, since the work of Stanley & Knight (see their review, 1941), of Bawden & Pirie (see Pirie 1945) and of others

---

* These conclusions were reached at substantially the same time in the author's genetic analyses of the normal 'scute' region in *Drosophila* and of the minute rearrangements to which it was found to be subject (Muller 1935c) and in the observations of Bridges (1935) on normal salivary gland chromosomes, in which 'repeat' regions were microscopically evident.

has shown that even the virus particles infecting some plants, which fulfil the definition of genes in being self-determining in their reproduction and capable of transmitting their mutations, are composed of nucleoprotein, in fact, of *nothing but* nucleoprotein. This at the same time lends encouragement to the conception, though it does not by itself actually prove it, that the most primitive forms of life consisted of nothing else than a gene: that is, a bit of substance of specific chemical composition, nucleoprotein or protonucleoprotein in composition (possibly much simpler than the present day material—see Muller 1926), which was capable of duplicating not merely itself, but even its mutations, and which lost this ability on being divided further. Other work has shown that nucleoprotein is also contained in viruses infecting animals and bacteria, and that it is contained in plastids, although in these cases there is other material present as well. However, the problems concerning how the nucleoprotein works in gene duplication and in other gene activities remain as yet almost an unworked field, so far as experimental evidence is concerned.

It has been misleading and unhelpful to refer to the self-duplication of the gene as 'autocatalytic,' and Troland's otherwise brilliant papers (1914, 1916, 1917) are marred by his insistence on this. For the term is a 'blanket' one, referring merely to the end result, that more material of a given kind is produced if some of that material is present to begin with than if it is not. There are many totally diverse mechanisms by which such a result is brought about, and an understanding of one of them seldom helps with another. For example, the formation of pepsin from pepsinogen by the action of pepsin itself, which, as Herriott (1938, 1939; Herriott, Bartz & Northrup 1938) has shown, digests away in non-specific fashion a certain part of the pepsinogen molecule, thus freeing the pepsin component, can have no relation to gene synthesis. Most 'autocatalytic' reactions are mere accelerations of those which would otherwise take place anyway, but more slowly; this, too, is untrue of gene duplication. The example most commonly cited, crystallization, fails in several essential respects to be analogous to gene duplication. The most distinctive features of the latter, not found in the former, are that complicated chemical changes are induced by the self-duplicating agent that would not take place otherwise, and that from a heterogeneous medium which may be alike for many different agents and in this sense non-specific, the given agent selects the components necessary for itself and arranges and combines them into a form modelled on its own. Accordingly, given a change (mutation) in the agent, a corresponding change is wrought in the identical product. It is this modifiability, leading to improvability, of the reaction, that allows it, unlike other so-called 'autocatalytic' reactions, to form the basis of biological evolution and so of life itself. If, as Troland for instance thought, this were a common property of matter, all matter would be potentially living.

In an ingenious attempt to account for the nature of this autosynthetic reaction, Delbrück (1941), elaborating upon a suggestion by the Russian Frank-Kamenetzky (1939), has recently given a detailed but admittedly hypothetical picture of how amino-acids might conceivably be strung together in the same pattern as that in a pre-existing gene model. On this scheme, peptide-precursors from the medium, having unfinished peptide links, become attached to the amino-acid residues corresponding to them in the polypeptide model, by means of a resonance occurring at the sites of the peptide links; this is followed by a finishing up of the peptide connexions, and associated undoing of the resonance attachments between the old and the newly formed polypeptides. Although this scheme deserves serious consideration, it is not yet known that gene synthesis really involves the making of peptide links. For it is conceivable that the chains were ready-made so far as the gene itself is concerned, and that the specific synthesis consists rather in the making of other connexions, as between 'R groups' of different parts of one or more chains, resulting in a special kind of folding and/or superstructure. It can be calculated, at any rate, that gene specificity can hardly reside merely in the linear arrangement of amino-acids in protamines like those which form the bulk of the material isolated from some fish sperm, since their paucity of kinds and regularity of order in the molecules investigated seems to be such as to allow less than a thousand possible combinations (author's calculation, unpublished). Such an objection would not necessarily apply, however, to protamines in a more polymerized condition (as they might be in the living material), or to other proteins of higher type (such as those reported by Stedman & Stedman 1943). How-

ever, it may be that an analogous mechanism of resonance could as well be used to explain the making of other connexions than peptide groups, in imitation of those in a pre-existing gene model.

In this as in previous proposals for explaining gene duplication, starting with Troland's, in 1914, the 'raw materials' in the medium are supposed to become attached to like parts of the pre-existing gene and so to arrive at the same arrangements as it has. As I pointed out in 1921, there is a known parallel for this in the phenomenon of synapsis, whereby like chromosome parts, that is, like individual genes, become attached to one another in two's, and one has only to suppose that this phenomenon may extend even to the parts of the gene as they are put together during the process of its duplication, to get an explanation of duplication in terms of this remarkable synaptic force. Delbrück and others also have adopted this view that the same principle is acting in gene synthesis and synapsis. Thus any knowledge of basic principles of synapsis, such as the fact that it appears to consist primarily of an attraction in twos, even when more than two like members are available, which would indicate a face-to-face attraction, may be expected to help, eventually, in the explanation of gene duplication also.

But whereas in Delbrück's and some other interpretations the like particles are in the first place brought into atomic distances of one another in the correct position purely by the accidents of thermal agitation—an agent the exactitude of which for the purpose is dubious—cytologists have long considered their observations to indicate that the attraction of like chromosome parts for like extends over microscopically visible distances. That the union is not merely due to a chance coming together of likes at some single point, followed by a zipper effect, is indicated, among other things, by some new observations of McClintock (1945). These show that in the meiosis of the fungus *Neurospora* the chromosomes regularly come into side by side contact while still in a condensed, closely coiled condition, and only later become extended. But there is the apparently insuperable difficulty in attraction at a distance, pointed out by various physicists, that no spatial pattern of attractive forces could retain its specificity over a distance many times its own diameter, for the lines of force from different parts would become too dispersed and mixed with one another. It would therefore seem necessary, for explaining a self-specific pattern capable of operating at a distance, to postulate that it is expressed in the form of a temporal fluctuation, that is, a vibrational effect of some sort, varying with each gene. This would not be subject to the same limitation of distance as a spatial pattern, for its special characteristics and direction would to a considerable extent withstand both distance and intermixture with other patterns, just as is true of simultaneous musical notes. And although two such mutually attracting bodies would have to be in the right phases with respect to one another they would automatically tend to impress each other into such phases, as their state of least energy. I will not go into the further possibilities of this proposal, which have been put forward independently by Jordan (1938, 1939), myself (1941), and Fabergé (1942), in rather different forms, following Lamb's (1908) proposal of such a scheme for centrosomes. It is true that objections have been raised to these ideas (Pauling & Delbrück 1940). In any case the effect would, if it exists, have to be derived from a peculiar dynamic condition within the gene, of a kind not hitherto touched upon by chemists.* Such an effect would demand detailed explanation, for it might well hold the secret of the uniqueness of the processes of both gene duplication and gene synapsis.

In most discussions of these phenomena it has been assumed that they are of a direct nature, involving the immediate union, or building, of like by like. But as Friedrich-Freksa (1940) pointed out, what *may* happen is the union of each part with an opposite or complementary part, which, serving as an intermediary, becomes in its turn attached to another component opposite or complementary to itself, and therefore like the first component. On this scheme, one of the two components is

---

* Since the presentation of the above passage, Bernal has stated (in a private communication) that he has inferred the existence of a vibrational attraction between viruses. In a paper by Bernal & Fankuchen (1941) evidence was given for long range attractions between viruses, caused by their ionic atmospheres according to the principles of Debye; granting these, it is only necessary to postulate regular periods for the size and/or shape of these atmospheres (cf. variable stars) to obtain temporally specified attractions.

the protein of the gene, electrically negative because of its basic arginine, while the component complementary to this is a structure of nucleic acids, with its positive charges so arranged as to match the negative ones of the protein. The nucleic acid is then pictured as tending to hold a like protein on each side of it. By ordering the placement of heterogeneously arranged parts into proper pattern next to the original gene this two-step process was supposed to serve not only in the synapsis but also in the synthesis of the genes. Generalizing further, the author observers: 'Die hier entwickelte Vorstellung von der identischen Verdoppelung von Nucleinproteinen lässt sich vergleichen mit den Vorgängen, die wir von technischen Vervielfältigungsverfahren gewohnt sind. Die Reproduktion geht nie unmittelbahr, sondern immer nur auf dem Umweg über etwas Gegensätzliches vor sich. Die Gegensätze erhoben und tief, hell und dunkel, Bild und Spiegelbild, positive und negative Ladungen, können zur Formwiederholung führen.' It may be noted that this scheme does not seem to be capable, in the way Friedrich-Freksa thought, of explaining, by itself, specific attractions extending over microscopic distances. However, it might conceivably be modified to do so, by having the charges go through periodical changes, with complementary temporal patterns characteristic of the individual parts. In another respect also the scheme is incomplete, as in the protein the positions of only the arginine (or of the hexone bases in general) are accounted for by the nucleotide arrangements. Possibly other extensions of the hypothesis could be devised to alleviate this deficiency.

Another scheme using complementary unions had been proposed, for synapsis, by Lindegren & Bridges in 1938. They treated the gene as an antigen, against which an antibody became formed, that joined on to it, and this antibody in turn then tended to join on to the other gene of the same kind as well, thus bringing the two like genes together indirectly. Again, the process does not give long range attractions. This year Emerson (1945), supported by Sturtevant (1944), has proposed essentially the same complementary mechanism for gene duplication. In this case, separate parts are not supposed to be collected from the medium since, following Pauling's idea, antibody formation involves only an appropriate folding of an already existing protein, to bring its parts into a relation complementary to the original in regard to surface shape and surface position of reactive groups. This corresponds most nearly to Friedrich-Freksa's category of 'erhoben und tief.' Here, then, by the moulding of a second antibody to the first antibody, the type of surface characteristic of the original antigen is supposed to be regained and the gene duplication is thus completed. It happens, however, that the 'anti-antibodies' hitherto examined in serology do not fulfil this condition, as they involve different parts of the surface from those participating in the original antigen-antibody reaction, and are probably too imperfect anyway. Moreover, antibodies against proteins native to cells have not hitherto been found to be produced by them at all. But here, as before, it is hardly fair to require of a preliminary suggestion, lacking any experimental evidence, a complete solution, and various accessory hypotheses could doubtless be proposed for taking care of these difficulties.

11. ON POSSIBLE ROLES OF THE NUCLEIC ACID

Finally, however, some remarkable experimental evidence having a further bearing on gene chemistry has appeared, and from an unexpected quarter, namely, in investigations on the bacteria of pneumonia. Griffith (1928), in England, had shown in 1927 that under certain conditions pneumococci of one variety become somehow transformed into another by growing them in the presence of killed bacteria of that other variety. Avery, MacCleod & McCarty (1944) have gone further, and have given evidence which they believe points to the conclusion that the effective substance in this treatment is the nucleic acid itself, of the variety to be imitated, in practically protein-free condition, and in fact that nucleic acid in its naturally polymerized form. If this conclusion is accepted, their finding is revolutionary, no matter whether, with these authors, one adopts the radical interpretation that a transformation of the genetic material in the treated organisms has been induced, converting it into material like that used in the treatment (cf. 'Kappa substances'?), or whether it is supposed that genetic material of the donor strain actually becomes implanted within the treated strain and multiplies there, or whether the material used in the treatment is regarded as merely exerting a selective action so as to favour the survival of such exceedingly rare spontaneous mutants as happen by accident to agree with the other variety.* There are certain obvious tests for distinguishing between these very different

_____

* Since the above was written, and just as this paper was about to be sent in, Delbrück, speaking at the Mutation Conference in New York City on 26 January 1946, reported an instance of apparent transformation of one strain of bacteriophage by another, though in respect of a portion only of its properties. In this case, the rate of change is so high as to exclude the hypothesis of spontaneous mutation and selection. In my opinion, the most probable interpretation of these virus and *Pneumococcus* results then becomes that of actual entrance of the foreign genetic material already there, by a process essentially of the type of crossing-over, though on a more minute scale. At the same conference, on 28 January, Mirsky gave reasons for inferring that in the *Pneumococcus* case the extracted 'transforming agent' may really have had its genetic proteins still tightly bound to the polymerized nucleic acid; that is, there were, .in effect, still viable bacterial 'chromosomes' or parts of chromosomes floating free in the medium used. These might, in my opinion, have penetrated the capsuleless bacteria and in part at least taken root there, perhaps after having undergone a kind of crossing-over with the chromosomes of the host. In view of the transfer of only a part of the genetic material at a time, at least in the viruses, a method appears to be provided whereby the gene constitution of these forms can be analysed, much as in the crossbreeding tests on higher organisms. However, unlike what has so far been possible in higher organisms, viable chromosome threads could also be obtained from these lower forms for *in vitro* observation, chemical analysis, and determination of the genetic effects of treatment.

possibilities. But, in any case, it would be proved that the substance used for treatment is able to assume highly individualized forms, capable of influencing cell metabolism in far-reaching and specific ways. And so, if this substance really is composed solely of nucleic acid, it would follow that the tetranucleotides, despite their relative uniformity, have richly varied and specific forms of polymerization that give them very diverse and distinctive properties.

The question thus becomes acute whether it is really the nucleic acid or, as has commonly been thought, the protein, which is primary in determining the differences between genes, or whether, as a third possibility, the two are of comparable value and, perhaps, complementary in some sense as Friedrich-Freksa proposed. The hope of further analysis of gene structure through extensions of this method rises high. It may perhaps provide us at the same time with the first instance of that Eldorado of geneticists, directed mutation, although the geneticist characteristically insists on being driven to this conclusion before accepting it.

Recently, until this *Pneumococcus* work, what has increasingly seemed to be the most plausible chemical role for the nucleic acid in the gene has lain in its possible contribution to the energetics of gene reactions, and it is very likely, even now, that this too will somehow fit into the picture. As the beautiful work of a considerable group of biochemists has been showing, some of the nucleotide fractions of nucleic acid, or close derivatives of them, are fundamental in those varied processes of the cell where it is necessary to effect a transfer of energy from a substance or group of substances relatively rich in energy to one relatively poor, with resulting synthesis, mechanical work, secretion against osmotic pressure, or other form of potential energy storage. Often, an energy-rich phosphate bond is used by the nucleotide for the transfer. Even amino-acids may be strung into polypeptides by such a mechanism, if a hypothesis of Lipmann's (1941) is correct. Thus it may be that nucleic acid in polymerized form provides a way of directing such a flow of energy into specific complex patterns for gene building or for gene reactions upon the cell.* But to what extent the given specificity depends on the nucleic acid polymer itself, rather than upon the protein with which it is ordinarily bound, must as yet be regarded as an open question.

12. ON THE NATURE OF GENE EFFECTS

Concerning the nature of the primary reactions of the gene, in producing its effects upon the cell, there is as yet almost as much ignorance, so far as real experimental evidence goes, as there is concerning gene duplication. It has usually been assumed, perhaps gratuitously, that the gene acts here through its protein component. De Vries (1899) long ago suggested that the primary product is much like the gene itself, and pro-

---

* Since the presentation of the above passage (which is unchanged from that of the original lecture) Spiegelman, at the Mutation Conference in New York in January 1946, independently proposed the same view for the role of the nucleic acid of the gene, and presented supporting evidence for it, derived from his experiments on yeast.

duced by a process akin to its duplication, and he thought that this product then migrates into the cytoplasm to do its work. Since Driesch (1894), also in the last century, suggested that the hereditary determiners act as 'ferments,' that is, enzymes, it has been easy to conceive of the different genes, or these supposed products that represent them, as carrying out all sorts of catalytic reactions by means of their surface structure, and since Troland's (1914, 1916, 1917) able advocacy of this idea, which he called the 'Enzyme Theory of Life,' it has been especially popular. On different variants of this view, the original gene itself, *in situ* in its chromosome, thus acts as a heterocatalytic enzyme, or products like itself and still surrounding it in the chromosome do this work, or products like it that migrate out. The apparently self-reproducing melibiozymase reported by Lindegren (1945) and Spiegelman (1945) in the cytoplasm of yeast, derived originally from a chromosomal gene, would seem to be an example of the last-mentioned type.

It is at least evident that if the primary gene products are *not* like (or complementary to) the gene itself, and so are not produced by a process akin to gene duplication, then the gene must certainly act as an enzyme in producing them, whatever they are, otherwise it itself would get used up. Since different enzymes can initiate or accelerate the most varied types of reactions, the primary products of gene activity would in that case be correspondingly varied, and would belong to no one category of substances. If, on the other hand, the primary gene action is to produce more molecules similar in composition (or complementary) to itself, or to a part of itself (but perhaps unable to reproduce themselves further), it would be taking too much for granted to assume that the process of their production was an enzymatic one in the usual sense. For, as has been seen, gene duplication probably involves a very special kind of mechanism, that it would be misleading casually to lump with that of heterocatalytic enzymes, at least until more is known about it. And the gene products themselves, in case they were produced by such an essentially autosynthetic process, would not necessarily have to act as enzymes either: that is, they might actually become used up in the reactions which they in turn take part in, inasmuch as the original gene would presumably be able to replace them by the same autosynthetic process as before. This is not to deny, however, that they *could* act as enzymes, in some cases at least, particularly if they represented the protein component of the gene.

Hence, despite the appeal of simplicity presented by the view in question, it seems too early to conclude that either the gene or its primary products do always, or usually, act as enzymes. One should be even more cautious in assuming, conversely, that most enzymes represent primary gene products. It has long been known, to be sure, that various identified enzymes in the cytoplasm are gene-determined, like everything else in it. And this method may be used, as in the remarkable studies of Beadle and his co-workers on *Neurospora*, to analyse in great detail, by the comparative chemistry of different mutants, how various complicated nets of chemical processes in the cytoplasm are carried out. For each step at least one enzyme usually seems to be required. However, the fact that an enzyme has been found to be lost, incapacitated, or changed as a result of the mutation of a given gene by no means proves that it has been directly derived from that gene, and further work may show that in some or most of these cases mutations in various other genes can affect the given enzyme as much.

Somewhat similar considerations apply concerning the relations between genes and proteins in general, including protein antigens. Only if the primary products of genes were autosynthetically produced would it seem probable that they were always proteins (if one overlooks the possibility of their representing only the polymerized nucleic acid component instead). And while any higher protein may serve as an antigen, such an autosynthetic relation would still be far from a proof of the converse proposition, that all (or most) protein antigens are primary gene products. The modifiability of the antigenic properties of proteins is in fact against this view. It is, of course, true that if a class of substances were found, such that each member of the class was never dependent for its production on more than one gene, these substances would in all probability be primary gene products. It has been thought that such evidence existed in the case of the antigens dealt with in intra- and interspecific blood and tissue transfusions and transplantations. However, the relatively few cases where one of these antigens has been found to arise only in the presence of a given combination of two or more different genes are enough to cast serious doubt on this conclusion. For the gene differences dealt with in any given test are usually but a small sample of all the genes in the genotype. And if, even in this small sample, two are found which interact in the production of an antigen, it can be calculated that, if the whole genotype could be considered, the number of antigens found to be produced by interaction would probably be far higher than the maximum number produced singly. It would accordingly seem necessary to have an enormous body of data of such a kind before it could be proved by this method that any given class of isolable substances usually constitute primary, unmodified gene products. And it may well be that new methods will be needed before the primary gene products can be identified as being such.

Whatever the primary or remote gene products, it seems likely on general considerations that there is a limited number of possible types of building blocks in the gene, and that genes differ only in the arrangements and numbers of these. Under differences in arrangement may here be included not only changes in the linear sequence of amino-acids in the polypeptide chains but also changes in shape, involving folding of the chains, and attachments between R groups, whereby their active surfaces would acquire very different chemical properties. In this way radically diverse organic compounds may be pictured as being formed under the influence of different genes in the cell.

Most of the mutations leading to the different compounds of a relatively simple type as compared with proteins (so-called 'extractives' and prosthetic groups, vitamins, the common monosaccharides, lipoids, etc.) seem to have occurred in the lengthy evolution that must have intervened between the early virus and the bacterial stage. For bacterial protoplasm is, fundamentally, surprisingly like ours in innumerable complicated details. A large part of the evolution beyond the unicellular stage, on the other hand, has involved changes in the time and space patterning of materials which themselves were much like those already present in one-celled organisms. Included here were alterations in the sizes and proportions of parts, relative rates of growth, placing and timing of ingrowths and outgrowths, of contractile or supporting fibres, etc. Moreover, most of these changes can be seen to have derived their selective advantage from the effects of these redistributions themselves, rather than from the chemical changes behind the latter. It is true that even the redistributions must have had a chemical basis, but it seems unlikely that most of them are due to the formation of new prosthetic groups or of other substances which, as compared with proteins, are relatively simple. Rather does it seem likely that largely the same stock outfit of these substances became modified in their rate, place and time of action by more subtle changes in the proteins, including the enzymes, with which the simpler substances interact, and which undoubtedly interact also with one another in the determination of such effects. The quantitatively variable character of protein-with-protein reactions renders them especially suitable for such adjustments in degree, rate, and time- and space-pattern of effects.

It must of course be the future task of biochemistry, combined with genetics, to unravel the whole complicated web of protoplasmic and bodily interactions, from the primary gene products to the last phenotypic effects. In this work, as I stated in an earlier paper (1933), the production of a change in any individual gene provides us with what is, 'in effect, a scalpel or injecting needle of ultramicroscopic nicety,' for 'experiments in which the finest, most fundamental elements of the body fabric are separately attacked.' Such studies as those on pigment production in flowers and in insect eyes, on the synthesis of amino-acids and vitamins in *Neurospora*, and on the trains of effects following given mutations in mice and in poultry, illustrate the progress which this general method is capable of making. It can, of course, be vastly extended, as when, for example, not only the substances that naturally diffuse out of cells but also those obtainable by extraction are employed in the tests of the effect of one strain on another, and when in addition to the processes known as 'autotrophic' the large number that have to do with higher conversions and utilizations are brought into the picture, as by analogous cross-tests of mutations that prove lethal even in the presence of media adequate for ordinary saprophytes. The tracing of the courses of reaction of the substances in question will need even more refined biochemical methods but, proceeding from step to step, and with the aid also of the tracer isotopes now becoming more available, it should provide a field which for a long time will be ever more fertile in results. Most difficult of all, it may be anticipated, will be the working out of the details of protein structure and reactions (and perhaps that of nucleic acid polymers). At present, such work with proteins themselves is in its earliest stages, and adequate methods have still to be worked out, but one can at least see that the field is there.

13. 'POSITION EFFECT'

In considering the way in which genes exert their effects, some evidence from genetics may be mentioned which indicates that even the same gene nucleoprotein may react very differently in the presence of unlike physical influences from other nucleoproteins nearby. Reference is made here to the peculiar finding that in *Drosophila* genes may be induced to undergo changes in their effects even without intrinsically conditioned gene mutations, merely by shifting their position in the chromosome and so giving them different gene neighbours. If many generations later replaced in their original positions they resume their previous mode of action. It is conceivable that these 'position effects' are due to localized chemical reactions between the products of nearby genes. But it seems more likely that they are caused by changes in shape of the gene, that give it a different amount or even direction of effectiveness (Muller 1935*d*). As the degree of change in the effect would necessarily depend also upon other cellular conditions, varying with the region of the body, it is not strange that positionally induced changes have been observed in the patterning of an effect as well as in its general quality or intensity.

Such positional changes in gene shape might conceivably arise directly from the synaptic forces exerted by the neighbour genes. For the same force which, when applied to like genes, causes a uniform attraction of all parts, might affect the parts of different genes unevenly, so as to warp them. Or, as Ephrussi & Sutton (1944) have plausibly suggested, and as some recent evidence tends to favour, the cause of the shape changes might in some cases lie in the degree or type of coiling of the chromosome thread, a condition which in its turn is influenced by synaptic forces. Considering the matter more broadly, the result further suggests that many actual gene mutations also may owe their effects to changes in the gene shape, but that, in those cases, the changes are permanent, inwardly conditioned ones.

If the extraordinary results reported by Noujdin (1936, 1938, 1944) are taken at their face value, results studied further by Prokofyeva-Belgovskaya (1945), there is in addition an intermediate class between permanent gene mutations and immediate position effects, involving semi-permanent changes of the gene inducible by a special type of positional influence, and capable of being repeated in gene duplication long after the influence has been withdrawn, but gradually subsiding. These cases, more especially, seem to be best understood as due to differences in coiling, if one takes into consideration in this connexion the work not only of Noujdin, and of Prokofyeva-Belgovskaya, but also of Ephrussi & Sutton (1944), and of Ris and Crouse (1945). At any rate, the changes lie in the degree of 'heterochromatization,' whatever this may finally turn out to mean. A peculiar case of semi-permanent change reported by Sonneborn (1943c) may lie in a different category.

14. THE RECOMBINING OF GENES

There are grounds for the inference that higher organisms represent the accumulation of millions of separate mutational steps. Each one of these was an exceedingly rare event, probably occurring, on the average, in considerably less than one in a million germ cells. Moreover, each such event had to be followed by a long process of selective multiplication of the mutated gene, extensive enough to overcome the original excessive majority against it. A second favourable mutation does not get a foothold in an asexually reproducing line until, on the average, as many individuals bearing the first mutation have already been produced as may be expected to include a favourable mutation again. Owing to the small selective advantage usually possessed even by a successful mutation, this will probably take thousands of generations. That may not be so bad for bacteria and forms below them, as they reproduce so rapidly and present so inordinately many individuals for the selective process. In addition, they benefit by not having so many directions in which change must be almost simultaneously co-ordinated before it can become decidedly advantageous. But for higher forms, where the opposite of all these conditions obtain, geological time would have been insufficient to bring their evolution to anything like its present state, had not a great innovation been instituted, namely, sexual reproduction.

Sexuality really opened the door of the germplasm not merely to one other individual, but through it to a congress of the population as a whole. After the establishment of sexual reproduction, mutations no longer had to wait their turn in line, but those occurring anywhere throughout a species could become multiplied simultaneously and could meanwhile become combined with one another. Moreover, multitudinous combinations could thus be tested out as combinations. This was a great co-operative genetic undertaking, in which the contributions of all were pooled for the common benefit, a circumstance which multiplied their mutational resources and the speed of their evolution over what it would otherwise have been.

For the sexual process to succeed in giving this genetic advantage in organisms with a complex genotype, not only some sort of fertilization process but also the essentials of the mechanism of meiosis, including some sort of synapsis, segregation, and whatever else is necessary for orderly recombination of genes, had to operate, for it is the final recombination that counts. And in this recombination it can be calculated that crossing-over must usually make a much larger contribution to the potential speed of evolutionary change than does the random assortment of genes lying in different chromosomes, contrary to some current opinion.

Crossing-over thus turns out to be the key process in sexual reproduction. The basis which made it possible for the chromosomal genes to have developed crossing-over lies not only in their linear attachment to one another but also in certain of their other fundamental properties. Among these are to be noted especially their specific auto-attraction, the essentially 'by twos' nature of this attraction, the ability of the chains of genes to develop tensions in spiralization, and the capacity of broken ends to reunite. And all of these properties, by the way, have come to light chiefly as a result of studies of crossing-over.*

One reason why it has been necessary for organisms to take so many mutational steps as to require crossing-over for their accumulation is because so many of the successful steps must be very minute. Their minuteness is indicated by the exceeding delicacy of adjustment of quantitative characters which often obtains, a degree of precision which must have been attained through many small whittlings, so to speak. The phenomenon of dominance of normal genes over most of the more frequently appearing mutants provides one line of evidence for a long and as it were invisible selective process, as Fisher (1928, 1930) first pointed out, and as has been accepted either in the original or a modified form by most geneticists. Analysis of species differences also shows that numerous small steps have usually been involved. Striking evidence of the meticulousness of selection is further given by the phenomenon in *Drosophila* called 'dosage compensation' (Muller 1932). This is too complicated to be explained here, but the conclusion to be drawn from it should nevertheless be stated. That is, that the degree of develop-

---

* If these are in fact very fundamental properties of linearly attached genes, it would not be so surprising that a process of crossing-over should occur even in bacteria and viruses, as suggested in the preceding footnote.

ment of the visible characters studied has commonly been determined through the action of past selection to a nicety considerably exceeding human powers of optical discrimination.

For the almost continuous range of genetic variation which this implies, the changes possible in proteins seem especially suited. It will, however, be realized that, as a result of any given character having been determined through such a multitude of variations, especially if these are mainly protein in the basis of their action, the analysis of the biochemical details involved in the production of just this one character becomes a task requiring an army of investigators, even at a stage of scientific development in which adequate methods for protein study have become available.

15. THE SURVIVAL OF MUTATED GENES

The process of gene mutation which made the attainment of our present stupendously complicated, integrated organizations possible, is still going on. In itself mutation is really a disrupting, disintegrating tendency, like the thermal agitation which seems mainly to cause it. For the overwhelming majority of mutations are bad, and it is only the Maxwell demon of selection inherent in gene duplication, that is, the differential multiplication of the mutations, which brings order out of mutation's chaos despite itself. Let the previous course of selection be relaxed through natural or artificial means, and the tendency to disorder and degeneration gains and lowers the level. That is, not merely the evolution of a species to its present state, but even its maintenance, requires a continuance of selection, for the gene population is always in a dynamic equilibrium between mutation and selection, not statically fixed. Any cross-breeding population always contains many harmful genes, most of them recessive and seldom manifesting themselves, inherited from tens and hundreds of generations ago. But they are ordinarily not increasing in number because an equilibrium has been reached at which they die off, through the death or failure to reproduce of some of the individuals that do manifest them, at a rate as fast as new ones appear through mutation, as Fisher (1930), Wright (1931), and Haldane (1927), more especially, have pointed out.

From some imperfect human data one may calculate (unpublished) that probably the great majority of persons possess at least one recessive gene, or group of genes, which, had it been inherited from both parents, would have caused the death of the given person between birth and maturity. (And this does not include those which kill off just prenatal stages.) That such genes are not still more abundant is only due to the fact that, in the past, the individuals getting the genes from both parents did die. If one could and would let them live and breed without limit, their number would creep up until finally it would be a case of treating everyone for everything—for thousands of things besides those ailments now known—and thus, in fact, making completely artificial men.

I am not arguing here against medicine and better conditions and against compensating for hereditary lacks by environmental benefits, otherwise I should be arguing against myself and my children and against all other persons and their children. Only one must recognize and somehow make the best of the inexorable rule that practically every mutation, even a 'small' and non-lethal one, with the rarest of exceptions, requires finally a genetic death, that is, a failure to live or to breed, somewhere along the line of its descent, if the population would remain genetically at par. For each mutation, then, a genetic death— except in so far as, by judicious choosing, several mutations may be picked off in the same victim. If new mutations occurred in one germ cell in ten, as perhaps in flies, then this (with the reservation just mentioned) would be the proportion of genetic deaths of individuals eventually required. If, as could easily happen, by reason of the prevalence of injudicious X-ray treatment, exposure to artificial radioactivity or to special chemicals, or unwise average age of parenthood, the mutation frequency became doubled or tripled, so as to be increased to one in five, or one in three, then the proportion of genetic deaths, and of those manifesting the mutant characters, would finally rise correspondingly, before attaining its equilibrium value.

It would, however, be many centuries before these effects were fully manifest, and it might be that before that time a way would have been discovered of identifying mutants in the germ-cell stage and selecting them out then, or of reducing the frequency of mutation, or even of directing mutations. Moreover, the most constructive task is not that of merely keeping undesirable mutant genes down to a reasonably low level, but of fostering that tiny minority of possible types having biologically progressive effects, in which lie all the genetic hope of the future. All this seems very utopian now, and quite out of harmony with the recent pronouncement of a most eminent and influential American scientist who has stated that the aim of 'biological engineering,' as he calls it, is the provision of 'better food and clothing' (perhaps for Americans?). It is also out of harmony with the strong negative correlation between education or 'I.Q.' rating and rate of reproduction, found in modern civilized communities generally, including the U.S.A. and even the U.S.S.R. (see Price 1939). But does it not seem proper to think in utopian terms in a year which has marked the greatest revolution of all time in man's powers over his physical environment and in which at the same time, as by a double miracle, the interrelations of man with man also are being consciously reorganized on a grand scale (though just now preeminently on this island as a model), in such a way as may yet be fitting to control for our common benefit these and other physical powers that transcend the rights of individual manipulation?

But before any kind of conscious guidance over our own genetic processess were attempted, it is to be hoped that education would lead us gradually to desire it, so that it became an entirely voluntary concern of all, and that increasing understanding and a better developed social consciousness would help us in revising our judgements as to what was really important and what was really desirable. As a part of all this there would be required a growth in our appreciation of the importance of the factors of the physical and especially of the social environment in the determination of human traits, since, in regard to his most salient characteristics, his psychological ones, man is by far the most plastic, somatically, of all organisms. So much is this the case that, in the realm of the psychological, most of the differences commonly regarded as due to genes are usually only characteristics that have been superimposed by training or by largely unwitting conditioning. The costly lesson taught us by the terrible Nazi perversion of genetics should help to make these facts better realized.

Surely, however, it will not be our desire, after understanding the manner of origin of the organization we have inherited, to remain for ever content with merely maintaining or even making the most of that inheritance which we already have, without adding to it. For mankind is cursed or blessed with what has been called 'the divine discontent,' which drives him ever further everywhere. With knowledge comes power, and as the use of power cannot for ever be denied, it will behove us in the biological as in the physical field to develop also wisdom, including that social attitude which is a part of wisdom. With power so used, indeed, the wisdom could be further increased, self-multiplied, by genetic in combination with other methods, to a degree which would seem to be unlimited. Thus the self-reproduction of the gene and the self-reproduction of intelligence would reinforce one another in an ascending curve.

The day seems far off for so using our· knowledge of the gene, but in these times of rapid movement in the physical and social realms, we might as well recognize this even more distant star to which our biological wagon is hitched. Meanwhile, for to-day, we must remove our gloves, and be content to work with our fingers in the protoplasmic mud, to get the heavy wheels of our science turning. And in this work, if we can cause even a little movement forward, that should be sufficient adventure for our own little lives.

REFERENCES

ALEXANDER, J. & BRIDGES, C. B. 1929 Some physiochemical aspects of life, mutation and evolution. *Colloid Chem.* **1**, 9–58.

ALTENBURG, E. 1946 The viroid theory in relation to plasmagenes, viruses, cancer, and plastids. *Amer. Nat.* (in the Press).

AUERBACH, C. & ROBSON, J. M. 1944 The production of mutations by allyl iso-thiocyanate. *Nature*, **154**, 81.

AVERY, O. T., MACLEOD, C. M. & MCCARTY, M. 1944 Studies on the chemical nature of the substance inducing transformation of *Pneumococcus* types. *J. Exp. Med.* **79**, 137–158.

BERNAL, J. D. & FANKUCHEN, I. 1941 X-ray and crystallographic studies of plant virus preparations. *J. Gen. Physiol.* **5**, 111–165.

BRIDGES, C. B. 1935 Salivary chromosome maps. With a key to the banding of the chromosomes of *Drosophila melanogaster. J. Hered.* **26**, 60–64.

DARLINGTON, C. D. 1935 The time, place and action of crossing-over. *J. Genet.* **31**, 185–212.

DARLINGTON, C. D. 1944 Heredity, development and infection. *Nature*, **154**, 164–169.

DELBRÜCK, M. 1935 Über die Natur der Genmutation und der Genstruktur. Dritter Teil: Atomphysikalisches Modell der Genmutation. *Nachr. Ges. Wiss. Göttingen*, Math.-phys. Kl., Biol., N.F., **1**, 223–234.

DELBRÜCK, M. 1941 A theory of auto-catalytic synthesis of polypeptides and its application to the problem of chromosome reproduction. *Cold Spr. Harb. Symp. Quant. Biol.* **9**, 122–126.

DELBRÜCK, M. 1942 Bacterial viruses. *Adv. Enzymol.* **2**, 1–32.

EMERSON, S. 1944 The induction of mutation by antibodies. *Proc. Nat. Acad. Sci., Wash.*, **30**, 179–183.

EMERSON, S. 1945 Genetics as a tool for studying gene structure. *Ann. Mo. Bot. Gdn*, **32**, 243–249.

EPHRUSSI, B. & SUTTON, E. 1944 A reconsideration of the mechanism of position effect. *Proc. Nat. Acad. Sci., Wash.*, **30**, 183–197.

FABERGÉ, A. C. 1942 Homologous chromosome pairing: the physical problem. *J. Genet.* **43**, 121–144.

FISHER, R. A. 1928 The possible modification of the response of the wild type to recurrent mutations. *Amer. Nat.* **62**, 115–126.

FISHER, R. A. 1930 *The genetical theory of natural selection.* Oxford: Clarendon Press.

FRANK-KAMENETZKY, D. A. 1939 Resonance theory of autocatalysis. *C.R. Acad. Sci. U.R.S.S.*, N.S., **25**, 669–70.

FRIEDRICH-FREKSA, H. 1940 Bei der Chromosomenkonjugation wirksame Kräfte und ihre Bedeutung für die identische Verdopplung von Nucleo-proteinen. *Naturwissenschaften*, **28**, 376–379.

GILMAN, A. & PHILIPS, F. S. 1946 The biological action and therapeutic application of the B-chloroethyl amines and sulfides. *Science* **103**, 409–415 & 436.

GRIFFITH, F. 1928 The significance of pneumococcal types. *J. Hyg., Camb.,* **27**, 113–159.

HALDANE, J. B. S. 1927 A mathematical theory of natural and artificial selection. *Proc. Camb. Phil. Soc.* **23**, 838–844.

HALDANE, J. B. S. 1935 The rate of spontaneous mutation of a human gene. *J. Genet.* **31**, 317–326.

HEITZ, E. & BAUER, H. 1933 Beweise für die Chromosomennatur der Kernschleifen in den Knäuelkernen von *Bibio hortulanus* L. (Cytologische Untersuchungen an Dipteren, I). *Z. Zellforsch.* **17**, 67–82.

HERRIOTT, R. M. 1938 Isolation, crystallization and properties of swine pepsinogen. *J. Gen. Physiol.* **21**, 501–540.

HERRIOTT, R. M. 1939 Kinetics of the formation of pepsin from swine pepsinogen and identification of an intermediate compound. *J. Gen. Physiol.* **22**, 65–78.

HERRIOTT, R. M., BARTZ, Q. R. & NORTHRUP, J. H. 1938 Transformation of swine pepsinogen into swine pepsin by chicken pepsin. *J. Gen. Physiol.* **21**, 575–582.

HOROWITZ, N. H. 1945 On the evolution of biochemical synthesis. *Proc. Nat. Acad. Sci., Wash.,* **31**, 153–157.

HUXLEY, J. 1942 *Evolution. The Modern Synthesis.* Harper and Bros.

JORDAN, P. 1938 Zur Frage einer spezifischen Anziehung zwischen Genmolekülen. *Phys. Z.* **39**, 711–714.

JORDAN, P. 1939 Zur Quanten-Biologie. *Biol. Zbl.* **59**, 1–39.

KOLTZOFF, N. K. 1934 The structure of the chromosomes in the salivary glands of *Drosophila. Science*, **80**, 312–313.

LAMB, A. B. 1908 A new explanation of the mechanism of mitosis. *J. Exp. Zool.* **5**, 27–33.

LINDEGREN, C. C. 1945 Mendelian and cytoplasmic inheritance in yeasts. *Ann. Mo. Bot. Gdn,* **32**, 107–123.

LINDEGREN, C. C. & BRIDGES, C. B. 1938 Is agglutination an explanation for the occurrence and for the chromomere-to-chromomere specificity of synapsis? *Science,* **87**, 510–511.

LIPMANN, F. 1941 Metabolic generation and utilization of phosphate bond energy. *Adv. Enzymol.* **1**, 99–162.

MCCLINTOCK, B. 1945 *Neurospora* I. Preliminary observations of the chromosomes of *Neurospora crassa. Amer. J. Bot.* **32**, 671–677.

MIESCHER, F. 1897 *Die histochemischen und physiologischen Arbeiten.* 2. Leipzig: Vogel.

MINCHIN, E. A. 1916 The evolution of the cell. *Amer. Nat.* **50**, 5–39, 106–119.

MULLER, H. J. 1916 The mechanism of crossing-over. *Amer. Nat.* **50**, 193–221, 284–305, 350–366, 421–434.

MULLER, H. J. 1918 Genetic variability, twin hybrids and constant hybrids, in a case of balanced lethal factors. *Genetics,* **3**, 422–499.

MULLER, H. J. 1921*a* Variations due to change in the individual gene. Read before Amer. Soc. Nat. Toronto, Dec. 29. *Amer. Nat.* **56**, 32–50 (1922).

MULLER, H. J. 1921*b* Mutation. Read before Int. Eugenics Congr., New York. Publ. in *Eugenics, Genetics, and the Family,* **1**, 106–112 (1923).

MULLER, H. J. 1926 The gene as the basis of life. *Proc. 4th Int. Congr. Plant Sci. (Ithaca),* **1**, 879–921 (publ. 1929).

MULLER, H. J. 1927 The problem of genic modification. *Ver. V. int. Kongr. Vererbungswiss. Z. indukt. Abstamm.-u. VererbLehre,* Suppl. **1**, 234–260 (publ. 1928).

MULLER, H. J. 1928 The measurement of gene mutation rate in *Drosophila*, its high variability, and its dependence upon temperature. *Genetics,* **13**, 279–357.

MULLER, H. J. 1932 Further studies on the nature and causes of gene mutations. *Proc. 6th int. Congr. Genet. (Ithaca),* **1**, 213–255.

MULLER, H. J. 1933 The effects of Roentgen rays upon the hereditary material. *The science of radiology,* pp. 305–318. Springfield, Ill.: Charles C. Thomas.

MULLER, H. J. 1935*a* On the dimensions of chromosomes and genes in Dipteran salivary glands. *Amer. Nat.* **69**, 405–411.

MULLER, H. J. 1935*b* The status of the mutation theory in 1935. Read at de Vries Memorial Meeting, Leningrad,

Nov. 1935. Publ. in *Pravda*, no. 6, pp. 40–50 (1936), and under title 'The present status of the mutation theory', in *Curr. Sci.* Special no., March, pp. 4–15 (1938).

MULLER, H. J. 1935*c* The origination of chromatin deficiencies as minute deletions subject to insertion elsewhere. *Genetica*, 17, 237–252.

MULLER, H. J. 1935*d* The position effect as evidence of the localization of the immediate products of gene activity. *Summ. Commun. XV Int. Physiol. Congr.* (Leningr.-Mosc.), 286–289; and *Proc. 15th Int. Physiol. Congr.* (Leningr.-Mosc.), pp. 587–589 (1938).

MULLER, H. J. 1941 Resumé and perspectives of the symposium on genes and chromosomes. *Cold Spr. Harb. Symp. Quant. Biol.* 9, 290–308.

MULLER, H. J. & ALTENBURG, E. 1919 The rate of change of hereditary factors in *Drosophila. Proc. Soc. Exp. Biol., N.Y.*, 17, 10–14.

MULLER, H. J. & PROKOFYEVA, A. A. 1934 Continuity and discontinuity of the hereditary material. *C.R. Acad. Sci. U.R.S.S.*, N.S., 4, 74–83. (Reprinted in revised form, 1935, under title: 'The individual gene in relation to the chromomere and the chromosome', *Proc. Nat. Acad. Sci., Wash.*, 21, 16–26.)

NOUJDIN, N. I. 1936 Influence of the Y-chromosome and of the homologous region of the *X* on mosaicism in *Drosophila. Nature*, 137, 319–320.

NOUJDIN, N. I. 1938 A study of mosaicism of the eversporting displacement type in *Drosophila melanogaster. Bull. Biol. Med. Exp. U.R.S.S.* 5, 548–551.

NOUJDIN, N. I. 1944 The regularities of the heterochromatin influence on mosaicism. (Russ. with Eng. sum.) *J. Gen. Biol.* (U.S.S.R.), 5 (no. 6), 357–389.

OLENOV, J. M. 1941 The mutational process in *Drosophila melanogaster* under avitaminous B_2 conditions. *Amer. Nat.* 75, 580–595.

OPARIN, A. I. 1938 *The origin of life.* New York: The Macmillan Co.

PAINTER, T. S. 1933 A new method for the study of chromosome rearrangements and the plotting of chromosome maps. *Science*, 78, 585–586.

PAULING, L. & DELBRÜCK, M. 1940 The nature of the intermolecular forces operative in biological processes. *Science*, 92, 77–79.

PIRIE, N. W. 1945 Physical and chemical properties of tomato bushy stunt virus and the strains of tobacco mosaic virus. *Adv. Enzymol.* 5, 1–30.

PRICE, B. 1939 An interpretation of differential birth-rate statistics. *Proc. 7th Int. Congr. Genet.* pp. 241–242.

PROKOFYEVA-BELGOVSKAYA, A. A. 1945 Heterochromatization as a change in the chromosome cycle. (Russ. with Eng. sum.) *J. Gen. Biol.* (*U.S.S.R.*), 6, no. 2, 93–124.

RIS, H. & CROUSE, H. 1945 Structure of the salivary gland chromosomes of Diptera. *Proc. Nat. Acad. Sci., Wash.*, 31, 321–327.

SACHAROV, W. W. 1939 The mutation process in ageing sperm of *Drosophila melanogaster* and the problem of the specificity of the action of the factors of mutation. (Submitted to 7th Int. Genet. Congr. 1939) mim. 1941, *Dros. Inf. Serv.* 15, 37–38.

SCHRÖDINGER, E. 1945 *What is life?* Camb. Univ. Press.

SINGH, R. B. 1940 The influence of age and prolongation of larval life on the occurrence of spontaneous mutations in *Drosophila.* Univ. of Edinburgh, Ph.D. thesis (Typed MS.).

SONNEBORN, T. M. 1943*a* Gene and cytoplasm. I. The determination and inheritance of the killer character in variety 4 of *Paramecium aurelia. Proc. Nat. Acad. Sci., Wash.*, 29, 329–338.

SONNEBORN, T. M. 1943*b* Gene and cytoplasm. II. The bearing of determination and inheritance of characters in *Paramecium aurelia* on the problems of cytoplasmic inheritance, *Pneumococcus* transformation, mutation and development. *Proc. Nat. Acad. Sci., Wash.,* **29**, 338–343.

SONNEBORN, T. M. 1943c Acquired immunity to a specific antibody and its inheritance in *Paramecium aurelia. Proc. Ind. Acad. Sci.* **52**, 190–191.

SONNEBORN, T. M. 1945a Gene action in *Paramecium. Ann. Mo. Bot. Gdn,* **32**, 213–221.

SONNEBORN, T. M. 1945b The dependence of the physiological action of a gene on a primer and the relation of primer to gene. *Amer. Nat.* **79**, 318–339.

SPIEGELMAN, S. 1945 The physiology and genetic significance of enzymatic adaptation. *Ann. Mo. Bot. Gdn,* **32**, 139–163.

STANLEY, W. M. & KNIGHT, C. A. 1941 The chemical composition of strains of tobacco mosaic virus. *Cold Spr. Harb. Symp. Quant. Biol.* **9**, 255–262.

STEDMAN, E. & STEDMAN, E. 1943 Distribution of nucleic acid in the cell. *Nature,* **152**, 503–504.

STURTEVANT, A. H. 1944 Can specific mutations be induced by serological methods? *Proc. Nat. Acad. Sci., Wash.,* **30**, 176–178.

TIMOFÉEFF-RESSOVSKY, N. W., ZIMMER, K. G. & DELBRÜCK, M. 1935 Über die Natur der Genmutation und der Genstruktur. Vierter Teil: Theorie der Genmutation und der Genstruktur. *Nachr. Ges. Wiss. Göttingen* (Math.-phys. Kl., Biol.), N.F., **1**, 234–241.

TROLAND, L. T. 1914 The chemical origin and regulation of life. *Monist,* **22**, 92–134.

TROLAND, L. T. 1916 The enzyme theory of life. *Cleveland Med. J.* **15**, 377–387.

TROLAND, L. T. 1917 Biological enigmas and the theory of enzyme action. *Amer. Nat.* **51**, 321–350.

DE VRIES, H. 1899 *Intracellulare Pangenesis.* Jena.

WILSON, E. B. 1896 *The cell in development and inheritance.* New York: Columbia Univ. Press.

WINGE, O. & LAUSTSEN, O. 1940 On a cytoplasmic effect of imbreeding in homozygous yeast. *C. R. Ser. Physiol.* **23**, 17–39.

WRIGHT, S. 1931 Evolution in Mendelian populations. *Genetics,* **16**, 97–159.

ZUITIN, A. I. & PAVLOVETZ, M. T. 1938 Age differences in spontaneous mutation in males of *Drosophila melanogaster* of different origin. *C.R. Acad. Sci. U.R.S.S.* N.S., **21**, 50–52.

BIBLIOGRAPHY OF OTHER RELEVANT LITERATURE

ALTENBURG, E. 1930 The effect of ultra-violet radiation on mutation. *Anat. Rec.* **47**, 383.

ALTENBURG, E. & MULLER, H. J. 1920 The genetic basis of truncate wing—an inconstant and modifiable character in *Drosophila. Genetics,* **5**, 1–59.

BERNAL, J. D. 1940 Structural units in cellular physiology. *The Cell and Protoplasm,* pp. 199–205. Science Press.

BJERKNES, F. V. 1900–2 *Vorlesungen über hydrodynamische Fernkräfte nach C. A. Bjerknes' Theorie,* 2 vols. Leipzig.

BLACKWOOD, O. 1931 X-ray evidence as to the size of a gene. *Phys. Rev.* **37**, 1698.

CASPERSSON, T. 1939 On the role of the nucleic acids in the cell. *Proc 7th Inter. Genet. Congr.* pp. 85–86.

DARLINGTON, C. D. 1932 *Recent advances in cytology.* With a foreword by J. B. S. Haldane. London: J. and A. Churchill Ltd. (2nd ed. 1937.)

DARLINGTON, C. D. 1937 The biology of crossing-over. *Nature,* **140**, 759–761.

DARLINGTON, C. D. 1939 *The evolution of genetic systems.* Cambridge University Press.

DARLINGTON, C. D. 1942 Chromosome chemistry and gene action. *Nature,* **149**, 66–68.

DARLINGTON, C. D. 1945 The chemical basis of heredity and development. *Discovery* (March no.), pp. 79–86.

DARLINGTON, C. D. & LA COUR, L. F. 1945 Chromosome breakage and the nucleic acid cycle. *J. Genet.* **46**, 180–267.

DELBRÜCK, M. 1944 Problems of modern biology in relation to atomic physics.

Lectures at Vanderbilt Univ. School of Medicine. (Mimeographed.)

DEMEREC, M. 1933 What is a gene? *J. Hered.* 24, 369–378.

DEMEREC, M. 1935 Role of genes in evolution. *Amer. Nat.* 69, 125–138.

DEMEREC, M. 1938 Eighteen years of research on the gene. *Publ. Carneg. Instn*, no. 501, pp. 295–314.

DRIESCH, H. 1894 *Analytische Theorie der organischen Entwicklung.* Leipzig.

ELLENHORN, J., PROKOFYEVA, A. A. & MULLER, H. J. 1935 The optical dissociation of *Drosophila* chromomeres by means of ultraviolet light. *C.R. Acad Sci. U.R.S.S.*, N.S., 1, 234–242.

ENGELHARDT, W. A. & LJUBIMOVA, M. N. 1939 Myosin and adenosinetriphosphatase. *Nature*, 144, 668–669.

GREENSTEIN, J. P. 1944 Nucleoproteins. *Adv. Prot. Chem.* 1, 209–287.

GREENSTEIN, J. P. & JENRETTE, W. V. 1941 Physical changes in thymonucleic acid induced by salts, tissue extracts, and ultraviolet irradiation. *Cold Spr. Harb. Symp. Quant. Biol.* 9, 236–254.

GULICK, A. 1938 What are the genes? I. The genetic and evolutionary picture. *Quart. Rev. Biol.* 13, 1–18.

GULICK, A. 1938 What are the genes? II. The physico-chemical picture; conclusions. *Quart. Rev. Biol.* 13, 140–168.

GULICK, A. 1941 The chemistry of the chromosomes. *Bot. Rev.* 7, 433–457.

GULICK, A. 1944 The chemical formulation of gene structure and gene action. *Adv. Enzymol.* 4, 1–39.

HALDANE, J. B. S. 1932 *The causes of evolution.* London: Harper and Bros.

HOLLANDER, A. 1939 Wave-length dependence of the production of mutations in fungus spores by monochromatic ultra-violet radiation. *Proc. 7th Int. Congr. Genet.* pp. 153–154.

HOROWITZ, N. H., BONNER, D., MITCHELL, H. K., TATUM, E. L. & BEADLE, G. W. 1945 Genic control of biochemical reactions in *Neurospora. Amer. Nat.* 79, 304–317.

KNAPP, E. & SCHREIBER, H. 1939 Quantitative Analyse der mutationsauslösen-

den Wirkung monochromatischen U.-V.-Lichtes in Spermatozoiden von *Sphaerocarpus. Proc. 7th Int. Congr. Genet.* pp. 175–176.

LEA, D. E. 1940 A radiation method for determining the number of genes in the X-chromosome of *Drosophila. J. Genet.* 39, 181–188.

LURIA, S. E. & DELBRÜCK, M. 1943 Mutations of bacteria from virus sensitivity to virus resistance. Genetics, 28, 491–511.

MARSHALL, W. W. & MULLER, H. J. 1917 The effect of long continued heterozygosis on a variable character in *Drosophila. J. Exp. Zool.* 22, 457–470.

MIRSKY, A. E. 1943 Chromosomes and nucleoproteins. *Adv. Enzymol.* 3, 1–34.

MÖGLICH, F. & SCHÖN, M. 1938 Zur Frage der Energiewanderung in Kristallen und Molekülkomplexen. *Naturwissenschaften*, 26, 199–200.

MORGAN, T. H. 1926 *The theory of the gene.* Yale Univ. Press.

MORGAN, T. H. & BRIDGES, C. B. 1919 The inheritance of a fluctuating character. *J. Gen. Physiol.* 1, 639–643.

MORGAN, T. H., STURTEVANT, A. H., MULLER, H. J. & BRIDGES, C. B. 1915 *The mechanism of Mendelian heredity* New York: Henry Holt and Co. (Rev. ed. 1923.)

MULLER, H. J. 1929 The method of evolution. *Sci. Mon.* 29, 481–505.

MULLER, H. J. 1932 Some genetic aspects of sex. *Amer. Nat.* 66, 118–138.

MULLER, H. J. 1934 Radiation genetics. (Abstr.) *Verh. 4 int. Kongr. Radiol. (Zürich)*, 2, 100–102.

MULLER, H. J. 1936 The need of physics in the attack on the fundamental problems of genetics. *Sci. Mon., N. Y.*, 44, 210–214.

MULLER, H. J. 1940 An analysis of the process of structural change in chromosomes of *Drosophila. J. Genet* 40, 1–66.

MULLER, H. J. 1946 Age in relation to the frequency of spontaneous mutations in *Drosophila. Yearb. Carneg.*

Instn. 1945 (in the Press). Amer. Phil. Soc. 1945, 150–153.

OLIVER, C. P. 1930 The effect of varying the duration of X-ray treatment upon the frequency of mutation. *Science*, **71**, 44–46.

PAINTER, T. S. 1934 A new method for the study of chromosome aberrations and the plotting of chromosome maps in *Drosophila melanogaster*. *Genetics*, **19**, 175–188.

PAULING, L., CAMPBELL, D. H. & PRESSMAN, D. 1934 The nature of the forces between antigen and antibody and of the precipitation reaction. *Physiol. Rev.* **23**, 203–219.

PLOUGH, H. H. 1941 Spontaneous mutability in *Drosophila. Cold Spr. Harb. Symp. Quant. Biol.* **9**, 127–136.

SCHULTZ, J. 1943 Physiological aspects of genetics. *Ann. Rev. Physiol.* **5**, 35–62.

SCHULTZ, J. 1944 The gene as a chemical unit. *Colloid chemistry; theoretical and applied*, **1**, 819–850, ed. by J. Alexander. Reinhold Publ. Corp.

STADLER, L. J. 1932 On the genetic nature of induced mutations in plants. *Proc. 6th Int. Congr. Genet.* **1**, 274–294.

STADLER, L. J. 1939 Genetic studies with ultra-violet radiation. *Proc. 7th Int. Congr. Genet.* pp. 269–276.

STURTEVANT, A. H. 1917 An analysis of the effect of selection. *Publ. Carneg. Instn*, no. 264, 68 pp.

TATUM, E. L. & BEADLE, G. W. 1945 Biochemical genetics of *Neurospora. Ann. Mo. Bot. Gdn*, **32**, 125–129.

TIMOFÉEFF-RESSOVSKY, N. W. 1934 The experimental production of mutations. *Biol. Rev.* **9**, 411–457.

TIMOFÉEFF-RESSOVSKY, N. W. 1935 Über die Natur der Genmutation und der Genstruktur. Erster Teil. Einige Tatsachen der Mutationsforschung. *Nachr. Ges. Wiss. Göttingen*, Math.-phys. Kl., Biol., N.F., **1**, 190–217.

TIMOFÉEFF-RESSOVSKY, N. W. 1937 Experimentelle Mutationsforschung in der Vererbungslehre. Beeinflussing der Erbanlagen durch Strahlung und andere Faktoren. *Wiss. Forsch. Ber. Naturw. Reihe*, **42**. Dresden and Leipzig: Theodor Steinkopff.

WADDINGTON, C. H. 1939 The physicochemical nature of the chromosome and the gene. *Amer. Nat.* **73**, 300–314.

WADDINGTON, C. H. 1939 *An introduction to modern genetics.* London.

WRIGHT, S. 1941 The physiology of the gene. *Physiol. Rev.* **21**, 487–527.

WRIGHT, S. 1945 Genes as physiological agents: general considerations. *Amer. Nat.* **79**, 298–303.

ZIMMER, K. G. 1935 Über die Natur der Genmutation under der Genstruktur. Zweiter Teil: Die Treffertheorie und ithre Beziehung zur Mutationsauslösing. *Nachr. Ges. Wiss. Göttingen*, Math.-phys. Kl., Biol., **1**, 217–223.

BIBLIOGRAPHY

Refer to the Preface for a list of classical papers included in other anthologies.

BABCOCK, ERNEST B. 1949. 'The development of fundamental concepts in the science of genetics.' *Portugaliae Acta Biologica.* Série A, Volume R. B. Goldschmidt. Pages 1–46.

BEADLE, G. W., and E. L. TATUM. 1941. 'Genetic control of biochemical reactions in Neurospora. *Proceedings of the National Academy of Sciences* 27: 499–506.

BEADLE, G. W. 1945. 'Biochemical genetics.' *Chemical Reviews* 37: 15–96.

BEADLE, G. W. 1946. 'The gene.' *Proceedings of the American Philosophical Society* 90: 422–431.

BEADLE, G. W. 1946. 'Genes and the chemistry of the organism.' *American Scientist* *34*: 31–53, 76.

BEADLE, G. W. 1951. 'Chemical genetics.' In *Genetics in the 20th Century*. Edited by L. C. Dunn. New York: Macmillan.

BEADLE, G. W. 1959. 'Genes and chemical reactions in Neurospora.' *Science 129*: 1715–1726. Nobel Prize Lecture.

BEADLE, GEORGE W. 1963. 'Genetics and modern biology.' Philadelphia: American Philosophical Society. *Memoirs, Volume 57*.

CARLSON, ELOF AXEL. 1966. *The Gene: A Critical History*. Philadelphia: W. B. Saunders.

DEMEREC, M. 1933. 'What is a gene?' *Journal of Heredity 24*: 369–378.

DEMEREC, M. 1955. 'What is a gene?—twenty years later.' *American Naturalist 89*: 5–20.

DEMEREC, M. 1967. 'Properties of genes.' In *Heritage from Mendel*. Edited by R. Alexander Brink. Madison: University of Wisconsin Press.

DUNN, L. C. 1969. 'Genetics in historical perspective.' In *Genetic Organization*. Volume 1. Edited by Ernst W. Caspari and Arnold W. Ravin. New York: Academic Press.

EAST, E. M. 1929. 'The concept of the gene.' *Proceedings of the International Congress of Plant Sciences. Ithaca, New York. August 16–23, 1926*. Volume *1*: 889–895. Menasha, Wis. George Banta Publishing Co. (Compare Muller's treatment of the same topic in the same volume.)

EPHRUSSI, B. 1942. 'Chemistry of "eye-color hormones" of Drosophila.' *Quarterly Review of Biology 17*: 327–338.

GARROD, A. E. 1908. *Inborn Errors of Metabolism*. London: Oxford University Press. See also Harris, 1963.

HALDANE, J. B. S. 1954. *The Biochemistry of Genetics*. London: Allen and Unwin.

HARRIS, H. 1963. *Garrod's Inborn Errors of Metabolism*. London: Oxford University Press.

LEWIS, E. B. 1967. 'Genes and gene complexes.' In *Heritage from Mendel*. Edited by R. Alexander Brink. Madison: University of Wisconsin Press.

MORGAN, THOMAS HUNT. 1926. *The Theory of the Gene*. New Haven: Yale University Press. Reprinted 1964 by Hafner, New York.

MULLER, H. J. 1929. 'The gene as the basis of life.' *Proceedings of the International Congress of Plant Sciences. Ithaca, New York. August 16–23, 1926*. Volume *1*: 897–921. Menasha, Wis.: George Banta Publishing Co.

MULLER, H. J. 1951. 'The development of the gene theory.' In *Genetics in the 20th Century*. Edited by L. C. Dunn. New York: Macmillan.

MULLER, H. J. 1967. 'The gene material as the initiator and the organizing basis of life.' In *Heritage from Mendel*. Edited by R. Alexander Brink. Madison: University of Wisconsin Press. Also in *American Naturalist 100*: 493–517 (1966).

RUSSELL, E. S. 1930. *The Interpretation of Development and Heredity*. London: Oxford University Press.

SONNEBORN, T. M. 1968. 'H. J. Muller, crusader for human betterment.' *Science 162*: 772–776.

STADLER, L. J. 1954. 'The gene.' *Science 120*: 811–819.

STURTEVANT, A. H. 1948. 'The evolution and function of genes.' *American Scientist 36*: 225–236.

STURTEVANT, A. H., and G. W. BEADLE. 1939. *An Introduction to Genetics*. Philadelphia. W. B. Saunders. Reprinted by Dover Books, New York.

TATUM, EDWARD L. 1959. 'A case history in biological research.' *Science 129*: 1711–1715. Nobel Prize Lecture.

WHITEHOUSE, H. L. K. 1969. *Towards an Understanding of the Mechanism of Heredity*. New York: St. Martin's Press.

WRIGHT, SEWELL. 1941. 'The physiology of the gene.' *Physiological Reviews 21*: 487–527.

7/The Substance of Inheritance

Classical genetics established the fact that the vast majority of the characteristics of organisms are determined by genes, which could be thought of as substances that are parts of chromosomes. The chromosomal phenomena that occur during fertilization, mitosis, and meiosis provide a formal explanation of the rules of inheritance. These rules were established and found to be of near-universal applicability without the chemical nature of the hereditary substance being known.

OSWALD AVERY AND GUNTHER S. STENT

The next important advances in genetics began when the chemical nature of genes was determined. In 1943 Oswald Avery and his associates made observations suggesting that the hereditary substance of bacteria is deoxyribonucleic acid, or DNA. This opened a whole new field of genetics, which was pursued with utmost vigor, and by the close of the 1960's the most intimate secrets of life were understood in molecular terms.

Two selections will describe these remarkable events. The first is part of a letter that Oswald Avery wrote to his brother Roy, describing his discovery. The second is an article by Gunther Stent presenting the events in historical and intellectual perspective.

248

May 17, 1943. Dr. Gasser and Dr. Rivers have been very kind and have insisted on my staying on—providing me an ample budget and technical assistance to carry on the problem that I've been studying. I've not published anything about it—indeed have discussed it only with a few—because I'm not yet convinced that we have as yet sufficient evidence. However, I did talk to Ernest about it in Washington and I hope he has told you first of all— I felt he should know because it bears directly on my coming eventually to Nashville. It is the problem of the transformation of pneumococcal types.

You will recall that Griffith in London some fifteen years ago described a technique whereby he could change one specific type into another specific type through the intermediate R form. For example: Type II → R → Type III. This he accomplished by injecting mice with a large amount of *heat-killed* Type III cells together with a small inoculum of a *living R* culture derived from Type II. He noted that not infrequently the mice so treated died and from their heart blood he recovered living encapsulated Type III pneumococci. This he could accomplish only by the use of mice. He failed to obtain transformation when the *same* bacterial mixture was incubated in broth. Griffith's original observations were repeated and confirmed both in our lab and abroad by Neufeld and others. Then you remember Dawson with us reproduced the phenomenon *in vitro* by adding a dash of anti-R serum to the broth culture. Later Alloway used *filtered extract* prepared from Type III cells and in

the absence of formed elements and cellular debris induced the R culture derived from Type II to become typical encapsulated Type III pneumococcus. This, you may remember, involved several and repeated transfers in serum broth—often as many as 5–6—before the change occurred. But it did occur and once the reaction was induced, thereafter without further addition of the inducing extract, the organisms continued to produce the Type III capsule; that is, the change was hereditary and transmissible in serum in plain broth thereafter. For the past two years, first with MacLeod and now with Dr. McCarty I have been trying to find out what is the chemical nature of the substance in the bacterial extract which induces this specific change. The crude extract (Type III) is full of capsular polysaccharide, C (somatic) carbohydrate, nucleoproteins, free nucleic acids of both the yeast and thymus type, lipids and other cell constituents. Try to find in that complex mixture the active principle! Try to isolate and chemically identify the particular substance that will by itself when brought into contact with the R cell derived from Type II cause it to elaborate Type III capsular polysaccharide, and to acquire all the aristocratic distinctions of the same specific type of cells as that from which the extract was prepared! Some job—full of headaches and heartbreaks. But at last *perhaps* we have it. The active substance is not digested by crystalline trypsin or chymotrypsin. It does not lose activity when treated with crystalline ribonuclease which specifically breaks down yeast nucleic acid. The

Professor Roy Avery of Vanderbilt University has kindly given permission to reprint this important letter.

Type III capsular polysaccharides can be removed by digestion with the specific Type III enzyme without loss of transforming activity of a potent extract. The lipids can be extracted from such extracts by alcohol and ether at $-12°C$. without impairing biological activity. The extract can be de-proteinized by Sevag method—shaking c̄ chloroform and amyl alcohol until protein-free and biuret-negative. When extracted, treated and purified to this extent, but still containing traces of protein, lots of C carbohydrate, and nucleic acids of both the yeast and thymus types are further treated by the dropwise addition of absolute ethyl alcohol, an interesting thing occurs. When alcohol reaches a concentration of about 9/10 volume there separates out a fibrous substance which on stirring the mixture wraps itself about the glass rod—like thread on a spool—and the other impurities stay behind as granular precipitate. The fibrous material is redissolved and the process repeated several times. In short, this substance is highly reactive and on elementary analysis conforms *very* closely to the theoretical values of pure *desoxyribosenucleic* acid (thymus type). Who would have guessed it? This type of nucleic acid has not to my knowledge been recognized in pneumococcus before—though it has been found in other bacteria.

Of a number of crude enzyme preparations from rabbit bone, swine kidney, dog intestinal mucosa, and *pneumococci*, and fresh blood serum of human, dog, and rabbit, only those containing active depolymerase capable of breaking down known authentic samples of desoxyribose nucleic acid have been found to destroy the activity of our substance—indirect evidence but suggestive that the transforming principle as isolated may belong to this class of chemical substance. We have isolated highly purified substance of which as little as 0.02 of a *microgram* is active in inducing transformation—in the reaction mixture (culture medium) this represents a dilution of one part in a hundred million—potent stuff that—and highly specific. This does not leave much room for impurities—but the evidence is not good enough yet. In dilution of 1:1000 the substance is highly viscous as are authentic preparations of desoxyribose nucleic acid derived from fish sperm. Preliminary studies with the ultracentrifuge indicate a molecular weight of approximately 500,000 = a highly polymerized substance.

We are now planning to prepare a new batch and get further evidence of purity and homogeneity by use of ultracentrifuge and electrophoresis. This will keep me here for a while longer. If things go well I hope to go up to Deer Isle, rest awhile—come back refreshed and try to pick up the loose ends in the problem and write up the work. If we are right, and of course that's not yet proven, then it means that nucleic acids are *not* merely structurally important but functionally active substances in determining the biochemical activities and specific characteristics of cells—and that by means of a known chemical substance it is possible to induce *predictable* and *hereditary* changes in cells. This is something that has long been the dream of geneticists. The mutations they induced by X-ray and ultraviolet are always unpredictable, random, and chance changes. If we prove to be right—and of course it is a big if—then it means that both the chemical nature of the *inducing stimulus* is known and the chemical structure of the *substance produced* is also known—the former being thymus nucleic acid—the latter Type III polysaccharides, and both are thereafter reduplicated in the daughter cells—and after innumerable transfers and without further addition of the inducing agent, the same active and specific transforming substance can be recovered far in excess of the amount originally used to induce the reaction —sounds like a virus—may be a gene. But with such mechanisms I am not now concerned—one step at a time—and the first step is, what is the chemical nature of the transforming principle? Someone else can work out the rest. Of course the problem bristles with implications. It touches the biochemistry of thymus type of nucleic acids which are known to constitute the major part of chromosomes but have been thought to be alike regardless of origin and species. It touches genetics, enzyme chemistry, cell metabolism, and carbohydrate synthesis, etc. But today it will take a lot of well-documented evidence to convince anyone that the sodium salt of desoxyribose nucleic acid, protein-free, could possibly be endowed with such biologically active and specific properties, and this evidence we are now trying to get. It's lots of fun to blow bubbles, but it's wiser to prick them yourself before someone else tries to. So there's the story, Roy—right or wrong it's been good fun and lots of work. This supplemented by war work and general supervision of other important problems in the lab has kept me busy as you can well understand. Talk it over with Goodpasture but don't shout it around until we're quite sure or at least as sure as present methods permit. It's hazardous to go off half-cocked, and embarrassing to have to retract later. I'm so tired and sleepy I'm afraid I have not made this very clear. But I want you to know—am sure that you will see that I cannot well leave this problem until we've got convincing evidence. Then I look forward and hope we may all be together—God and the war permitting —and live out our days in peace.

The following article by Gunther S. Stent, of the University of California, Berkeley, was published in 1970.

DNA
The Golden Jubilee

In September of 1950 the Genetics Society of America held a symposium at Ohio State University to celebrate the fiftieth anniversary of the rediscovery of the work of Gregor Mendel, founder of the scientific study of heredity. The proceedings of that symposium were published under the title *Genetics in the 20th Century*, in the form of twenty-six essays written by some of the most eminent geneticists of the time.[1]* In his introduction

* Superscript numbers refer to notes at the end of this article.

This essay is dedicated in grateful affection to my teacher André Lwoff.

From Gunther Stent, 'DNA.' *Daedalus*, Journal of the American Academy of Arts and Sciences, Volume *99*, Number *4*, 1970. Reprinted by permission.

to these essays the editor, L. C. Dunn, of Columbia University, writes that their 'primary purpose was to survey the progress of the first fifty years of genetics and to exemplify the status of some of its problems today.' These problems, Dunn points out, extended beyond the confines of pure and applied biological science to the arena of the then ascendant Cold War: 'no one in 1950 can be unaware of the fact that the principles upon which genetics rest have been declared politically unacceptable in Russia and that the other communist countries generally have followed suit.' However, its alleged conflict with the tenets of dialectical materialism notwithstanding, 'genetics has become a many-sided body of knowledge and method dealing with questions which are recognized as of central importance in all efforts to understand living matter— how it perpetuates itself through reproduction, how it changes and adapts itself to its environment. Many of its principles have turned out to have a general character, so that not only do the rules apply to plants, as Mendel first found, but to animals of all kinds, to man himself, and to the whole world of microorganisms, bacteria and viruses, revealed since Mendel's time.' Finally, Dunn observes that 'in spite of its evident diversification, genetics has fortunately retained the essential unity given to it by [Mendel's] discovery of a fundamental element of heredity, the gene, so that varied problems can be stated in a common language which is becoming more generally understood.'

The achievements which the distinguished essayists celebrated form a body of knowledge that is nowadays generally referred to as *classical*

genetics. And it is precisely in the gene concept, which had provided its common language (and which had given such offense to primitive ideologues such as Trofim Lysenko), that classical genetics is to be differentiated from the *molecular genetics* that was to follow in its wake. The fundamental unit of classical genetics is an indivisible and abstract gene. In contrast, the fundamental unit of molecular genetics is a concrete chemical molecule, the nucleotide, with the gene being relegated to the role of a secondary unit aggregate comprising hundreds or thousands of such nucleotides.*

---

* The draft of this paper referred to the classical gene as being not only an indivisible and abstract unit but also a *transcendental* unit. In the Bellagio conference discussion my use of all three of these adjectives was criticized. So far as 'transcendental' is concerned, I have now eliminated it, as a possible source of confusion, even though I still think that its common (rather than Kantian) meaning, namely possessing attributes so fantastic as to be beyond ordinary comprehension, *is* applicable to the classical gene. As far as the other two adjectives are concerned, Curt Stern expressed the view that the gene was 'indivisible' only because geneticists did not *know* how to divide it, and that it was not really 'abstract' because geneticists were perfectly aware that it had a material basis, genes having in fact been shown to reside on chromosomes. To me it appears, however, that both indivisibility and abstractness of the classical gene were fundamental epistemological qualities, which took their origin in the kind of operations by means of which the gene was then studied. From the classical purview, it would have been *meaningless* to divide the gene and *futile* to endow it with a concrete physical identity. For instance,

As perusal of the Golden Jubilee essays shows, the gene concept had remained largely devoid of any material content for the fifty years following the rediscovery of Mendel's work. Besides not having fathomed its physical nature, classical geneticists had been unable to explain how the gene manages to preside over specific cellular physiological processes from its nuclear throne, or how it manages to achieve its own faithful replication in the cellular reproductive cycle. Herman J. Muller of Indiana University, then one of the elder statesmen of genetics, epitomizes that condition of classical genetics in his essay 'The Development of the Gene Theory' in these terms: 'the real core of genetic theory still appears to lie in the deep unknown. That is, we have as yet no actual knowledge of the mechanism underlying that unique property which makes a gene a gene—its ability to cause the synthesis of another structure like itself, in which even the mutations of the original gene are copied . . . What must happen is that just that precise reaction is *selectively* caused to occur, out of a virtually infinite series of possible reactions, whereby materials taken from a common medium become synthesized into a pattern just like that of the structure which itself guides the reaction. We do not know of such things yet in

I remember that in 1949 there appeared in *Life* magazine an electron micrograph of an ultra-thin chromosome section, which purported to show the first picture of a gene. The patent speciousness of this claim at the time illustrates the basic cognitive difficulty that existed as recently as twenty years ago: how could one recognize a gene as a gene even if one happened to lay eyes on it?

chemistry.' But for the classical geneticist study of the detailed nature and physical identity of the gene, though undoubtedly of great intellectual interest, is not really an essential part of his work. His theories on the mechanics of heredity and the experimental predictions to which these theories lead are largely formal, and their success does not depend on the knowledge of structures at the submicroscopic, or molecular, level where the genes lie. Application of the adjective 'classical' to that first phase of genetic research is rather analogous to its well-established use in 'classical physics.' The fundamental and indivisible conceptual unit of nineteenth-century classical physics was the atom, which despite its unfathomed nature, had allowed very far-reaching insights into the macroscopic properties of matter. 'Atomic' twentieth-century physics later succeeded in explaining the nature of the atom in terms of subatomic phenomena, just as molecular genetics was to succeed in explaining the molecular nature of the gene in terms of subgenic phenomena.*

---

* At the Bellagio conference Erik Erikson wondered what is actually meant by 'classical.' He received a variety of responses to his query, such as that 'classical,' means 'of no concern to me,' and hence 'embalmed'; 'what you learn in school'; 'perfect in content and form'; 'simple and perfect in form'; and that it pinpoints dogmas from which young people are encouraged to make heretical departures. Léon Rosenfeld stated that for Bohr 'classical physics' meant perfection in physical description, and that the later atomic physics is 'nonclassical' in the sense that the language of everyday macroscopic experience can give only an imperfect description of the microscopic world of atoms. Dr. Rosenfeld's remark

One Gene–One Enzyme

Among the twenty-six Golden Jubilee essays there are only three which deal to any extent with matters that were of immediate relevance for the then-nascent molecular genetics. One of these essays, 'Genetic Studies With Bacteria,' was written by Joshua Lederberg, at the time a twenty-five-year-old professor-prodigy in the University of Wisconsin. Lederberg reviews his recent work on bacterial sexuality, which he and Edward L. Tatum had discovered just four years earlier.[2] His discussion of bacterial genetics is still entirely 'classical,' in that the gene is treated as the fundamental unit responsible for the phenomena under study. Indeed, Lederberg takes some pains to make clear that the genetic mechanisms of the lowly bacteria can be understood from the classical viewpoint developed through study of higher forms. His essay pertains to molecular genetics only, insofar as it describes what was to become one of the chief 'molecular' experimental systems, the sexually fertile bacterium *Escherichia coli* K12 isolated at Stanford University 'in the fall of 1922 from the stools of a diphtheria convalescent.'

George W. Beadle, then Thomas Hunt Morgan's successor as chairman of the Biology Division at the California Institute of Technology, contributed an essay entitled 'Chemical

makes the correspondence between classical genetics even closer than I had previously thought. For the macroscopic character differences on which the concept of the Mendelian gene was based are likewise closer to our everyday experience than the microscopic nucleotide unit of molecular genetics.

Genetics,' which traces the development of the concept that genes preside over cellular function by controlling the chemical reactions of cell metabolism. Beadle describes how in the 1930's he and Boris Ephrussi (who also contributed an essay, concerned mainly with extranuclear inheritance) worked on the genetic control of the formation of the eye color of the *Drosophila* fruit fly and how he finally became discouraged over the difficulties encountered with that material. In 1940 he and E. L. Tatum turned their attention to a more favorable organism, the bread mold *Neurospora*. According to Beadle: 'With the new organism our approach could be basically different. Through control of the constituents of the culture medium we could search for mutations in genes concerned with the synthesis of already known chemical substances of biological importance. We soon found ourselves with so many mutant strains unable to synthesize vitamins, amino acids and other essential components of protoplasm that we could not decide which ones to work on first.' Beadle reports that in the intervening ten years he, Tatum, and their collaborators managed to analyze the genetic and biochemical characteristics of enough *Neurospora* mutants to lend strong support for the 'one-gene–one-enzyme,' hypothesis.[3] (In his essay Beadle actually prefers the name 'one-gene–one-function' hypothesis, but it was the former name which became popular in the community of geneticists.) This hypothesis states that each gene has only one primary function, which in most or all cases is to direct the synthesis of one and only one enzyme, and thus to control one single chemical reaction catalyzed by that one enzyme. Though (as Beadle emphasizes) the idea that genes control single functions was not really original with him or Tatum, there can be little doubt that their clear formulation and strong experimental evidence for that hypothesis had a profound impact on subsequent thought about the gene. It must be noted, however, that the gene of Beadle's essay is still the classical, indivisible, abstract unit. But in promulgating the belief that each gene is doing only one thing, the one-gene–one-enzyme hypothesis gave hope of ultimately being able to find out how that thing is done.

The third essay touching on molecular genetics was one by Alfred E. Mirsky of the Rockefeller Institute for Medical Research, entitled 'Some Chemical Aspects of the Cell Nucleus.' Chemical study of the cell nucleus was begun in the 1860's by Mendel's contemporary, the Swiss chemist Friedrich Miescher,[4] before the notion had emerged that the cell nucleus is the seat of heredity. Miescher undertook an analysis of cells, such as pus cells and salmon sperm, in which he knew the nucleus to represent a large fraction of the total cell mass. These analyses revealed that the nucleus contains a hitherto unknown, phosphorous-rich, acid substance, to which a later worker gave the name *nucleic acid*.

By the turn of the century the ubiquitous presence of nucleic acid in plant and animal cells had been demonstrated, and the German biochemist Albrecht Kossel had identified the nucleic acid building blocks (Figure 1): the four nitrogenous bases *adenine, guanine, cytosine,* and *uracil* (the former two belonging to the class *purines* and the latter two to the class *pyrimidines*), a five-carbon sugar, and phosphoric acid. Further analytical work, largely by P. A. Levene[5] and by W. Jones in the 1920's, showed that there exist two fundamentally different kinds of nucleic acid, which are now called *ribonucleic acid, or RNA,* and *deoxyribonucleic acid,* or *DNA.* RNA contains *ribose*, whereas DNA contains *deoxyribose* as its five-carbon sugar. DNA, furthermore, does not contain the pyrimidine uracil; instead of uracil it contains 5-methyl uracil, or thymine. Nitrogenous base, sugar, and phosphoric acid were found to be linked to form a *nucleotide*. By the 1930's it had been shown that nucleic acid molecules contain several such nucleotides linked through phosphate diester bonds between their ribose or deoxyribose sugar molecules (Figure 1). Another ten years had to elapse before the enormously high chain length of nucleic acids was finally appreciated. As we now know, nucleic acid molecules represent *polynucleotide* chains of thousands, and sometimes millions of nucleotides in continuous chemical linkage.

Mirsky's essay is concerned mainly with the biological significance of the DNA type of nucleic acid. It sets forth how the invention in 1924 by R. Feulgen[6] of a specific color stain for DNA made it possible to show that the cell DNA is located almost exclusively in the *chromosomes* of the nucleus, whereas the cell RNA, despite its being a 'nucleic acid,' is located mainly in the cytoplasm. Since the chromosomes are the cell organelles in which the genes were known to reside, it did not seem farfetched to imagine that DNA plays some important role in hereditary processes. But as the chromosomes contain even

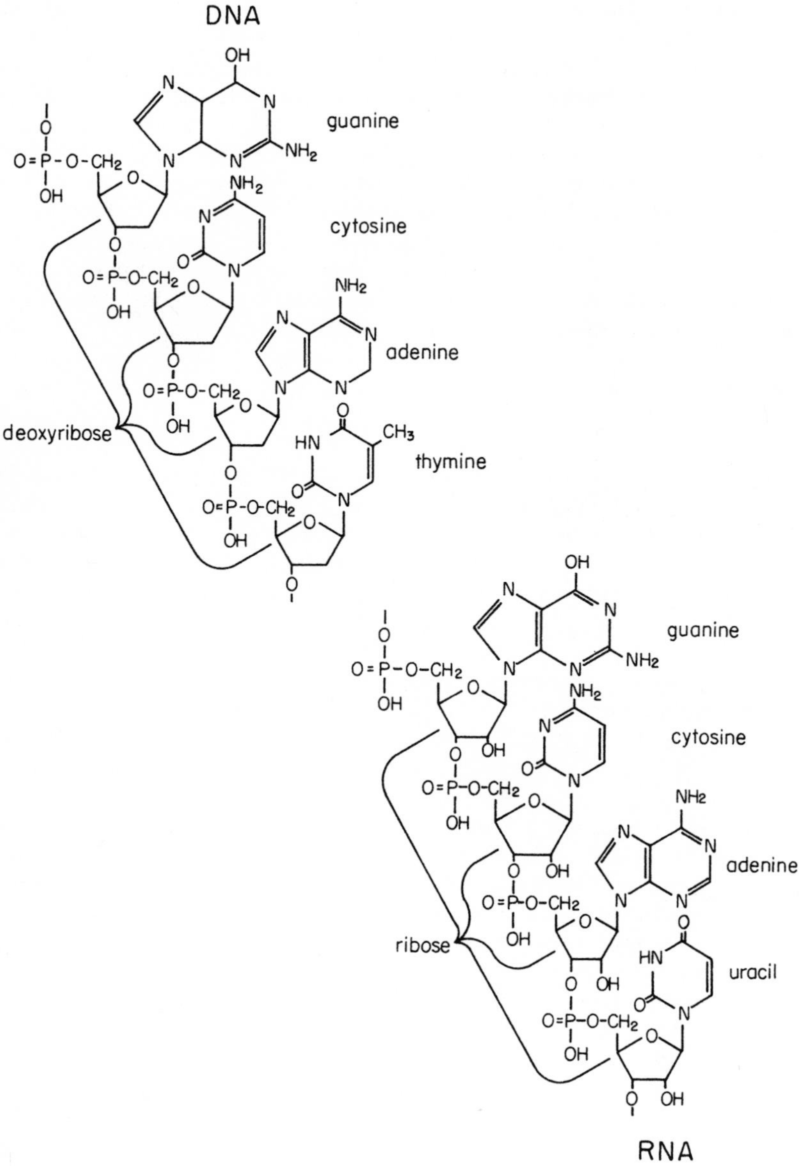

FIG. 1. The polynucleotide chains of deoxyribonucleic acid, or DNA, and of ribonucleic acid, or RNA.

more protein than DNA, it was not necessary to infer that the genes actually contain any DNA. In support of the view that 'DNA is part of the gene substance,' however, Mirsky cites his own observations that 'in the different cells of an organism the quantity of DNA for each haploid set of chromosomes is constant . . . constancy per cell is certainly an unusual characteristic for chemical components of cells . . . Even a substance, such as RNA, present in all cells, varies greatly in amount in different cells.'

As far as the chemistry of DNA is concerned, Mirsky announces the recent demise of the *tetranucleotide theory*, 'according to which a polynucleotide containing one of each of the four bases was considered to be a fundamental unit in DNA. Using chromatographic procedures for analysis [Erwin] Chargaff and his colleagues have shown that the four bases [adenine, guanine, cytosine, and thymine] are not present in equimolar proportions [as had been previously believed], and this has removed whatever experimental basis there was for the tetranucleotide theory.' Furthermore, 'Chargaff and his colleagues have analyzed the DNA's prepared from a number of different sources and have obtained results showing that in preparations from different organisms the ratios of the bases are different, although they are the same in preparations from different tissues of the same organism. It is highly probable, therefore, that the proportion of the four bases differ in DNA's of various organisms.'

Mirsky does not, however, mention another seemingly less important finding which Chargaff presented in the very paper[7] cited in the essay. This finding is the DNA compositional equivalence rule, which states that although the proportion of the four bases differs in the DNA's of various organisms, it is nevertheless true that the molar proportion of adenine is always equal to that of thymine and the molar proportion of guanine is always equal to that of cytosine. Three years later, this rule was to provide a crucial clue for divining the structure of DNA.

Transforming Principle

In 1944, or six years before the Golden Jubilee, work had been published (and was certainly well known to most of the essayists) which proved that DNA is not only 'part of the gene substance' but *is* the gene substance. Yet Mirsky's essay is the only one of the twenty-six in which the implications of that work are discussed. (Lederberg's Golden Jubilee essay refers to it briefly as a promising development in bacterial genetics, and Beadle's essay devotes two sentences to it, saying that it 'has certainly introduced another chapter in genetics, and one that promises to be among the most exciting. It has given chemists new incentive to learn about the nucleic acids, compounds which everyone recognizes to be extremely important biologically and about which so little is known.')

This work had been carried out by Oswald T. Avery[8] and his collaborators at the Rockefeller Institute and represented the identification of the active principle of *bacterial transformation*, a phenomenon first observed in 1928 by the British bacteriologist F. Griffith.[9] Avery could show that upon addition of purified DNA extracted from normal *donor* bacteria to mutant *recipient* bacteria, which differ from the donor bacteria in one mutant gene, some of the recipient bacteria are transformed hereditarily into the donor type. Thus the normal donor gene must have entered the transformed recipient bacterium in the form of a donor DNA molecule and there displaced its homologous mutant gene. Hence it followed that the bacterial DNA represents the bacterial genes. In 1944 this conclusion seemed so radical that even Avery himself was reluctant to accept it until he had buttressed his experiments with the most rigorous controls.

Avery's controls were evidently not rigorous enough for most contemporary geneticists, including his Rockefeller Institute colleague Mirsky, who was of the opinion that 'it is quite possible that DNA, and nothing else, is responsible for the transforming activity, but this has not been demonstrated conclusively. In purification of the active principle more and more of the protein attached to DNA is removed . . . It is difficult to eliminate the possibility that the minute quantities of protein that probably remain attached to DNA, though undetectable by the tests applied, are necessary for activity.' But Mirsky concedes that 'it can be regarded as established that DNA is at least part of the active [transforming] principle.' As we shall see presently, Avery's work, just as Mendel's, had been too far ahead of its time, so that for some eight years it had very little impact on genetic research. Although, in contrast to Mendel, Avery and his discovery were well known, the genetic role of DNA

had to be rediscovered in 1952 through work with bacteriophages.*

---

* In the Bellagio conference discussion Charles Weiner stated that he 'always recognizes it as a danger signal when someone says 'so and so was too far ahead of his time,' a discovery or a man. This is a method of obscuring what went on . . . to say that it was ahead of its time is, essentially, to take a shortcut through history.' Of course, pronouncing a discovery to have been ahead of its time just *because* it was not immediately appreciated is to make an empty tautology. But I am appealing here to another criterion of premature discovery that ought to illuminate rather than obscure what went on. By this criterion a discovery is 'ahead of its time' if the inferences to which it leads cannot be connected by a series of simple logical steps with contemporary canonical knowledge. Thus, Curt Stern's essay shows that Mendel was 'ahead of his time' because the statistical laws which Mendel found to govern heredity could not be connected with the then known aspects of anatomy and physiology; and I am trying to show in the following just why it was difficult in 1944 to connect DNA with the gene. Lest it be concluded that the judgment of prematurity can be rendered only with hindsight, I shall provide an example of a recent discovery that falls within the purview of this essay and can be judged as being ahead of the *present* time. Three or four years ago there appeared reports purporting to have shown that the memory of a task learned by a trained donor animal can be transferred to a naïve recipient animal through the vehicle of nucleic acid molecules. Now whereas it is quite generally appreciated that the possibility of such memory transfer would, in fact, constitute a fact of capital importance for our understanding of the higher nervous system, these reports have so far remained without heuristic effect on brain research. For

Bacteriophages, or *phages* as they are called in the trade, are subcellular parasites of bacteria that were discovered in 1917.[10] They occupy only about one-thousandth of the volume of their bacterial host cell and hence are so small that they cannot be seen in ordinary microscopes using visible light. Phages were first seen in 1940, in the wake of the development of the electron microscope, and they were found to be tadpole-shaped particles having a head and a tail. (It was eventually established that the tail is composed of protein and that the head represents a stuffed bag whose casing is protein and whose stuffing is DNA.)

In 1938 the physicist Max Delbrück, then a postdoctoral fellow at the California Institute of Technology, started experimenting with phages, in the ex-

pectation that study of their self-replication might throw light on what Muller's essay described as 'that unique property which makes a gene a gene—its ability to cause the synthesis of another structure like itself, in which even the mutations of the original gene are copied.' Delbrück's work began with designing the *one-step-growth experiment*, in which he used as his experimental material a phage active on *Escherichia coli*.[11] The one-step-growth experiment showed that each phage particle infecting an *E. coli* bacterium gives rise to some hundred phage progeny particles after a brief half-hour latent period. Thus this experiment brought clearly into focus the central problem of self-replication: how does the parental phage particle manage to produce its crop of a hundred progeny during that half hour? Two years later Delbrück met Salvador Luria, then a recently arrived refugee from war-torn Europe, and Alfred Hershey, of Washington University in St. Louis. This meeting brought into being the American Phage Group, whose collective memory has been preserved in a series of autobiographical essays.[12] The members of this group were united by a single common goal—the desire to reach what Muller referred to as 'the real core of genetic theory,' or to extend the frame of reference of genetics beyond the billiard ball gene.

Although the intellectual foundations were laid by the Phage Group for the edifice of molecular genetics during the next dozen years, the first real breakthrough came only in 1952 with an experiment by Hershey and his young assistant, Martha Chase.[13] Hershey and Chase showed by use of phage particles labeled with radio-

---

there is no chain of reasonable inferences by means of which our present, albeit very imperfect, view of the functional organization of the brain can be reconciled with the possibility of its acquisition, storage, and retrieval of experiential information through nucleic acid molecules. Thus for the community of neurobiologists there is no point in paying serious attention to these claims, or even to spend any time on checking whether they are true or false. This attitude may, at first glance, appear to be 'unscientific,' but it is, in fact, the very way in which science has to operate. Or, as Arthur Eddington advised his fellow scientists, it is not a good policy to put overmuch confidence in facts until they have been proven by theory. This appears to be also the view taken by Michael Polanyi (*Science*, 141 [1963], 1010) who showed that a premature discovery of his own was understandably ignored for forty years because during all that time it could not be connected with the then orthodox theories of physical chemistry.

phosphorus ^{32}P in their DNA and with radiosulfur ^{35}S in their protein that at the outset of phage infection of the *E. coli* cell only the DNA of the phage actually enters the cell; the protein of the phage remains outside, devoid of any further function in the reproductive drama about to ensue within. Thus it could be concluded that the genes of the parent phage responsible for directing the synthesis of progeny phages reside in its DNA. This second demonstration that DNA is the hereditary material had an immediate and profound impact on genetic thought.

Why did Avery's announcement that DNA is the genetic material have in its day so much less effect on genetic thought than the Hershey-Chase experiment proving the same point eight years later (with much less compelling evidence, it might be noted)? First, it was only in the late 1940's that, thanks to the pioneering efforts of Luria and Delbrück,[14] bacteria and phages came to be accepted as genuine genetic organisms to which the gene concept could be legitimately applied (Lederberg's Golden Jubilee essay still has the flavor of a missionary effort to spread that gospel). So, in 1944 many people regarded bacterial transformation as some bizarre metamorphosis without relevance to hereditary processes in higher forms. Further, and more important, the teranucleotide theory of DNA structure still held sway, so that it was very difficult to imagine how a DNA molecule made up of monotonously repeating units containing one each of the four bases *could* be the carrier of *genetic information*. Thus even those persons who were prepared to accept in the mid-1940's the conclu-

sions that bacterial transformation is a truly hereditary process and that DNA is really the transforming principle were wont to consider the phenomenon as a case of gene *mutation*. That is to say, they imagined that DNA is a chemical capable of causing specific or directed mutations rather than being the gene substance itself. But the demise of the tetranucleotide theory announced in Mirsky's essay now meant that the four types of bases can follow each other in any arbitrary order in the polynucleotide chain. Since the base composition was found to be different in DNA samples obtained from different organisms, it could be supposed at the time of the Hershey-Chase experiment that any given DNA molecule harbors its genetic information in the form of a precise sequence of the four bases along its polynucleotide chain.

With the acceptance of the parental phage DNA as the genetic material and the birth of the notion that genetic information is encoded into DNA as a nucleotide base sequence, the fundamental problem of biological self-replication could now be restated in terms of two DNA functions, *autocatalytic* and *heterocatalytic*. By means of the autocatalytic function, the phage DNA replicates its own precise nucleotide base sequence a hundredfold to generate the genes with which its progeny phages are to be endowed. And by means of the heterocatalytic function, the phage DNA directs, or presides over, the synthesis of the phage-specific proteins that are to furnish the body of its progeny. The successful elucidation of these two DNA functions was to be the work of the next decade.

The Double Helix

In 1951, one year after the Golden Jubilee, Linus Pauling, of the California Institute of Technology, discovered the basic structure of protein molecules.[15] Proteins, we might recall here briefly, are long chain molecules, built up of a sequence of twenty different kinds of *amino acids*. These amino acids are joined to each other through a chemical linkage called the *peptide* bond. The length of different kinds of protein chains present in living cells varies considerably, but on the average these chains contain about three hundred amino acids linked end-to-end. Pauling found that the three-dimensional conformation of the amino acid chain is a helix, to which he gave the name α *helix*. The discovery of the α helix represented the first great achievement in the attempt to work out protein structure by the methods of X-ray crystallography.

Pauling's success inspired James D. Watson, a twenty-two-year-old member of the Phage Group and a pupil of Luria's, to abandon the genetic and physiological experiments on phage reproduction that he had been carrying out. Watson instead decided to try to work out the three-dimensional structure of the DNA molecule, which, a few months after his decision, the Hershey-Chase experiment showed to be the carrier of the phage genes. To gain the necessary skills in X-ray crystallography, Watson joined John C. Kendrew of the University of Cambridge, who, like Pauling, was studying protein structure. In Cambridge, Watson met Francis Crick, a Ph.D. student, to whom it had also occurred that the three-dimensional structure of DNA would be likely to provide important insights into the nature of the gene. Watson and Crick then began a collaboration whose story is now so well known through Watson's famous autobiographical account[16] that there is little need to retell it here. Suffice it to say that in the spring of 1953 Watson and Crick discovered that the DNA molecule is a *double helix*, composed of two intertwined polynucleotide chains (Figure 2). The DNA double helix is self-complementary, in that to each adenine nucleotide on one chain there corresponds a thymine nucleotide on the other chain, and that to each guanine nucleotide on one chain there corresponds a cytosine nucleotide on the other chain. The specificity of this complementary relation devolves from hydrogen bonds formed between two opposite nucleotides, adenine-thymine and guanine-cytosine, at each step of the double helical molecule (Figure 3). The complementary base-pairing relation of the two chains thus accounted for the base composition equivalence rule noticed by Chargaff three years earlier.

On first sight, Watson's and Crick's discovery of the double-helical structure of DNA resembled Pauling's then two-year-old discovery of the α helix, particularly since the formation of specific hydrogen bonds also plays an important role in Pauling's structure. But, on second sight, the promulgation of the DNA double helix emerges as an event of a qualitatively different heuristic character. First, in working out the structure of the double helix Watson and Crick had for the first time introduced genetic reasoning into structural determination, by demanding that the evidently highly regular three-dimensional struc-

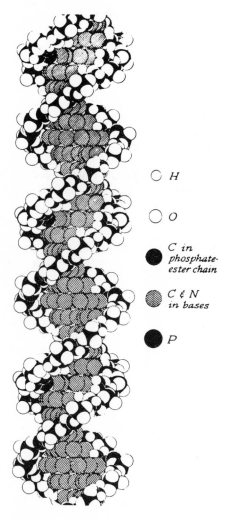

○ *H*

○ *O*

● *C in*
phosphate-
ester chain

◐ *C & N*
in bases

● *P*

FIG. 2. A space-filling model of the Watson-Crick double helix structure of DNA. (From G. S. Stent, *Molecular Biology of Bacterial Viruses*, San Francisco: W. H. Freeman. Copyright © 1963. Reprinted by permission.)

ture of DNA must be able to accommodate the informational aspect of an arbitrary nucleotide base sequence along the two polynucleotide strands.

Second, the discovery of the DNA double helix opened up enormous vistas to the imagination. It was to provide the high road to 'the real core of genetic theory' which, according to Muller, was still lying in the deep unknown at the time of the Golden Jubilee. Molecular genetics had now become a going concern.

Watson and Crick had concluded their letter to *Nature,*[17] in which they first described the DNA double helix, with a sentence that can surely lay claim to being one of the most coy statements in the literature of science: 'It has not escaped our notice that the specific [nucleotide base] pairing we have postulated immediately suggests a possible copying mechanism of the genetic material.' In a second letter to *Nature*[18] they soon told what it was that had not escaped their notice: the DNA molecule can achieve its autocatalytic function upon separation of the two helically intertwined, complementary polynucleotide strands (Figure 4). Each of the two parent strands would then serve as a *template* for the ordered assembly of its own complementary daughter strand, by having every nucleotide on the parent strand attract and line up for polynucleotide synthesis a free nucleotide carrying the complementary purine or pyrimidine base. Thus in the case of phage reproduction, the DNA of the infecting phage particle would undergo successive rounds of unwinding and complement addition. In this way an intrabacterial pool of replica phage DNA molecules would be built up, identical in nucleotide base sequence to the DNA of the parent phage; this pool would provide the genes for the offspring phages.

It took about five more years to

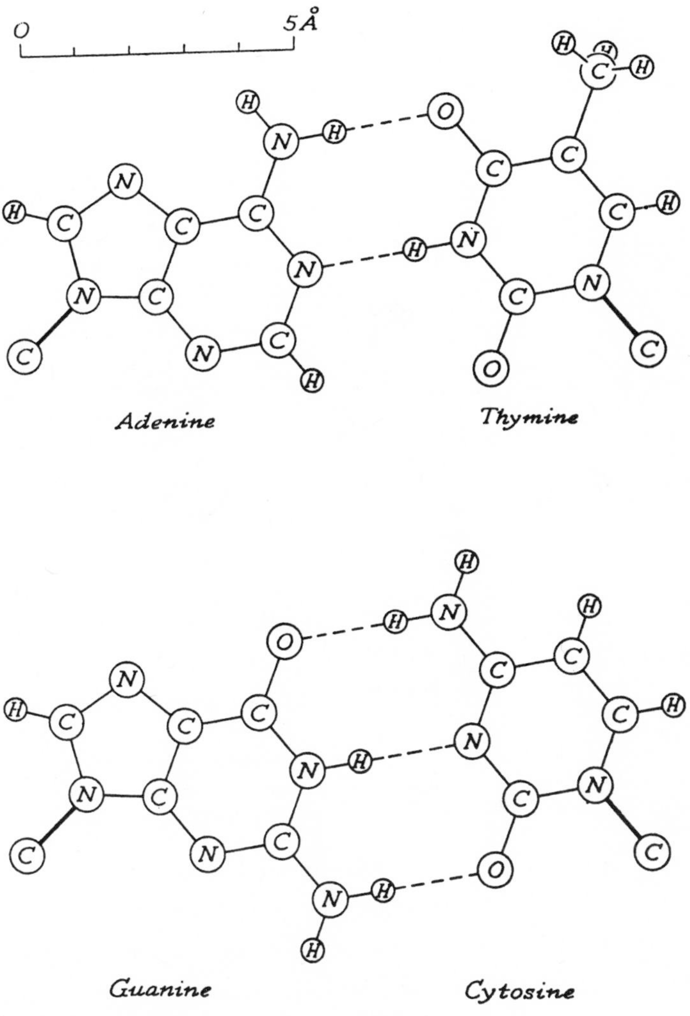

FIG. 3. The complementary base pairing of adenine and thymine, and of guanine and cytosine in the double-helical DNA molecule. Hydrogen bonds are shown as dashed lines. The carbon atom of the deoxyribose to which each base is attached is also shown. (From G. S. Stent, *Molecular Biology of Bacterial Viruses,* San Francisco: W. H. Freeman. Copyright © 1963. Reprinted by permission.)

prove that the solution to the problem of the autocatalytic function proposed by Watson and Crick is essentially correct. The main proof depended on the prediction[19] that under that proposal the atoms of the parental double helix ought to become distributed in a *semi-conservative* manner. That is to say, each of the two daughter molecules generated by fis-

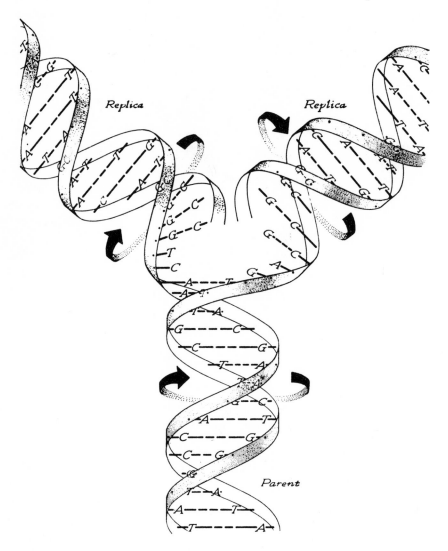

FIG. 4. Replication of the DNA double helix, according to the mechanism of Watson and Crick. (From G. S. Stent, *Molecular Biology of Bacterial Viruses*, San Francisco: W. H. Freeman. Copyright © 1963. Reprinted by permission.

sion of the parental DNA double helix should contain one polynucleotide chain of parental provenance and one chain synthesized *de novo*. In 1958, Matthew Meselson and Frank W. Stahl, one a graduate student of Pauling's and the other a postdoctoral student in Delbrück's laboratory at Caltech, managed to demonstrate that the DNA of *E. coli* is replicated semi-conservatively.[20] For the purpose of this demonstration, Meselson had adapted the method of *equilibrium density gradient sedimentation* to the detection of minute density differences of large molecules, particularly for measuring the density increase produced by the replacement of a common atomic species, such as nitrogen ^{14}N, by its rare heavier isotope, such as ^{15}N, in the molecules. This method was to become one of the most widely used techniques in later molecular biological research.

Reform of the Gene

Understanding of the heterocatalytic function of DNA, which from the very outset of its formulation appeared as a more complex problem than the autocatalytic function, required a rather greater effort and a somewhat longer time. First of all, it proved necessary to reform the concept of the classical gene. Today it seems impossible to make an accurate reconstruction of the historical course taken by this reform and to ascertain to whom each facet can be attributed. Many of the essential ideas were first proposed in informal discussions on both sides of the Atlantic and were then quickly broadcast to the cognoscenti by private international bush-telegraph. Months and often years elapsed before a new idea was committed to print, and then very often it was not by the person who had first thought of it. Nevertheless, it can hardly be doubted that here too Watson and Crick played a dominant role, as well as Seymour Benzer, then at Purdue University.

One of the points of departure of this reform was Beadle's and Tatum's one-gene–one-enzyme theory. By the early 1950's the analytical work of Frederick Sanger,[21] of the University of Cambridge, had lent direct support to the credence that a given species of enzyme represents a homogeneous class of protein molecules in which a definite number of the twenty different kinds of amino acids are assembled in a unique sequence. From this credence developed the so-called *sequence hypothesis*, which states that the exact spatial conformation of a protein molecule, and hence the specificity of its biologic function, is *wholly determined* by that unique amino acid sequence from which it is built. At the time that this hypothesis was advanced there was no proof whatsoever of its validity. But it came to be embraced immediately as molecular-genetic dogma, because it allowed rephrasing of the one-gene–one-enzyme theory in more concrete terms: the gene directs the synthesis of one enzyme by directing the assembly of the twenty kinds of amino acids into a protein molecule of given amino acid sequence. But since it was already taken for granted that the genetic information is held in the form of a particular nucleotide base sequence in DNA, it now became clear that the *meaning* of the particular nucleotide base sequence making up a sector of DNA corresponding to a gene could be nothing other than the specification of an amino acid sequence of the corresponding protein molecule. By means of incisive *fine structure* genetic experiments on a single gene of one of the *E. coli* phages, Benzer was able to gather convincing support for the notion that the gene is, in fact, a linear array of DNA nucleotides which determines a linear array of protein amino acids. It was Benzer more than anyone else who showed that the fundamental genetic unit is the DNA nucleotide base.[22]

Genetic Code

This reform of the gene concept led directly to the belief that there must exist a *genetic code* that relates the nucleotide base sequence in the DNA polynucleotide chain to the amino acid sequence of the corresponding enzyme protein. An obvious consideration quickly revealed that this code could be no simpler than one involving the specification of each amino acid in the protein molecule by at least *three* successive nucleotide bases in the DNA. Certainly there cannot exist a *one-to-one* correspondence between nucleotide bases in the DNA and amino acids in the protein molecule, because the four kinds of nucleotide bases taken *one* at a time could specify only one out of four, not one out of twenty, kinds of amino acids. Nor would it suffice that *two* adjacent nucleotide bases specify one amino acid, since in that case only $4 \times 4 = 16$ kinds of amino acids could be coded for by the four kinds of nucleotide bases. But four kinds of nucleotide bases taken *three* at a time provide $4 \times 4 \times 4 = 64$ different code words, or *codons*, and hence each of the twenty kinds of protein amino acids could be represented by at least one such codon in the genetic code, with the extra number of codons allowing for the possibility that each kind of amino acid is represented by more than a single codon.

These a priori insights had certainly been reached within a few months of the discovery of the DNA double helix and were first published in 1954 by the physicist-cosmologist George Gamow.[23] But it was not until 1961 that it was finally proven that the genetic code does involve a language in which a triplet of successive nucleotide bases in the DNA polynucleotide chain stands for one protein amino acid. That proof came from purely formal genetic experiments carried out by Crick[24] and his collaborators on the same phage gene which had figured in Benzer's earlier reform of the gene concept. As we shall see presently, the code was unexpectedly broken that same year.

Central Dogma

It was one thing to have formulated the general principles according to which genetic information is stored and replicated in the DNA. But it was quite another to work out the molecular mechanisms of the heterocatalytic function through which that information is realized as protein molecules. Here Watson and Crick also played a dominant role by formulating in the years 1953–1955 what came to be known as the *central dogma*. According to that dogma the heterocatalytic function is a *two-stage* process, in which the other type of nucleic acid, RNA, also becomes involved. In the first stage, the DNA molecule serves as the template for the synthesis of an RNA polynucleotide chain onto which the sequence of nucleotide bases in the DNA chain is *transcribed*. In the second stage, the RNA chain is then *translated* by the cellular machinery for protein synthesis into protein molecules of amino acid sequence specified via the genetic code. It is to be noted that an essential feature of the central dogma is the one-way flow of information,

DNA → RNA → protein,

a flow which is never reversed.

In order to study the processes envisaged by the central dogma it became necessary to employ the methods of biochemistry to open the black box containing the cellular machinery which actually effects the transcription-translation drama of the central dogma. One of the first insights then provided by the application of biochemical methods was the identification of the *ribosome* as the *site* of cellular protein synthesis.[25] The ribosome is a small particle present in vast numbers in all living cells (one *E. coli* bacterium contains about 15,000 ribosomes); its mass is composed of about one-third protein and two-thirds RNA. But how is the information for specific amino acid sequences encoded in the gene made available to the ribosome in its protein assembly process? In answer to this question it was proposed in 1961 by François Jacob and Jacques Monod,[26] of the Pasteur Institute, that the RNA onto which, according to the central dogma, the genetic nucleotide base sequence is first transcribed is a molecule of *messenger RNA* (Figure 5). This messenger RNA molecule is picked up by a ribosome, on whose surface then proceeds the translation of RNA nucleotide sequence into protein amino acid sequence, codon by codon. In this translation process, the messenger RNA chain runs through the ribosome as a tape runs through a tape recorder head. While the tail of a messenger RNA molecule is still running through one ribosome, its head may already have been picked up by another ribosome, so that a single molecule of messenger RNA can actually service several ribosomes at the same time.

How the amino acids are actually assembled into the correct predetermined sequence by the messenger RNA as it runs through the ribosome had been envisaged by Crick[27] before the concept of the messenger RNA had even been clearly formulated. Crick thought it unlikely (working from first principles, as was his wont) that the twenty different amino acids could interact in any specific way directly with the nucleotide base triplet on the RNA template chain. He therefore proposed the idea of a nucleotide *adaptor*, with which each amino acid is outfitted prior to its incorporation into the polypeptide chain. This adaptor was thought to contain a nucleotide base triplet, or *anticodon*, complementary (in the Watson-Crick nucleotide base pairing sense) to the nucleotide triplet codon that codes for the particular amino acid to which the adaptor is attached. The anticodon nucleotides of the adaptor would then form specific hydrogen bonds with their complementary codon nucleotides on the messenger RNA and thus bring the amino acids bearing the adaptor into the proper, predetermined alignment on the ribosome surface.

No sooner had the adaptor hypothesis been formulated than students of the biochemistry of protein synthesis began to encounter an ensemble of specific reactions and enzymes that gradually resembled more and more the a priori postulates of that hypothesis.[28] First, a special type of small RNA molecule, the

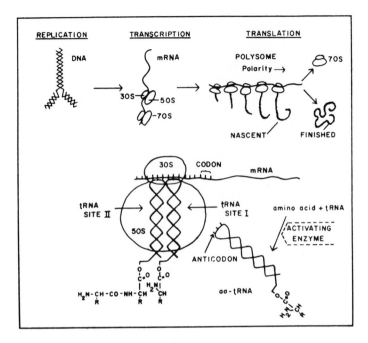

FIG. 5. A summary diagram of the autocatalytic (replication) and heterocatalytic (transcription and translation) functions of DNA. *Upper left:* The DNA double helix replicates according to the semiconservative Watson-Crick mechanism. *Upper center:* The DNA nucleotide sequence has been transcribed onto a molecule of messenger RNA, or mRNA. The mRNA molecule is engaged by ribosomes, which are composed of two subunits, a smaller one called 30S and a larger one called 50S; the two subunits together constitute the intact, or 70S, ribosome. *Upper right:* Each of several ribosomes working in tandem on the same mRNA (a 'polysome') translates the mRNA nucleotide sequence into the corresponding polypeptide chain. When a ribosome has translated the entire nucleotide sequence corresponding to a gene, the completed protein chain is released and the ribosome is free to attach itself to another mRNA molecule. *Lower part:* Details of the process of amino acid assembly. The nascent protein chain (here consisting merely of two amino acids) is attached to that molecule of transfer RNA, or tRNA, which figured as adaptor of the last amino acid to be added into the chain. This molecule of tRNA is in turn held to site II of the 50S ribosomal subunit. The next amino acid to be incorporated into the nascent protein chain is specified by that nucleotide triplet codon of the mRNA which faces site I of the 50S subunit. Into site I can fit only a molecule of tRNA whose anticodon matches the codon displayed by the mRNA and to which the appropriate amino acid has become attached, thanks to the recognition effected by the activating enzyme. Once the tRNA has entered site I, its amino acid is brought into juxtaposition with the last amino acid of the nascent protein chain, and the next peptide bond can be formed. When the chain has thus been elongated by one amino acid residue, mRNA and tRNA molecules move over the ribosome from right to left, to translocate the tRNA now carrying the nascent chain from site I to site II and to display the next codon at site I. (From H. K. Das, A. Goldstein, and L. C. Kanner, *Molecular Pharmacology,* 2, 1966, 158. Reprinted by permission.)

transfer RNA, was discovered, which contains about 80 nucleotides in its polynucleotide chain. Each cell contains several dozen distinct species of transfer RNA, each species being capable of combining with one and only one kind of amino acid. This transfer RNA turned out to be Crick's postulated adaptor, since that transfer RNA species which accepts any given amino acid contains the anticodon nucleotide triplet in its polynucleotide chain which is complementary to the codon representing that same amino acid in the genetic code.

Second, a set of enzymes, the *amino acid activating enzymes*, was discovered, each of whose members is capable of catalyzing the combination of one kind of amino acid with its cognate transfer RNA molecule. Thus the set of activating enzymes which matches each amino acid with its proper transfer RNA adaptor (by means of which the amino acid is recognized in protein synthesis) evidently represents the *dictionary of heredity*, the cellular agency that 'knows' the genetic code.

Breaking the Code

The actual deciphering of the genetic code began with a discovery made by the young biochemist Marshall Nirenberg at the National Institutes of Health. In the spring of 1961 Nirenberg had managed to develop a 'cell-free' system capable of linking amino acids into protein molecules. This system contained ribosomes, transfer RNA, and amino acid activating enzymes extracted from *E. coli*. Though Nirenberg was by no means the first to reassemble *in vitro* the cellular machinery for protein forma-

tion, his system had one very important advantage over its predecessors: here protein synthesis depended on the addition of messenger RNA to the reaction mixture. Thus it became feasible to direct the *in vitro* formation of specific proteins by introducing into this system specific types of messenger RNA. Now when Nirenberg introduced a synthetically produced *monotonous* RNA containing *only* the uracil nucleotide (instead of the four types of nucleotide bases present in natural messenger RNA), he obtained a dramatic result. Addition of the artificial, monotonous messenger RNA resulted in the *in vitro* formation of an equally monotonous 'protein,' namely a 'protein' containing only one kind of amino acid— phenylalanine.[29] This result could have only one meaning: in the genetic code the uracil-uracil-uracil (or in the DNA, the equivalent thymine-thymine-thymine) nucleotide triplet represents the amino acid phenylalanine.

Nirenberg announced his identification of the first codon in August 1961, at the International Congress of Biochemistry in Moscow, where it caused a sensation. (Crick later wrote that he was 'electrified.') Thus at one stroke the breaking of the genetic code had become accessible to direct chemical experimentation, because now the effect of introducing various synthetically produced types of messenger RNA of known composition into the cell-free protein-synthesizing system could be examined. The Moscow announcement set off a code-breaking race, sometimes called the Code-War of the U3 Incident, which culminated in deciphering the definite, or at least probable, meaning of many of the 64 codons by 1963.

In 1964 Nirenberg made a second experimental breakthrough in the deciphering of the genetic code by means of his cell-free system for protein synthesis.[30] At that time he discovered that it is possible to detect in his reaction mixture the specific *binding* to ribosomes of molecules of transfer RNA carrying their cognate amino acids. In particular, he found that addition to his reaction mixture of a very short polynucleotide chain consisting of only three nucleotides, instead of messenger RNA, promoted the *specific binding* to ribosomes of those and only those transfer RNA molecules that carry the anticodon complementary in the Watson-Crick base-pairing sense to the nucleotide triplet added to the reaction mixture. Protein formation does not, of course, occur under these conditions, since the short nucleotide triplet cannot serve as the template for directing the assembly of many amino acids. By means of this new technique Nirenberg found, in confirmation of his earlier identification of the codon representing the amino acid phenylalanine, that addition of the uracil-uracil-uracil nucleotide triplet promotes the binding of phenylalanine transfer RNA to ribosomes. And since it was relatively easy to prepare by chemical methods all 64 possible nucleotide triplets, the entire code could be worked out by means of this binding method within little more than a year.

The results of Nirenberg's work, which were presently supported or confirmed by other methods, are now generally presented in the form of a table, the arrangement of which was conceived by Crick (Figure 6). It has been suggested that this table of the code represents for biology what the periodic table of the elements represents for chemistry. The important features of the genetic code are these:

(1) The code contains synonyms, in that many amino acids are coded for by more than one kind of codon. For instance, the nucleotide base triplets uracil-uracil-uracil *and* uracil-uracil-cytosine are synonyms, in that they both code for phenylalanine.

(2) The code has a definite structure, in that synonymous codons representing the same amino acid are nearly always in the same 'box' (see Figure 6) of the table. That is, the synonymous codons general differ from each other only in the third of their three nucleotides. An explanation for this aspect of the code was provided by Crick in terms of the geometry of the hydrogen bonds involved in the recognition of the messenger RNA codon by the transfer RNA anticodon at the ribosomal site of protein formation.

(3) The code is very nearly universal. Though most of the code was deciphered by use of the protein-synthesizing machinery of *E. coli*, later tests showed that the results are very much the same whether the transfer RNA and amino acid activating enzymes (the agency that 'knows' the code) are obtained from bacterial, plant, or animal sources (including mammals). The universality of the code among contemporary living forms shows that the code has remained unchanged over a very long period of organic evolution. One explanation which has been offered for the, at first sight surprising, evolutionary permanence of the code is that any genetic mutation engendering a change in the code—by means of

THE GENETIC CODE

1st↓ 2nd→	U	C	A	G	↓3rd
U	PHE	SER	TYR	CYS	U
	PHE	SER	TYR	CYS	C
	LEU	SER			A
	LEU	SER		TRP	G
C	LEU	PRO	HIS	ARG	U
	LEU	PRO	HIS	ARG	C
	LEU	PRO	GLUN	ARG	A
	LEU	PRO	GLUN	ARG	G
A	ILEU	THR	ASPN	SER	U
	ILEU	THR	ASPN	SER	C
	ILEU	THR	LYS	ARG	A
	MET	THR	LYS	ARG	G
G	VAL	ALA	ASP	GLY	U
	VAL	ALA	ASP	GLY	C
	VAL	ALA	GLU	GLY	A
	VAL	ALA	GLU	GLY	G

FIG. 6. The table of the genetic code. In this table the letters U, C, A, and G represent the four kinds of nucleotides, containing the bases uracil, cytosine, adenine, and guanine. The three- and four-letter abbreviations represent the twenty kinds of protein amino acids. The codon corresponding to any given position on this table can be read off according to the following rules: The base of the first nucleotide of the codon is given by the capital letter on the left, which defines a horizontal row containing four lines. The base of the second nucleotide is given by the capital letter on the top, which defines a vertical column containing sixteen codons. The intersection of rows and columns defines a 'box' of four condons, all of which carry the same bases in their first and second nucleotides. The base of the third nucleotide is given by the capital letter on the right, which defines one line within any given horizontal row. The triplets UAA, UAG, and UGA are 'no sense' codons, to which there corresponds no amino acid.

which evolution of the code would have had to proceed—would necessarily be lethal to the organism in which it took place: such a mutational alteration of the code would cause a sudden change in the amino acid sequence of *all* of the protein molecules of the mutant organism. And such a wholesale change would be most unlikely to be compatible with survival of the organism. Another possible explanation for the evolutionary permanence of the code is that there exists some as yet unfathomed geometrical or stereochemical relation between the anticodon nucleotide triplet and the amino acid which it represents. Indeed, if such a relation exists, it would be bound to hold one of the keys to understanding the origin of life.

The Operon

The developments recounted so far have revealed the *qualitative* aspect of the heterocatalytic function: how the nucleotide base sequence carried by the DNA is finally translated into the predetermined amino acid sequence. However, there must pertain to the heterocatalytic function also a *quantitative* aspect: how the DNA manages to govern the synthesis of appropriate amounts of the different proteins whose structure is encoded into it. For it is an easily ascertainable fact that in *E. coli* the number of protein molecules read off one gene per generation can exceed by more than ten-thousandfold the corresponding number read off another gene in the same cell. Furthermore, the rate of translation of any given gene is not always the same, being very high

under some physiological conditions and very low under others.

The framework for understanding this quantitative aspect, which was not really covered by the original formulation of the central dogma, was finally provided in 1961 by the *operon theory* of Jacob and Monod[26] (Figure 7). In order to explain the functional regulation of the genes, this theory envisages that a group of genes, or an *operon*, is subject to coordinate control. The genes belonging to the same operon occupy contiguous sectors of the DNA; that is, they are closely linked and share a common, special regulatory segment of DNA, their *operator*. This operator, which is located at one extreme of the operon group of genes, can exist in two states: open or closed. As long as the operator is open, messenger RNA can be transcribed from all the genes of the operon, and hence synthesis of the protein species encoded into these genes may proceed on the ribosomes into which these messenger RNA molecules are fed. As soon as the operator is closed, transcription of the messenger RNA, and hence synthesis of the corresponding protein species, ceases. Thus the rate of translation of the genes belonging to any operon depends on the fraction of the total time during which their operator happens to be open. Now whether the operator is open or closed in turn depends on whether it has interacted with a *repressor* protein, itself the product of a special regulatory gene. Combination of the operator segment with its related repressor inhibits messenger RNA transcription of the genes of that operon and therefore closes the operator. Hence, in the last analy-

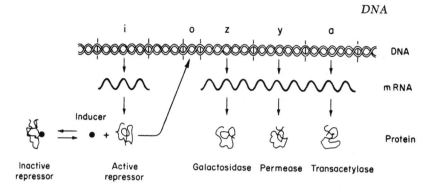

FIG. 7. The operon theory of Jacob and Monod, as applied to the lactose fermentation genes of *E. coli*. Three contiguous genes, *z, y,* and *a,* code for the protein structure of three enzymes, galactosidase, permease, and transacetylase, and share a common operator gene *o*. A gene *i* codes for the protein structure of the repressor, which can attach to and thus 'close' gene *o*. Combination with an inducer, such as lactose, inactivates the repressor, thus preventing it from 'closing' *o*. (From G. S. Stent, in *The Neurosciences,* New York: Rockefeller University Press, 1967. Reprinted by permission.)

sis, the rate of translation of any gene is a function of the intracellular concentration of active repressor capable of combining with the operator segment of DNA to which that gene is linked.

Thus, by the mid-1960's the general nature of both autocatalytic and heterocatalytic functions of the DNA was understood.[31] Through formation of complementary hydrogen bonds DNA achieves both functions by serving as a template for the synthesis of replica polynucleotide chains, making DNA chains for the autocatalytic and RNA chains for the heterocatalytic function. RNA, in turn, completes the heterocatalytic function by formation of complementary hydrogen bonds with the anticodons of transfer RNA molecules

in the amino acid assembly processes. The real core of genetic theory was lifted from the deep unknown in which Muller had found it to lie only fifteen years earlier. We now *do* have actual knowledge of the mechanism underlying that unique property which makes a gene a gene: Formation of complementary hydrogen bonds seems to be all there is to how like begets like.

Alas, the very success of molecular genetics in explaining one of the most profound mysteries of life in terms of workaday chemical reactions altered the spiritual qualities of this field. Molecular genetics now presents an integral canon of biological knowledge which must be preserved and passed on to succeeding generations in the academies. Moreover, as a subject for scholarly research it is far from exhausted, and its practical exploitation in such biotechnological domains as agriculture and eugenics has, as yet, barely begun. But its appeal as an arena for heroic strife against the Great Unknown is gone. And in contrast to classical genetics, which at the apogee of its development still harbored the enigma of the gene as a skeleton in the closet, molecular genetics seems to have left no transcendant legacy. So the would-be explorer of uncharted territory must direct his attention elsewhere. One of the most formidable unsolved problems of biology recommending itself to such romantic types is embryology. Understanding the processes responsible for the orderly development of the fertilized egg into a complex and highly differentiated multicellular organism still seems to boggle the imagination. The recent course of embryological research, however, suggests that it may be just more of the same old molecular genetics, albeit at a much more complicated level.

There now seems to remain really only one major frontier of biological inquiry which still offers some romance of research: the higher nervous system. Its fantastic attributes continue to pose as hopelessly difficult and intractably complex a problem as did the gene a generation ago. Indeed, the higher nervous system presents the most ancient and deepest biological mystery in the history of human thought: the relation of mind to matter. Increasing numbers of veteran molecular geneticists are now turning their attention to the higher nervous system in the hope of finding relief from jejune genetic investigations along more or less clearly established lines. They have good reason to hope that, unlike the quest for fathoming the gene, the quest for understanding the brain will not soon reach a disappointingly workaday denouement. For since the mind-matter mystery is not likely to be amenable to scientific analysis, the most interesting attribute of life may never be explained.[32]

I am indebted to Curt Stern and Alfred E. Mirsky for helpful criticisms of the manuscript of this essay.

REFERENCES

1. L. C. DUNN, ed., *Genetics in the 20th Century* (New York: Macmillan, 1951).
2. E. L. TATUM and J. LEDERBERG, 'Gene Recombination in the Bacterium *Escherichia Coli,' Journal of Bacteriology*, 53 (1947), 673–684.
3. G. W. BEADLE, 'Genes and the Chemistry of the Organism,' *American Scientist*, 34 (1946), 31–53.
4. F. MIESCHER, 'Ueber die chemische Zusammensetzung der Eiterzellen,' *Hoppe-Seyler Medizinisch Chemische Untersuchungen*, 4 (1871), 441.
5. P. A. LEVENE and L. W. BASS, *Nucleic Acids* (New York: The Chemical Catalog Company, 1931).
6. R. FEULGEN and H. ROSSENBECK, 'Mikroskopisch-chemischer Nachweis einer Nukleinsäure vom Typus der Thymonucleinsäure und die darauf beruhende elektive Färbung von Zellkernen in mikroskopischen Präparaten,' *Zeitschrift für Physiologische Chemie*, 135 (1924), 203–248.
7. E. CHARGAFF, 'Chemical Specificity of Nucleic Acids and Mechanisms of Their Enzymatic Degredation,' *Experientia*, 6 (1950), 201–209.
8. O. T. AVERY, C. M. MACLEOD, and M. MCCARTY, 'Studies on the Chemical Nature of the Substance Inducing Transformation of Pneumococcal Types,' *Journal of Experimental Medicine*, 79 (1944), 137–157.

9. F. GRIFFITH, 'Significance of Pneumococcal Types,' *Journal of Hygiene*, 27 (Cambridge, Eng., 1928), 113.

10. Readers interested in obtaining an overview of the contribution of phage research to modern biology are referred to G. S. Stent, *Molecular Biology of Bacterial Viruses* (San Francisco: W. H. Freeman, 1963).

11. E. L. ELLIS and M. DELBRÜCK, 'The Growth of Bacteriophage,' *Journal of General Physiology*, 22 (1939), 365.

12. J. CAIRNS, G. S. STENT, and J. D. WATSON, eds., *Phage and the Origins of Molecular Biology* (New York: Cold Spring Harbor Laboratory for Quantitative Biology, 1966).

13. A. D. HERSHEY and M. CHASE, 'Independent Function of Viral Protein and Nucleic Acid in Growth of Bacteriophage,' *Journal of General Physiology*, 36 (1952), 39–56.

14. S. E. LURIA and M. DELBRÜCK, 'Mutations of Bacteria from Virus Sensitivity to Virus Resistance,' *Genetics*, 28 (1943), 491.

15. L. PAULING, R. B. COREY, and H. R. BRANSON, 'The Structure of Proteins: Two Hydrogen-Bonded Helical Configurations of the Polypeptide Chain,' *Proceedings of the National Academy of Sciences*, 37 (Washington, D. C., 1951), 205.

16. J. D. WATSON, *The Double Helix* (New York: Atheneum, 1968).

17. J. D. WATSON and F. H. C. CRICK, 'A Structure for Deoxyribose Nucleic Acid,' *Nature*, 171 (1953), 737.

18. J. D. WATSON and F. H. C. CRICK, 'Genetic Implications of the Structure of Deoxyribonucleic Acid,' *Nature*, 171 (1953), 964.

19. M. DELBRÜCK and G. S. STENT, 'On the Mechanism of DNA Replication,' W. D. McElroy and B. Glass, eds., *The Chemical Basis of Heredity* (Baltimore: Johns Hopkins Press, 1957), p. 699.

20. M. MESELSON and F. W. STAHL, 'The Replication of DNA in *Escherichia Coli*,' *Proceedings of the National Academy of Sciences*, 44 (Washington, D. C., 1958), 671.

21. F. SANGER, in D. E. Green, ed., *Currents in Biochemical Research* (New York: Interscience, 1956).

22. S. BENZER, 'The Elementary Units of Heredity,' W. D. McElroy and B. Glass, eds., *The Chemical Basis of Heredity* (Baltimore: Johns Hopkins Press, 1957), p. 70.

23. G. Gamow, 'Possible Relation Between Deoxyribonucleic Acid and Protein Structures,' *Nature*, 173 (1954), 318.

24. F. H. C. CRICK, L. BARNETT, S. BRENNER, and R. J. WATTS-TOBIN, 'General Nature of the Genetic Code for Proteins,' *Nature*, 192 (1961), 1227.

25. K. MCQUILLEN, R. B. ROBERTS, and R. J. BRITTEN, 'Synthesis of Nascent Proteins by Ribosomes in *Escherichia Coli*,' *Proceedings of the National Academy of Sciences*, 45 (Washington, D. C., 1959), 1437.

26. F. JACOB and J. MONOD, 'Genetic Regulatory Mechanisms in the Synthesis of Proteins,' *Journal of Molecular Biology*, 3 (1961), 318.

27. F. H. C. CRICK, 'On Protein Synthesis,' *The Biological Replication of Macromolecules*, Symposium of the Society for Experimental Biology, XII (London: Cambridge University Press, 1958), p. 138.

28. M. B. HOAGLAND, P. C. ZAMECNIK, and M. L. STEPHENSON, 'Intermediate Reactions in Protein Biosynthesis,' *Biochimica et Biophysica Acta*, 24 (1957), 215. M. B. Hoagland, M. L. Stephenson, J. F. Scott, R. J. Hecht, and P. C. Zamecnik, 'A Soluble Ribonucleic Acid Intermediate in Protein Synthesis,' *Journal of Biological Chemistry*, 231 (1958), 241.

29. M. NIRENBERG and J. H. MATTHEI, 'The Dependence of Cell-Free Protein Synthesis in *Escherichia Coli* Upon Naturally Occurring or Synthetic Polyribonucleotides,' *Proceedings of the National Academy of Sciences*, 47 (Washington, D. C., 1961), 1588.

30. M. NIRENBERG and P. LEDER, 'RNA

Codewords and Protein Synthesis: The Effect of Trinucleotides Upon Bringing sRNA to Ribosomes,' *Science*, 145 (1964), 1399.

31. By that time the Cold War had eased, both in its strategic and its genetic aspects. Molecular genetics had become definitely an in-subject in the Soviet Union, thanks in part to the final fall from favor of Lysenko after the dismissal of N. Khrushchev. But it seems likely that in any case the 'materialist' concrete DNA gene of molecular genetics (and the possibility of inheritance of 'acquired' characters through phenomena such as the DNA-mediated bacterial transformation) is more palatable to dialectical-materialist thought than the 'idealist' abstract gene of classical 'Mendelism-Morganism.' See Zh. Medvedev, *The Rise and Fall of Lysenko* (New York: Columbia University Press, 1969).

32. To readers interested in the development of genetic thought, I strongly recommend A. H. Sturtevant, *A History of Genetics* (New York: Harper and Row,

1965). Some of the philosophical issues attending the rise of molecular genetics are treated in M. Delbrück, 'A Physicist Looks at Biology,' *Transactions of the Connecticut Academy of Sciences*, 38 (1949), 173 (reprinted, see note 12); in G. S. Stent, 'That Was the Molecular Biology That Was,' *Science*, 160 (1968), 390; and in D. Fleming, "Emigré Physicists and the Biological Revolution,' in D. Fleming and B. Bailyn, eds., *The Intellectual Migration: Europe and America, 1930-1960* (Cambridge, Mass.: Harvard University Press, 1969), 152. Professional introductions to molecular genetics are provided by J. D. Watson, *The Molecular Biology of the Gene* (New York: Benjamin, 1965), and G. S. Stent, *Molecular Genetics* (San Francisco: W. H. Freeman, 1970). A fuller treatment of the notions touched on in the closing paragraphs of this essay can be found in G. S. Stent, *The Coming of the Golden Age: A View of the End of Progress* (New York: Natural History Press, 1969).

BIBLIOGRAPHY

There is an enormous literature in the field of molecular genetics. References to some of the key data papers are given in *Heredity and Development* (Suggested Readings, Chapters 7 and 8). The references listed below are, for the most part, general and usually they have extensive bibliographies.

BECKWITH, JONATHAN R., and DAVID ZIPSER. 1970. *The Lactose Operon*. Cold Spring Harbor, N.Y.: Cold Spring Harbor Laboratory.

BOREK, ERNEST. 1965. *The Code of Life*. New York: Columbia University Press.

BRAUN, W. 1953. *Bacterial Genetics*. Philadelphia: W. B. Saunders.

BRINK, R. ALEXANDER. 1967. *Heritage from Mendel*. Madison: University of Wisconsin Press. Many individual articles on modern genetics.

CAIRNS, J., G. S. STENT, and J. D. WATSON. Editors. 1966. *Phage and the Origins of Molecular Biology*. Cold Spring Harbor, N.Y.: Cold Spring Harbor Laboratory.

CARLSON, ELOF AXEL. 1966. *The Gene: A Critical History*. Philadelphia: W. B. Saunders.

CARLSON, ELOF AXEL. 1967. *Modern Biology. Its Conceptual Foundations*. New York: George Braziller. An anthology including many papers pertinent to the topics of this chapter.

CARLSON, ELOF AXEL. 1967. *Gene Theory.* Belmont, Calif.: Dickenson Publishing Co. Fourteen articles, mostly on molecular biology.

CARLSON, ELOF AXEL. 1971. 'An unacknowledged founding of molecular biology: H. J. Muller's contributions to gene theory, 1910–1936.' *Journal of the History of Biology 4*: 149–170.

CASPARI, ERSNT W. and ARNOLD W. RAVIN. 1969. *Genetic Organization. A Comprehensive Treatise.* Volume 1. New York: Academic Press.

COHEN, DAVID. 1965. *The Biological Role of the Nucleic Acids.* New York: American Elsevier.

Cold Spring Harbor Symposia on Quantitative Biology. Much of the progress of modern genetics is recorded in this series. The more relevant volumes are.
1971. Volume 35. 'Transcription of genetic material.'
1969. Volume 34. 'Mechanism of protein synthesis.'
1968. Volume 33. 'Replication of DNA in microorganisms.'
1966. Volume 31. 'The genetic code.'
1961. Volume 26. 'Cellular regulatory mechanisms.'
1958. Volume 23. 'Exchange of genetic material.'
1956. Volume 21. 'Genetic mechanisms.'
1953. Volume 18. 'Viruses.'
1951. Volume 16. 'Genes and mutation.'
All published by the Cold Spring Harbor Laboratory, Cold Spring Harbor, N.Y.

COBURN, ALVIN F. 1969. 'Oswald Theodore Avery and DNA.' *Perspectives in Biology and Medicine 12*: 623–630.

DAVIDSON, J. N. 1968. 'Nucleic acids—the first hundred years.' *Progress in Nucleic Acid Research and Molecular Biology 8*: 1–6.

DEBUSK, A. GIB. 1968. *Molecular Genetics.* New York: Macmillan.

DE ROBERTIS, E.D.P., W. W. NOWINSKI, and F. A. SAEZ. 1970. *Cell Biology.* Fifth Edition. Philadelphia: W. B. Saunders.

DUNN, L. C. Editor. 1951. *Genetics in the 20th Century.* New York: Macmillan. A series of papers that surveyed the field just before the dawn of Watson and Crick.

DUPRAW, ERNEST J. 1968. *Cell and Molecular Biology.* New York: Academic Press.

DUPRAW, ERNEST J. 1970. *DNA and Chromosomes.* New York: Holt, Rinehart & Winston.

FLEMING, DONALD. 1968. 'Emigré physicists and the biological revolution.' *Perspectives in American History 2*:152–189.

HARRIS, HARRY. 1970. *The Principles of Human Biochemical Genetics.* New York: American Elsevier.

HARTMAN, PHILIP E., and SIGMUND R. SUSKIND. 1969. *Gene Action.* Second Edition. Englewood Cliffs, N.J.: Prentice-Hall.

HERSHEY, A. D., and others. 1971. *The Bacteriophage Lambda.* Cold Spring Harbor, N.Y.: Cold Spring Harbor Laboratory.

HERSKOWITZ, IRWIN. 1967. *Basic Principles of Molecular Genetics,* Boston: Little, Brown.

HESS, EUGENE L. 1970. 'The origins of molecular biology.' *Science 168*: 664–669.

INGRAM, VERNON M. 1963. *The Hemoglobins in Genetics and Evolution.* New York: Columbia University Press.

JACOB, F., and J. MONOD. 1961. 'On the regulation of gene activity.' *Cold Spring Harbor Symposia on Quantitative Biology 26*: 193–209.

JACOB, F., and J. MONOD. 1961. 'Genetic regulatory mechanisms in the synthesis of proteins.' *Journal of Molecular Biology 3*: 318–356.

JUKES, THOMAS H. 1966. *Molecules and Evolution*. New York: Columbia University Press.

KENDREW, JOHN C. 1966. *The Thread of Life. An Introduction to Molecular Biology*. Cambridge: Harvard University Press.

LECHEVALIER, H.A., and M. SOLOTOROVSKY. 1965. *Three Centuries of Microbiology*. New York: McGraw-Hill.

LIVINGSTONE, FRANK B. 1967. *Abnormal Hemoglobins in Human Populations*. Chicago: Aldine.

LURIA, S. E. 1970. 'Molecular biology: past, present, future.' *Bioscience 20*: 1289–1293, 1296.

MCELROY, W. D., and BENTLEY GLASS. Editors. 1957. *The Chemical Basis of Heredity*. Baltimore: Johns Hopkins University Press.

OLBY, ROBERT. 1968. 'Before *The Double Helix*.' *New Scientist 38*: 679–681.

OLBY, ROBERT. 1970. 'Francis Crick, DNA, and the central dogma.' *Daedalus*. Fall 1970: 938–987.

PAULING, LINUS, and ROGER HAYWARD. 1964. *The Architecture of Molecules*. San Francisco: W. H. Freeman.

PAULING, LINUS. 1970. 'Fifty years of progress in structural chemistry and molecular biology.' *Daedalus*. Fall 1970: 988–1014.

RAVIN, ARNOLD. 1965. *The Evolution of Genetics*. New York: Academic Press.

SAGER, RUTH, and FRANCIS J. RYAN. 1961. *Cell Heredity*. New York: Wiley.

SCHRÖDINGER, ERWIN. 1945. *What is Life? The Physical Aspects of the Living Cell*. New York: Macmillan.

Scientific American. 1968. *The Molecular Basis of Life. An Introduction to Molecular Biology*. San Francisco: W. H. Freeman. Reprints of articles from *Scientific American*.

Scientific American. 1970. *Facets of Genetics*. San Francisco: W. H. Freeman. Reprints of articles from *Scientific American*.

STAHL, FRANKLIN W. 1969. *The Mechanics of Inheritance*. Second Edition. Englewood Cliffs, N.J.: Prentice-Hall.

STENT, GUNTHER S. 1963. *Molecular Biology of Bacterial Viruses*. San Francisco: W. H. Freeman.

STENT, GUNTHER S. 1968. 'That was the molecular biology that was.' *Science 160*: 390–395.

STENT, GUNTHER S. 1969. 'The 1969 Nobel Prize for Physiology or Medicine.' *Science 166*: 479–481.

STENT, GUNTHER S. 1969. *The Coming of the Golden Age. A View of the End of Progress*. Garden City, N.Y.: Natural History Press.

STENT, GUNTHER S. 1971. *Molecular Genetics. An Introductory Narrative*. San Francisco: W. H. Freeman.

TAYLOR, J. H. Editor. 1965. *Selected Papers on Molecular Genetics*. New York: Academic Press.

WAGNER, R. P., and H. K. MITCHELL. 1964. *Genetics and Metabolism*. Second Edition. New York: Wiley.

WHITEHOUSE, H. L. K. 1969. *Towards an Understanding of the Mechanisms of Heredity*. New York: St. Martin's Press.

WATSON, J. D. 1970. *Molecular Biology of the Gene*. Second Edition. New York: W. A. Benjamin.

WATSON, J. D. 1968. *The Double Helix*. New York: Atheneum. A personal account of the discovery of DNA.

WOESE, CARL R. 1967. *The Genetic Code*. New York: Harper and Row.

YCAS, M. 1969. *The Biological Code*. New York: Wiley-Interscience.

ZAMECNIK, P. C. 1960. 'Historical and current aspects of the problem of protein synthesis. *Harvey Lectures 54*:256–281.

ZUBAY, GEOFFREY L. 1968. *Papers in Biochemical Genetics*. New York: Holt, Rinehart & Winston. An anthology.

8 / The Genetics of Man

There are no principles of genetics that apply solely to man and there is no important principle of genetics that was discovered by studying man's inheritance. But geneticists, being human, have always been interested in what their science had to say about themselves. Not surprisingly this has been one of the most interesting fields of biology for the layman.

During the first half-century after the discovery of Mendel's paper—from 1900 to 1950—the major triumphs of human genetics consisted of finding that Mendel's First Law applies to ourselves. Things are very different today: the new techniques of cytology and molecular biology have resulted in a more rapid rate of discovery. Much is being learned, not only about the formal genetics of man, but also about the genetics of disease, race, individual characteristics, birth defects, and the consequences of variations in chromosome number.

One matter that has always been of interest is the extent to which man can use genetic information to improve his lot and control his future. In Book V of Plato's *Republic*, Socrates notes how breeders of domestic animals are careful to breed only from the most desirable parents and suggests that the State should apply similar principles to man.

JOSHUA LEDERBERG

Two and a half millennia later Joshua Lederberg, Nobel Laureate at Stanford University, has some interesting and important things to say on this same topic.

EXPERIMENTAL GENETICS AND HUMAN EVOLUTION

JOSHUA LEDERBERG
Department of Genetics, Stanford University School of Medicine, Palo Alto, California

Planning based on informed foresight is the hallmark of organized human intelligence, in every theater from the personal decisions of domestic life to school bond elections to the world industrial economy. One sphere where it is hardly ever observed is the prediction and modification of human nature. The hazards of monolithic sophistocratic rationalization of fundamental human policy should not be overlooked, and medicine is - wisely dedicated to the welfare of individual patients one at a time. However, though lacking machinery for global oversight, we must still find ways to cope with the population explosion, environmental pollution, clinical experimentation, the allocation of scarce resources like kidneys (transplant or artificial), even a convention on when life begins and ends, which confounds discussion of abortion and euthanasia. Concern for the biological substratum of posterity, i.e., eugenics, is divided by the same cross-purposes. Nevertheless, whether or not he dares to advocate concrete action, every student of evolution must be intrigued by what is happening to his own species (what else matters?), and especially the new evolutionary theory needed to model a self-modifying system that makes imperfect plans for its own nature.

Repeated rediscovery notwithstanding, the eugenic controversy started in the infancy of genetic science. More recently, the integration of experimental genetics and biochemistry has provoked a new line of speculation about more powerful techniques than the gradual shift of gene frequencies by selective breeding for the modification of man. This article will first recapitulate a widely held skepticism about the criteria for the 'good man' who is the aim of eugenic policy. . . .

The debate needed to ventilate these issues has started in a few conferences: *Man and his future* (G. Wolstenholme, ed.) Ciba Foundation Symposium, 1962; *Control of human heredity and evolution* (T. M. Sonneborn, ed.) Macmillan, 1965; and *Biological aspects of social problems* (Meade & Parkes, eds.) Plenum Press, 1965, which document many other ideas and references to primary literature. I would refer especially to Dobzhansky (1962) and Harris (1964) for outlines of the philosophical and technical foundations of the discussion.

. . . The outlook of this article is unavoidably culture-bound; many of my allusions pertain to academic life in the United States and might seem utterly absurd to the vast majority of the world's population, of which we are hardly an unbiased sample. The

From *American Naturalist 100*: 519–531. 1966. Reprinted by permission.

futility of discussing the patterns of human evolution without fairer representation of its actual components is the most cogent criticism of any simplicistic definition of eugenic goals.

1) Human culture has grown so rapidly that the biological evolution of the species during the last hundred generations has only begun to adjust to it. Microscopic processes of human evolution go on, but the instability of the historical milieu obscures any coherent pattern of biological adaptation since the paleolithic. Cultural cohesion tends to mute strident biological innovation, by the exclusion of deviants (whether 'positive' or 'negative'). But of course it generates its own biases, with many short-term fluctuations in the selective value of different genotypes and the long term-cancellation of many advantages irrelevant to civilized life.

2) Even on the time scale of the cultural revolution, we must acknowledge a singularity in the history and evolution of the planet: the emergence of scientific insight and technological power in the present era. In one lifetime, the parish has become the solar system.

3) The historical examples of the application of technology such as armaments and the population explosion are premonitions of the future. The hazards of imbalance as between technical power and social wisdom are well advertised, but technology itself is out-of-balance. For example, the technology of arms control has only recently attracted a fraction of the scientific attention devoted to its politics.

4) There has been considerable discussion of the supposed hazard to the human gene pool from the sheltering of the tacitly 'unfit' by medicine or social welfare. Not so widely understood is the futility of negative-eugenic programs: most deleterious genes are represented and maintained in the population mainly by normal (conceivably sometimes supernormal?) heterozygotes. If we attack the heterozygotes as well as overtly afflicted homozygotes, almost no human being will qualify. In addition, many well-established institutions, such as the comfort of the automobile, and of heated shelters, war, and inheritance of unearned wealth or power, are equally suspect as dysgenic. It is very difficult to see how we can reconcile any aggressive negative eugenic program with humanistic aspirations for individual self-expression and the approbation of diversity. Positive eugenic programs can be defended roughly in proportion to their ineffectiveness: applied on a really effective scale they would state the same dilemmas. At present the main hazard of these proposals is the oblique even if unintended weight they may appear to give to the enforcement of negative eugenics on outcast groups.

Genetic counseling can nevertheless play an important role within the framework of personal decision and forèsight for the immediate family. It can offer grave negative cautions about inbreeding and recurrence of genetic disease; it might also encourage optimists to look for compatibility or complementarity of positive attainments as a factor in mating preference. However, the public advertisement of 'superior germ plasm' (sperm banks) is open to so many distortions—like most manipulations of mass taste— that its implementation would probably run very differently from its sponsors' hopes. As in adoption proceedings, the anonymity of third parties can be set aside only at great risk to the stability of family life.

5) The cultural revolution has begun its most critical impact on human evolution, having generated technical power which now feeds back to biological nature. The last decade of molecular biology has given us a mechanistic understanding of heredity, and an entry to the same for development. These are just as applicable to human nature as they are to microbial physiology. Some themes of biological engineering are already an inevitable accompaniment of scientific and medical progress over the next five to 20 years.

The sharpest challenges to our pretensions about human nature are already in view—and may be overlooked by too farsighted focussing on more sophisticated possibilities, like 'chemical control of genotype.' (To save repeating a phrase, let me call this genetic alchemy, or *algeny*). Algeny is diversionary, not because I doubt its eventual realization, but because the obvious difficulties provide a too convenient refuge for evading sooner anxieties. Perhaps I might point to some analogous history. Some years ago, I suggested that the genetics of somatic cells of mammals could be worked out most directly by exploiting precedented interactions of cells in fusion, coalescence of karyotypes, and segregation. This was already being brilliantly realized, but much more energy has still been spent in vain pursuit of DNA- and virus-mediated transfer of genes in mammalian cells. These algenical visions still dominate the imagination of most of my colleagues, and may of course ultimately succeed.

The realization of applied biology is, simply, medicine; a more effective slogan on which to focus an alternative to eugenics is 'euphenics,' whose meaning should be transparent to readers of this journal. Euphenics then means all the ameliorations of genotypic maladjustment, including liability to any disease, that could be brought about by treatment of the affected individual, more efficaciously, the earlier in his development. Disease is any deficit relative to a desired norm, and with its shots to accelerate brain growth, the next generation or two will surely have an even more dismal clinical appreciation of our intellectual capacity than we as students did for our professors.

. . . The next few years will see the development of tissue and organ transplantation on a large scale. It would be a mistake to think of this as merely the repair of catastrophic defects in kidney or heart. Many more of us have slighter imbalance in our homeostats, muscles, teeth, stomachs or scalps, whose amenability to exchange will add up more weightily for standards of human performance. These implants will compete with their mechanical counterparts, which already prove the eminence of the trivial. The automobile is evolving into an all-purpose exoskeleton now augmented not only with locomotors but also a variety of sensors, effectors, and communicators. As it can also be equipped with auxiliary blood-pumps gas-bubblers, and a laundry (kidney), much of the effort that goes into making these medical devices implantable may be already irrelevant to contemporary man.

Embedded in molecular biology are the crucial answers to grave and basic questions about aging, the major degenerative diseases, and cancer; and it seems an easy gamble that very consequential changes in life-span and the whole pattern of life are in the offing, provided only that the momentum of existing scientific effort is sustained. Quite apart from the glimpses of the bizarre that mechanical and transplanted organs may offer, this is a general issue of the utmost importance to the fabric of human relationships; we have hardly begun to face it.

It is already a very heavy burden on the conscience of our physicians that the ebbing of life is a gradual process; that the spontaneous beating of the heart is no longer the uncontrollable axiom of human life; indeed that many a 'person' could be maintained indefinitely as an organ culture if there were any motive for it. Biological science already has a great deal to say and more questions to ask about the foundations of personality and its temporal continuity, which we have not begun to apply to the disposition of our own lives. The whole issue of self-identification needs scientific reexamination before we apply infinite effort to preserve a material body, many of whose molecules are transient anyhow. Inevitably, biological knowledge weighs many human beings with personal responsibility for decisions that were once relegated to divine Providence. In mythical terms, human nature began with the eating of the fruit of the Tree of Knowledge. Curiously, Genesis correlates this with the pain of childbirth, an insight that the growth of man's brain has gone beyond the safe and comfortable. However, the expulsion from Eden only postponed our access to the Tree of Life.

If the limit to a brain volume of 1500 ml is dictated by the proportions of the female pelvis—and obstetrics testifies how marginal the adaptation is—the simple practice of cesarean section could set us on a new evolutionary track. Very little is now known of the embryological homeostat for size and complexity of the brain. However, the few hints from early effects of some hormones, and the 'NGF' regulator of the sympathetic ganglia warrant the expectation of prophylactic control of the development of the growing brain. As such techniques become available, the responsibility for their administration can no more be evaded than for sending a child to school. Unfortunately, there are bound to be serious risks on both sides of the equation.

The elaboration of euphenics is, however, not the main purpose of a discussion of human evolution, except for the one point—the added difficulties it creates for any measure of human value. If this subject were not at the heart of the eugenic controversy, it would be arrogant to insist on the discussion of it.

Reconsider how we must reevaluate the cumulative score of a human genotype regarded over a lifetime, and for its contribution to the human future. Besides present perplexities, look to future perturbations:

1) Durability. The mere extension of lifespan alters the scores. Performance must be measured over the whole term of life, not based only on youthful precocity.

2) The euthenic context. Educational opportunity and practice are changing rapidly. Consider

a) Recognition of individual diversity. Educators have begun to learn, and exercise the knowledge, that children vary widely in the details of their information-processing machinery, e.g., the relative acuity of their sensory modalities. Many 'dull' children must be reclassified as overspecialized; we might well make virtue out of necessity in enabling each child to exploit his inherent skills. This can be accomplished realistically by

b) Computer-assisted teaching. The computer display is perhaps just an extension of the printed book, but we need a much more versatile adjustment of the information channels to the subtle requirements and performance of each child. This can hardly be achieved where each teacher must deal with any number of children simultaneously. In fact, the apparent value of a genotype will fluctuate according to the current status and the availability of the teaching programs and their relevance to the values of the community! Human teachers remain indispensable for developing and guiding these programs and for their insights into motivational and social sides of their students' behavior. In the long run, the individuation of the euthenic environment can only accentuate the importance of genotypic variation.

c) Within the United States and in other ways throughout the world, we observe an unprecedented experiment in equality of opportunity without regard to race. The uncontrollability of environments has left no room for the scientist to embrace any conclusions whatever about the genetic basis of differences in racial performance. Community attitudes have made genes for dark skin handicaps to academic achievement, often overriding superior brains. Many other genes play on the interaction of child and community, and ultimate human performance, just as deviously. As our knowledge of, and more to the point, the community's response to, these idiosyncrasies evolves, there will be a corresponding revision of the value equation.

d) Job skills change. Neat handwriting and mental arithmetic once crucial for white collar work are now obsolete. Tolerance for assembly line tedium is following muscular power onto the wasteheap of redundant skills. Social skills, leadership, and esthetic breadth are becoming the criteria of job success in many fields as machines take over the more routine tasks, in which logical rigor may soon be encompassed.

e) Western culture and its limited population is being succeeded by a much broader world culture. Is there much point in setting eugenic standards relevant only to a small minority of the world's population even as we watch the unprecedented breakdown of intercultural barriers? The jet airplane has already had an incalculably greater effect on human population genetics than any conceivable program of calculated eugenics.

3) The world situation. The central problem for the species must bias any momentary evaluation. Until recently, this was perceived as agricultural efficiency. Hunger still haunts the earth, but we might just manage to marshal the technical resources to assuage it. The specter of the industrialized world is suddenly nuclear suicide, and this has already led to some concern as to the biological adaptation of the species most appro-

priate to an age dominated by nuclear power. Political institutions are likely to change course much more rapidly than any biological response. As has been pointed out repeatedly, adaptability is man's unique adaptation.

This begs the question how to anticipate future needs, how far adaptability can be generalized, and how well it can compete, in any well-defined microniche, with more rigorous specialization. To put it another way, how do we identify the most adaptable genotypes now living and what is the price, to the detriment in special skills, of this adaptability?

4) Response to euphenics. The medico-technological context of human performance is more predictable than the socio-political. We are already committed to the attempted eradication of infectious agents like malaria, tuberculosis, cholera, variola, and poliovirus. In consequence, any breakdown of public health services can be catastrophic by exposing large, imperfectly immunized populations to these parasites. If the interplay of Hemoglobin S and malaria is a useful model, genetic adaptations to a germ-free environment are taking place too; chemical pollution might replace germs as a major selective factor except that its cumulative impact on adults is less cogent than acute infanticide. The context of modern man, in fact, includes steadily increasing reliance on medicine, i.e., euphenics, from ovulation onwards. It makes as little sense to decry genetic adaptations to this as to other components of civilized life. The quality of a genotype cannot now be evaluated in terms of a hypothetical state of nature (wherein we would quickly grunt in chilly displeasure at our unfurred skins),

but must match the pragmatic expectations of the milieu of the individual and his descendants. In fact selection is so slow, especially for rare genes, as to make this a theoretical issue for some time. It would be a tour-de-force to demonstrate any change in the frequency of a specific deleterious gene in a human population that could be unambiguously traced to a relaxation of natural selection against it. In comparison to the pace of medical progress, these exigencies are trivial.

As medical practice evolves so does the evaluation of health and vigor. What has happened to pancreatic diabetes is happening to phenylketonuria, and is bound to happen to many other biochemical and developmental diseases. Indeed, it would be no surprise to find compensating advantages, in certain contexts, for some of these genotypes.

The availability of transplants and prosthetics is an extension of the social process which relaxes the demands placed on a single genotype. We can imagine the systematic use of chimerism as another way of merging the best that each of a variety of genotypes can offer.

Recall that the most successful exercises in plant breeding have not established pure lines of vigorous individuals. Instead, somewhat overspecialized strains are nurtured and the latent resources of individually unpromising parents are merged in vigorous hybrid offspring. (A good farmer has learned how common sense conflicts with reality when he tries to use ears of hybrid corn as seed for another generation.)

5) Social adjustment. We are on the shakiest ground trying to sort out the genetic basis of such social dis-

eases as crime and delinquency. In any case we have a long way to go in elucidating how nature and nurture interact in this field; e.g., what penalty the species would suffer by extirpating every gene that might in some environment contribute to crime and rebelious behavior.[1] Instability of family life, the estrangement of the generations, and the shallowness of human communication are more prevalent and cumulatively more serious diseases than violent crime, and must be given equal account in any effort to define the 'good man,' or in any lament of human deterioration.

Who will toss the first stone?

6) The sexual dimorphism. Most genic discussions have been overwhelmingly male-oriented, as is academic life. Western culture is more paradoxical than ever in its assignment of roles to women, and thereby in the design of their education and the advertised criteria of feminine success, stressed by conflicting demands for decoration and utility, dependence and initiative. The lack of useful occupation for many older women is a premonition of the leisure society where 'work may become the prerogative of a chosen elite.' Half the beneficiaries of eugenic design will be women. Will their creativity and happiness be augmented in a genotype that recombines XX and a set of male-oriented autosomes? Or shall we bypass the dimorphism and evolve a race where this does not matter? To shout 'Vive la différence' and then ignore it is hypocrisy.

Occupational discrimination by sex has been outlawed as a byproduct of the civil rights movement in the United States, which raises nice biological questions. The sexual dimorphism is one of the most primitive of genetic differentials. Yet, in forthcoming attempts to enforce and evade the law, we shall see how thin the scientific groundwork is to answer how far the statistics of female performance in industrial society are biologically vs. socioculturally determined. In some ways this may be even harder to answer objectively than for the racial counterpart, since we are even less able to perform a meaningful experiment. What finesse it will take to design genotypes optimized for both sexes, i.e., properly rechanneled by the developmental switch with respect to the full set of desiderata, besides the primary sex characteristics!

7) The leisure society. This discussion has been dominated by criteria of performance at work. The whole framework may be obsolescent on the time scale of a few generations. As machines come to do almost all of the work, and this must include managerial and inventive tasks as well as clerical and manual, what are the relevant human values? Will not boredom be the most pernicious disease, and a zest for life without the compulsion of labor the rare essential for the species? Play rather than work will be the substratum of human activity, and the transmutation of play into cultural progress will replace the underpinning by industrial and military technology of its superstructure of basic science.

Perhaps the scientist who works for his joy in it is the most nearly pre-adapted for that topsy-turvy world, obviously an impeccable criterion for eugenic choice.

---

[1] Professor Walter Bodmer proposes labelling this concept 'the social load.'

This leads us finally to algeny. Man is indeed on the brink of a major evolutionary perturbation, but this is not algeny, but *vegetative propagation*. (No one will be surprised that Haldane had anticipated this reasoning years ago).

For the sake of argument, suppose we could mimic with human cells what we know in bacteria, the useful transfer of DNA extracted from one cell line to the chromosomes of another cell. Suppose we could even go one step further and sprinkle some specified changes of genotype over that DNA. What use could we make of this technology in the production as opposed to the experimental phase?

Repair genetic-metabolic disease? Indeed, if a diffusible hormone or enzyme were involved; but the same virtues are more readily available by transplantation. The advantage is consequential only if some nondiffusible product or irreversible developmental commitment (like a neuronal pattern) were involved. However, it is utterly unreasonable to anticipate the correct reprogramming of every treated cell. Then we must perform the algeny on gamete or zygote, but in so doing we face the difficulty of testing the consequences of the intervention! If the purpose is a better human being, by any standard, we would need 20 years to prove that the developmental perturbation was the intended, or in any way a desirable one. And if it were, we would face the same hazards generation after generation. The premise of this argument is that the inherent complexity of the system precludes any merely prospective experiment in algeny. It is bound to fail a large part of the time, and possibly with dis-astrous consequences if we slip even a single nucleotide.

To recapitulate, if the desired effect is achieved by modifying some somatic cells, the same end is available by transplanting cells already known to have these properties. In general this should be much easier than systematically changing the existing ones. If the zygote or a gamete needs to be altered, the operation is bound to have an uncertain outcome, and needs some kind of retrospective test. This ability to manipulate zygote nuclei should depend on prior capacity for nuclear transplantation and vegetative proliferation of the involved cells—both as part of the operation, and for the experimental calibration of the results.

If we have efficacious methods for testing and selecting new genotypes, do we have much need for algeny? Would not recombination and mutation give ample material for test? Perhaps for some time. But I would credit the possibility of designing a useful protein from first premises, replacing evolution by art. It would then be requisite to implant a specified nucleotide sequence into a chromosome. This would still be useless without retrospective inspection and approval of the result, e.g., in a clone of somatic cells. What to do with the mishaps needs to be answered before we can believe that these risks will be undertaken in the fabrication of humans. But, during an experimental phase, algeny may be as useful for the generation of designed genotypes, especially if they can be verified in cell culture, as other combinatorial tricks in the geneticists' repertoire.

Vegetative reproduction, once we are reminded that it is an indispensable facet of experimental technique in the microbial analogy, cannot be so readily dismissed. In fact there is ample precedent for it, and not only throughout the plant and microbial kingdoms, but in many lower animals. Monozygotic twins in man are accidental examples. Experimentally, we know of successful nuclear transplantation from diploid somatic as well as germline cells into enucleated amphibian eggs. There is nothing to suggest any particular difficulty about accomplishing this in mammals or man, though it will rightly be admired as a technical tour-de-force when it is first implemented or will this sentence be an anachronism before it is published?) Indeed I am more puzzled by the rigor with which apogamous reproduction has been excluded from the vertebrate as compared to the plant world, where its short-run advantages are widely exercised. If the restriction is accidental from the standpoint of cell biology, nevertheless a phylum that was able to fall into this trap might be greatly impeded in its evolutionary experimentation towards creative innovation.

Vegetative or clonal reproduction has a certain interest as a investigative tool in human biology, and as an indispensable basis for any systematic algenics; but other arguments suggest that there will be little delay between demonstration and use. Clonality outweighs algeny at a much earlier stage of scientific sophistication, primarily because it answers the technical specifications of the eugenicists in a way that Mendelian breeding does not. If a superior individual (and presumably then genotype) is identified, why not copy it directly, rather than suffer all the risks of recombinational disruption, including those of sex. The same solace is accorded the carrier of genetic disease: why not be sure of an exact copy of yourself rather than risk a homozygous segregant; or at worst copy your spouse and allow some degree of biological parenthood. Parental disappointment in their recombinant offspring is rather more prevalent than overt disease. Less grandiose is the assurance of sex-control; nuclear transplantation is the one method now verified.

Indeed, horticultural practice verifies that a mix of sexual and clonal reproduction makes good sense for genetic design. Leave sexual reproduction for experimental purposes; when a suitable type is ascertained, take care to maintain it by clonal propagation. The Plant Patent Act already gives legal recognition to the process, and the rights of the developer are advertised 'Asexual Reproduction Forbidden.'

Clonality will be available to and have significant consequences from acts of individual decision . . . given only community acquiescence or indifference to its practice. But here this simply allows the exercise of a minority attitude, possibly long before its implications for the whole community can be understood. Most of us pretend to abhor the narcissistic motives that would impel a clonist, but he (or she) will pass just that predisposing genotype intact to the clone. Wherever and for whatever motives close endogamy has prevailed before, clonism and clonishness will prevail.

Apogamy as a way of life in the plant world is well understood as an evolutionary cul-de-sac, often associated with hybrid luxuriance. It can be an unexcelled means of multiplying a rigidly well-adapted genotype to fill a stationary niche. So long as the environment remains static, the members of the clone might congratulate themselves that they had outwitted the genetic load; and they have indeed won a short-term advantage. In the human context, it is at least debatable whether sufficient latent variability to allow for any future contingency were preserved if the population were distributed among some millions of clones. From a strictly biological standpoint, tempered clonality could allow the best of both worlds—we would at least enjoy being able to observe the experiment of discovering whether a second Einstein would outdo the first one. How to temper the process and the accompanying social frictions is another problem.

The internal properties of the clone open up new possibilities, e.g., the free exchange of organ transplants with no concern for graft rejection. More uniquely human is the diversity of brains. How much of the difficulty of intimate communication between one human and another, despite the function of common learned language, arises from the discrepancy in their genetically determined neurological hardware? Monozygotic twins are notoriously sympathetic, easily able to interpret one another's minimal gestures and brief words; I know, however, of no objective studies of their economy of communication. For further argument, I will assume that genetic identity confers neurological similarity,[1] and that this eases com-

_____

[1] A far cry from the duplication of personality!

munication. This has never been systematically exploited as between twins, though it might be singularly useful in stressed occupations—say a pair of astronauts, or a deep-sea diver and his pump-tender, or a surgical team. It would be relatively more important in the discourse between generations, where an older clonont would teach his infant copy. A systematic division of intellectual labor would allow efficient communicants to have something useful to say to one another.

The burden of this argument is that the cultural process poses contradictory requirements of uniformity (for communication) and heterogeneity for innovation). We have no idea where we stand on this scale. At least in certain areas—say soldiery—it is almost certain that clones would have a self-contained advantage, partly independent of, partly accentuated by the special characteristics of the genotype which is replicated. This introverted and potentially narrowminded advantage of a clonish group may be the chief threat to a pluralistically dedicated species.

Even when nuclear transplantation has succeeded in the mouse, there would remain formidable restraints on the way to human application, and one might even doubt the further investment of experimental effort. However several lines are likely to become active. Animal husbandry, for prize cattle and racehorses, could not ignore the opportunity, just as it bore the brunt of the enterprises of artificial insemination and oval transplantation. The dormant storage of human germ plasm as sperm will be replaced by the freezing of somatic tissues to save potential donor nuclei. Experiments on the efficacy of human nuclear transplantation will continue on a somatic basis, and these tissue clones used progressively in chimeras. Human nuclei, and individual chromosomes and genes of the karyotype, will also be recombined with cells of other animal species—these experiments now well under way in cell culture. Before long we are bound to hear of tests of the effect of dosage of the human 21st chromosome on the development of the brain of the mouse or the gorilla. Extracorporeal gestation would merely accelerate these experiments. As bizarre as they seem, they are direct translations to man of classical work in experimental cytogenetics in Drosphila and in many plants. They need no further advance in algeny, just a small step in cell biology.

My colleagues differ widely in their reaction to the idea that anyone could conscientiously risk the crucial experiment, the first attempt to clone a man. Perhaps this will not be attempted until gestation can be monitored closely to be sure the fetus meets expectations. The mingling of individual human chromosomes with other mammals assures a gradualistic enlargement of the field and lowers the threshold of optimism or arrogance, particularly if cloning in other mammals gives incompletely predictable results.

What are the practical aims of this discussion? It might help to redirect energies now wasted on naive eugenics and to protect the community from a misapplication of genetic policy. It may sensitize students to recognize the significance of the fruition of experiments like nuclear transplantation. Most important, it may help to provoke more critical use of the lessons of history for the direction of our future. This will need a much wider participation in these concerns. It is hard enough to approach verifiable truth in experimental work; surely much wider criticism is needed for speculations whose scientific verifiability falls in inverse proportion to their human relevance. Scientists are by no means the best qualified architects of social policy, but there are two functions no one can do for them: the apprehension and interpretation of technical challenges to expose them for political action, and forethought for the balance of scientific effort that may be needed to manage such challenges. Popular trends in scientific work towards effective responses to human needs move just as slowly as other social institutions, and good work will come only from a widespread identification of scientists with these needs.

The foundations of any policy must rest on some deliberation of purpose. One test that may appeal to skeptical scientists is to ask what they admire in the trend of human history. Few will leave out the growing richness of man's inquiry about nature, about himself and his purpose. As long as we insist that this inquiry remain open, we have a pragmatic basis for a humble appreciation of the value of innumerable different approaches to life and its questions, of respect for the dignity of human life and of individuality, and we decry the arrogance that insists on an irrevocable answer to any of these questions of value. The same humility will keep open the options for human nature until their consequences to the legacy momentarily entrusted to us are fully understood. These concerns are entirely consistent with the rigorously mechanistic formulation of life which has been the systematic basis of recent progress in biological science.

Humanistic culture rests on a definition of man which we already know to be biologically vulnerable. Nevertheless the goals of our culture rest on a credo of the sanctity of human individuality. But how do we assay for *man* to demarcate him from his isolated or scrambled tissues and organs, on one side, from experimental karyotypic hybrids on another. Pragmatically, the legal privileges of humanity will remain with objects that look enough like men to grip their consciences, and whose nurture does not cost too much. Rather than superficial appearance of face or chromosomes, a more rational criterion[1] of human identity might be the potential for communication with the species, which is the foundation on which the unique glory of man is built. . . .

These are not the most congenial subjects for friendly conversation, especially if the conversants mistake comment for advocacy. If I differ from the consensus of my colleagues it may be only in suggesting a time scale of a few years rather than decades. Indeed, we will then face two risks, (1) that our scientific position is extremely unbalanced from the standpoint of its human impact, and (2) that precedents affecting the long-term rationale of social policy will be set, not on the basis of well-debated principles, but on the accidents of the first advertised examples. The accidentals might be as capricious as the

nationality, batting average, or public esteem of a clonont, the handsomeness of a parahuman progeny, the private morality of the experimenters, or public awareness that man is part of the continuum of life.

LITERATURE CITED

DOBZHANSKY, T. 1962. Mankind evolving. Yale Univ. Press, New Haven, Conn.
HARRIS, M. 1964. Cell culture and somatic variation. Holt, Rinehart and Winston, Inc., New York.
MEADE, J. E., and A. S. PARKES (eds.). 1965. Biological aspects of social problems. Plenum Press, New York.
SONNEBORN, T. M. (ed.). 1965. Control of human heredity and evolution. The Macmillan Co., New York.
WOLSTENHOLME, G. (ed.). 1962. Man and his future. Ciba Foundation Symposium. Little, Brown & Co., Boston, Mass.

GENETIC ENGINEERING, OR THE AMELIORATION OF GENETIC DEFECT
JOSHUA LEDERBERG

Few subjects pose as many difficulties for rational discussion as does the bearing of genetic research on human welfare. . . . The most sophisticated geneticist today is baffled by challenges such as Huntington's disease. Will the son of an afflicted father be afflicted later in life? What can he do to assure that his own children will not have it?

Perhaps some year soon we will know enough at least to recognize the genotype before neuronal degeneration has been irreversibly set in mo-

---

[1] A suggestion upon which it would be arrogant to insist against competing views of the essence of human nature.

From *The Pharos* of Alpha Omega Alpha *34*, 1:9–12. 1971.

tion. But our failure to be able to provide significant help today is a humbling reality next to the effusive (though justifiable) predictions about future accomplishments.

What then of the bold claims for a brave new world of genetic manipulation? Their substance is grounded on the recent solution of many fundamental mysteries of genetic biochemistry. Many of the obstacles to genetic engineering, apart from the moral and political questions that may be posed, are technological; which is to say that their solution is consistent with our basic scientific knowledge of the gene. But this is as if to say that 'merely technical obstacles' prevent building a land bridge from San Francisco to Honolulu. . . . We still hear the most absurd generalization, such as 'whatever is technically feasible tends to get done.'

Anyone who has actually labored to 'do' *anything* knows that the more appropriate slogans are: 'almost nothing ever gets done, especially if it costs money'; or, 'when a need is generally perceived, articulately formulated and wisely analyzed, the technical problems will be surmounted. But this will happen much sooner if a mass advertising campaign can be built around it.' . . .

Where then does the scientist fit into such a discussion?

He can fairly justify his life and work in terms of fundamental knowledge about nature. Studies on the implantation of nuclei into eggs of different genotypes are a rewarding approach to learning how genes function and how this relates to how the egg develops. Were they done for the purported purpose of learning the technology of cloning in man, we

would then be obliged to set a priority (positive or negative) on it from the standpoint of the human values that might justify or repudiate the investment for the solution of grave problems that encumber human welfare. Then we must and usually do insist that the problems are real ones and that technical solutions are credible. What is more often obscured is the need to examine all the side effects, to inhibit the premature exploitation of new cures that may be far worse than the disease, to assure that as much sophistication goes into looking for the side effects as was eagerly purchased for the primary solution. . . . Pesticide-poisoning and air pollution have been figured as technological jinn. It would be fairer to lay the blame on technological idiocy and the refusal to make the economic investments needed to develop all the science required for the safe and healthful utilization of the new tricks.

What then are the *problems* to which genetic science can be applied? Some may think of rescuing man from the prospect of nuclear annihilation by recasting the genes for aggression—or acquiescence—that are supposed to predestine a future of territorial conflict. Even if we postulate for sake of argument that we know the genetics of militarism, we have no way to apply it without solving the political problem that is the primary difficulty to begin with. If we could agree upon applying genetic (or any other effective) remedies to global problems in the first place, we probably would need no recourse to them in the actual event. . . .

So much for the grand designs of genetic engineering. There remain the very real tragedies of genetic disease.

The societal interest in preventing or ameliorating mental retardation and other forms of congenital malformation is obvious. . . . It is also entirely congruent with the needs of the family and, if we believe in the nobility of man and the worth of *human* life, also of the afflicted child as well.

The most effective avenues of preventing genetic disease include (1) the primary prevention of gene mutations, and (2) the detection and humane containment of the DNA lesions once introduced into the gene pool. The 'natural' mutation process in man results in the introduction of a new bit of genetic misinformation once in every ten gametes. Most of the human cost of this 'mutational load' is paid during early stages of fertilization and pregnancy, where it makes up a fair part of the total fetal wastage. But about 2 per cent of newborns suffer from a recognizable discrete genetic defect. This is just the tip of the iceberg; the heritability of many common diseases suggests that from one-fourth to one-half of *all* disease is of genetic origin, for there are important variations in susceptibility to the frankest of environmental insults.

Not all of this health deficit can be attributed to recurrent mutations. An unknown proportion results from the selective advantage that is paradoxically associated with the heterozygous state of many genes, even some with lethal effect in the homozygote (such as sickle-cell hemoglobin). Nevertheless, a significant part of medicine— much more than most practitioners overtly recognize—is in fact directed to lesions that are inherently preventable, if we could control the mutation process in the background.

About a tenth of the 'natural muta-

tion rate' can be attributed to background radiation—from cosmic rays and from radioactive potassium and other isotopes in our natural environment. Therefore, doubling the background, which would correspond to the 'maximum permissible standards' now advocated by federal agencies, would add another 10 per cent to the existing mutation rate: one-ninth rather than one-tenth of our gametes would carry deleterious mutations. This is an enormous impact in absolute terms; a modest increase in relative terms. We must, nevertheless, pay careful attention to the benefits that would be connected with this level of radiation exposure to be sure we are getting a fair bargain.

It must be pointed out that industrial nuclear energy activities today add less than 10 per cent to the average background (hence, less than 1 per cent to the mutation rate); medical x-rays add 50 per cent and 5 per cent, respectively. The same question applies: the now-more-prevalent standards for the judicious and cost-effective use of diagnostic x-rays do not necessarily or automatically excuse the dispensable residue.

A significant portion of 'spontaneous' mutations must be attributed to environmental chemicals, many of which are clearly established as mutagens in laboratory experiments (for example, the peroxy compounds that characterize smog). The extent to which such materials reach the germ cells is absolutely unknown at present. There are good reasons to believe, however, that (1) the inductions of mutations, in germ cells and of cancer and in somatic cells, are fundamentally similar processes—most chemical carcinogens being also mutagenic when properly tested; and (2) a large part of the incidence of cancer is of chemical-environmental origin, cigarette-smoking being only the best-known and best-advertised example. It therefore follows that environmental mutagenesis is equally prevalent. If the relative effects of radiation in the two systems are any hint, the cryptic penalties of the mutations are likely, in the long run, to exact even a larger price in human misery than the short run cancers. . . .

Once a mutation has been allowed to occur in a gamete, and this then participates in fertilization and the production of a new individual, we face a much more difficult problem in any effort at genetic hygiene. For now we must deal with the destinies of human individuals, not merely the chemistry of an isolated segment of DNA. Our problem, seen in the large, is compounded by every humanitarian effort to compensate for a genetic defect, insofar as this shelters the carrier from natural selection. So it must be accepted that medicine, even prenatal care (which may permit the fragile fetus to survive), already intrudes on the question 'Who shall live?,' the challenge so often thrust at rational discussions of policies that might influence the frequencies of deleterious genes. It is so difficult to do only good in such matters that we are best off putting our strongest efforts into the prevention of mutations, so as to minimize the heavy, moral and other, burdens of decision once the gene pool has been seeded with them.

We still cannot evade an evolutionary legacy of genetic damage that would remain with us for generations, even if all new mutation could be stopped by fiat. Our fundamental resources remain very feeble: in a few cases, we can diagnose the heterozygous carriers of recessive mutations, and the genetic counselor can then advise the prospective parents of the odds that they will have affected children. Where voluntary childlessness is unacceptable, it is also sometimes possible to monitor a pregnancy by sampling cells from the amniotic fluid. This can then enable the mother to proceed with confidence, or to request an elective abortion, on the basis of firm knowledge of the genotype of the fetus. We can expect a rapid extension of technical facilities for such diagnoses. At present, they are limited to examination of the chromosomes (for gross chromosomal abnormalities, such as Down's syndrome), and to enzyme assays on cultured cells, which can diagnose a few dozen rare diseases with varying degrees of reliability. We will surely be learning, during the next decade, how to use much more sophisticated approaches to the structure of the DNA and RNA of such cells for more basic diagnostic methods.

In many cases, a deeper understanding of the causal chain by which a DNA alteration leads to pathology may help us devise new forms of therapy to compensate for the genetic defect. This may be as crude as the use of insulin in diabetes, or as subtle as the use of controlled diets in phenylketonuria. (Both approaches are valuable; neither is entirely satisfactory.)

Another approach to constructive therapy, which may mitigate a variety of diseases, is an extension of the existing uses of specific virus strains. At present, their role in medicine is confined to their use as vaccines, for the provocation of immunity against related, wild viruses. This is a specialized example of the modification of cell metabolism by inoculated DNA, discovered empirically by Jenner, and still quite imperfectly understood (our ignorance being concealed by the conceptualizations of clinical virology which still fail to explain just how a vaccine works—e.g., to state just which cells of the vaccinated individual are carrying the viral genetic information, and in what form). We can visualize the engineering of other viruses so that they will introduce compensatory genetic information into the appropriate somatic cells, to restore functions that are blanked out in a given genetic defect. As with vaccine viruses, this presumably will leave the germ cell DNA unaltered, and therefore does not attack the defective gene as such. If we can cope with the disease, should we bother about the gene? Or may we not leave that problem to another generation? . . .

As to cloning: There is no urgent social problem to be addressed by such a technique. It does serve as a metaphor to indicate that future generations will have infinitely more powerful ways than we do to deal with whatever they perceive as [the] socially urgent issues of human nature. We can therefore focus, more confidently, on dealing with the distress of individual human beings in the immediate generation. The cloning issue shows that intrusive genetic engineering, if it is pursued for any other reason, will have plenty of policy problems to digest even before the 'technology' has reached the point of detailed synthesis of genotypes by design.

Finally, medical scientists in general well appreciate and usually respond to ethical concerns about the application of new techniques in man, by contrast to experimental animals. For a long time, it has been known that one could operate on the brain, in such 'interesting' ways as dividing the corpus callosum, with the possibility of the development of autonomous 'intellects' in the two hemispheres. It would be unthinkable to apply such surgical technology to man without the persuasion and conviction that it would be for the benefit of the patient-subject. We will not be given the benefit of the doubt in public discussions of such questions; there are many influential people who really believe that 'anything feasible will be done,' and we may have to restate the obvious many times in reviewing the ethical constraints on possible experimentation.

To return to the 'clone-a-man' metaphor: in my view, we simply do not know enough about the question, at either a technical or an ethical level (and these are intertwined), to dogmatize about whether or not it should ever be done. Certainly it cannot be thought of, within the framework of our generally-accepted standards of medical ethics, unless (1) we can make and communicate a reasonably confident prediction of the outcome, and, more important, (2) it has the informed consent, and serves a reasonable humanitarian purpose, of and for the individuals who are involved. In genetic matters, this must include the interests of the prospective newborn, as well as of his parents, and of the community. If we demand that he be represented in person, then no one could reasonably be allowed to be born, whether by 'natural' sexual fertilization, by the design of his parents, or otherwise. The specific question of 'cloning-a-man' is almost the least important one I can think of; the one it opens up—who must be held to account for the next generation, and how—may be the most.

BIBLIOGRAPHY

This list of references emphaszes recent publications on the genetics of man but there are a few key references from the earlier years. At the very end there is a supplementary bibliography on bioengineering.

ALLEN, GARLAND E. 1970. 'Biology and culture: science and society in the eugenic thought of H. J. Muller.' *Bioscience 20*: 346–353.

AMERICAN EUGENICS SOCIETY. 1961. 'Statement of the eugenic position.' *Eugenics Quarterly 8*:181–184.

AUERBACH, CHARLOTTE. 1961. *The Science of Genetics*. New York: Harper and Row.

BARR, MURRAY L. 1959. 'Sex chromatin and phenotype in man.' *Science 130*: 679–685.

BARTALOS, MIHALY. Editor. 1968. *Genetics in Medical Practice*. Philadelphia: Lippincott.

BATESON, W. 1906. 'An address on mendelian heredity and its application to man.' *Brain 29*: 157–179.

BLACKER, C. P. 1945. *Eugenics in Prospect and Retrospect*. London: Hamish Hamilton.

BLACKER, C. P. 1952. *Eugenics: Galton and After*. London: Duckworth.

BOYER, SAMUEL H. IV. Editor. 1963. *Papers on Human Genetics*. Englewood Cliffs, N.J.: Prentice-Hall. An anthology of important papers.

BROWN, W. M. COURT. 1967. *Frontiers of Biology: Human Population Cytogenetics*. New York: American Elsevier.

CARSON, HAMPTON L. 1963. *Heredity and Human Life*. New York: Columbia University Press.

CARTER, C. O. 1962. *Human Heredity*. Baltimore: Penguin Books.

CASTLE, W. E. 1916. *Genetics and Eugenics*. Cambridge: Harvard University Press. Revised edition, 1924.

CAVALLI-SFORZA, L. L. 1967. 'Human populations.' In *Heritage from Mendel*. Edited by R. A. Brink. Madison: University of Wisconsin Press. Pages 309–331.

CLARKE, C. A. 1964. *Genetics for the Clinician*. Philadelphia: F. A. Davis.

Cold Spring Harbor Symposia in Quantitative Biology. 1964. 'Human Genetics.' Volume 29.

CONKLIN, EDWIN GRANT. 1915. *Heredity and Environment in the Development of Man*. Princeton: Princeton University Press. Fifth edition, 1923.

CREW, F.A.E. 1966. *The Foundations of Genetics*. New York: Pergamon. Pages 155–188.

CROW, JAMES F. 1967. 'Genetics and Medicine.' In *Heritage from Mendel*. Edited by R. A. Brink. Madison: University of Wisconsin Press. Pages 351–374.

DARLINGTON, C. D. 1969. *Genetics and Man*. New York: Schocken Books.

DAVENPORT, CHARLES B. 1911. *Heredity in Relation to Eugenics*. New York: Henry Holt.

DOBZHANSKY, THEODOSIUS. 1962. *Mankind Evolving*. New Haven: Yale University Press.

DOBZHANSKY, THEODOSIUS. 1964. *Heredity and the Nature of Man*. New York: Harcourt, Brace & World.

DUNN, L. C. 1962. 'Cross currents in the history of human genetics.' *American Journal of Human Genetics 14*: 1–13.

EMERY, ALAN E. H. 1968. *Heredity, Disease, and Man. Genetics in Medicine*. Berkeley: University of California Press.

FUHRMANN, WALTER, and FRIEDRICH VOGEL. 1969. *Genetic Counseling. A Guide for the Practicing Physician*. New York: Springer.

GALTON, FRANCIS. 1883. *Enquiries into Human Faculty*. London: Macmillan.

GALTON, FRANCIS. 1889. *Natural Inheritance*. New York: Macmillan.

GALTON, FRANCIS. 1914. *Hereditary Genius: An Inquiry Into Its Laws and Consequences*. Second Edition. London: Macmillan.

GATES, REGINALD RUGGLES. 1946. *Human Genetics*. Two Volumes. New York: Macmillan.

GERMAN, JAMES. 1970. 'Studying human chromosomes today.' *American Scientist 58*: 182–206.

GLASS, BENTLEY. 1943. *Genes and the Man*. New York: Teachers College, Columbia University.

GLASS, DAVID C. Editor. 1968. *Biology and Behavior. Genetics*. New York: Rockefeller University Press and Russell Sage Foundation.

HALLER, MARK H. 1963. *Eugenics: Hereditarian Attitudes in American Thought*. New Brunswick: Rutgers University Press.

HAMERTON, J. L. Editor. 1962. *Chromosomes in Medicine*. London: Heineman.

HANDLER, PHILIP. Editor. 1970. *Biology and the Future of Man*. New York: Oxford University Press. Chapter 20.

HARRIS, HARRY. 1970. *The Principles of Human Biochemical Genetics*. New York: American Elsevier.

HOGBEN, LANCELOT. 1939. *Nature and Nurture*. London: Allen and Unwin.

HOLMES, S. J. 1923. *Studies in Evolution and Eugenics*. New York: Harcourt, Brace.

KING, JAMES C. 1971. *The Biology of Race*. New York: Harcourt Brace Jovanovich.

KNUDSON, ALFRED G. JR. 1965. *Genetics and Disease*. New York: McGraw-Hill (Blakiston Division).

LERNER, I. MICHAEL. 1968. *Heredity, Evolution and Society*. San Francisco: W. H. Freeman.

LEVITAN, MAX, and ASHLEY MONTAGU. 1971. *Textbook of Human Genetics*. New York: Oxford University Press.

LUDMERER, KENNETH M. 1969. 'American geneticists and the eugenics movement: 1905–1935.' *Journal of the History of Biology 2*: 337–362.

MCKUSICK, VICTOR A. 1964. *On the X Chromosome of Man*. Washington: American Institute of Biological Sciences. See also *Quarterly Review of Biology 37*: 69–175.

MCKUSICK, VICTOR A. 1969. *Human Genetics*. Second Edition. Englewood Cliffs, N.J.: Prentice-Hall.

MCKUSICK, VICTOR A. 1971. *Mendelian Inheritance in Man. Catalog of Autosomal Dominant, Autosomal Recessive, and X-linked Phenotypes*. Third Edition. Baltimore: Johns Hopkins University Press.

MCKUSICK, VICTOR A. 1971. 'The mapping of human chromosomes.' *Scientific American*. April 1971. Pages 104–113.

MEDAWAR, P. B. 1960. *The Future of Man*. The BBC Reith Lectures, 1959. London: Methuen.

MILLER, ORLANDO J. 1964. 'The sex chromosome abnormalities.' *American Journal of Obstetrics and Gynecology 90*: 1078–1139.

MITTWOCH, URSULA. 1967. *Sex Chromosomes*. New York: Academic Press.

MOODY, P. A. 1967. *Genetics of Man*. New York: W. W. Norton.

MORGAN, T. H. 1924. 'Human inheritance.' *American Naturalist 58*: 385–409.

MOTULSKY, A. G. Editor. 1969. *Genetic Prognosis and Counseling*. New York: Harper and Brothers.

MULLER, H J., and others. 1939. 'The Geneticist's manifesto.' *Journal of Heredity 30*: 371–373.

NEEL, JAMES V., and WILLIAM J. SCHULL. 1954. *Human Heredity*. Chicago: University of Chicago Press.

NEEL, JAMES V., MARGERY SHAW, and WILLIAM J. SCHULL. Editors. 1965. *Genetics and Epideminology of Chronic Diseases*. Washington: U. S. Department of Public Health, Education and Welfare. Public Service Publication No. 1163.

NEWMAN, HORATIO HACKETT. 1921. *Evolution, Genetics and Eugenics*. Chicago: University of Chicago Press. Second Editor, 1925.

NEWMAN, HORATIO H., FRANK N. FREEMAN, and KARL J. HOLZINGER. 1937. *Twins: A Study of Heredity and Environment*. Chicago: University of Chicago Press.

OSBORN, FREDERICK. 1968. *The Future of Human Heredity. An Introduction to Eugenics in Modern Society*. New York: Weybright and Talley.

PENROSE, LIONEL S. 1963. *Outlines of Human Genetics*. London: Heinemann.

PENROSE, LIONEL S. 1963. *The Biology of Mental Defects*. New York: Grune & Stratton.

RACE, R. R., and R. SANGER. 1962. *Blood Groups in Man*. Oxford: Blackwell.

RAYNER, STURE. 1966. 'Julian Huxley and his views on eugenics in evolutionary perspective.' *Hereditas 56*: 207–212.

REED, SHELDON C. 1963. *Counseling in Medical Genetics*. Philadelphia: W. B. Saunders.

ROBERTS, J. A. FRASER. Editor. 1967. *An Introduction to Medical Genetics*. Fourth Edition. New York: Oxford University Press.

ROSLANSKY, JOHN D. Editor. 1966. *Genetics and the Future of Man*. New York: Appleton-Century-Crofts.

SHIELDS, JAMES. 1962. *Monozygotic Twins Brought Up Apart and Brought Up Together*. New York: Oxford University Press.

SMITH, DAVID. 1964. 'Autosomal abnormalities.' *American Journal of Obstetrics and Gynecology 90*: 1055–1077.

SNYDER, L. H. 1959. 'Fifty years of medical genetics.' *Science 129*: 7–13.

STERN, CURT. 1960. *Principles of Human Genetics*. Second Edition. San Francisco: W. H. Freeman.

STURTEVANT, A. H. 1954. 'Social implications of the genetics of man.' *Science 120*: 405–407.

STURTEVANT, A. H. 1965. *A History of Genetics*. New York: Harper and Row. Chapter 20.

SUTTON, H. ELDON. 1965. *An Introduction to Human Genetics*. New York: Holt, Rinehart & Winston.

THOMPSON, J. S., and M. W. THOMPSON. 1966. *Genetics in Medicine*. Philadelphia: W. B. Saunders.

TURPIN, RAYMOND, and JÉRÔME LEJEUNE. 1969. *Human Afflictions and Chromosomal Aberrations*. New York: Pergamon.

VANDENBERG, STEVEN G. Editor. 1965. *Methods and Goals in Human Behavior Genetics*. New York: Academic Press.

VOLPE, E. PETER. 1971. *Human Heredity and Birth Defects*. New York: Pegasus.

WEYL, NATHANIEL, and STEFAN T. POSSONY. 1963. *The Geography of Intellect*. Chicago: Henry Regnery.

WHITTINGHILL, MAURICE. 1965. *Human Genetics and Its Foundations*. New York: Reinhold.

WILLIAMS, ROGER J. 1953. *Free and Unequal. The Biological Basis of Individual Liberty*. Austin: University of Texas Press.

WORLD HEALTH ORGANIZATION. 1964. *Human Genetics and Public Health*. New York: Columbia University Press.

YUNIS, JORGE J. Editor. 1965. *Human Chromosome Methodology*. New York: Academic Press.

There has been a resurgence of interest in the question of inherited differences in intelligence among the 'races' of man. Should you be interested in this topic the following references will introduce you to the literature:

JENSEN, ARTHUR R., and others. 1969. 'Environment, heredity, and intelligence.' *Harvard Educational Review*. Reprint Series No. 2.

JENSEN, ARTHUR R., RICHARD C. LEWONTIN, and EUGENE RABINOWITCH. 1970. 'The Jensen thesis: three comments.' *Bulletin of the Atomic Scientists*. May 1970. Pages 17–26.

Genetic Engineering, or the application of some of the newer techniques of genetics and developmental biology to control man's future, is discussed in the following books and articles:

AUGENSTEIN, LEROY G. 1968. *Come, Let Us Play God*. New York: Harper & Row.

DARLINGTON, C. D. 1958. 'The control of evolution in man.' *Eugenics Review 50*. Number 3.

DAVIS, BERNARD D. 1970. 'Prospects for genetic intervention in man.' *Science 170*: 1279–1283.

HUISINGH, DONALD. 1969. 'Should man control his genetic future?' *Zygon 4*: 188–199.

LEACH, GERALD. 1970. *The Biocrats*. New York: McGraw-Hill.

LEDERBERG, JOSHUA. 1970. 'Genetic engineering and the amelioration of genetic defect.' *Bioscience 20*: 1307–1310.

LURIA, S. E. 1969. 'Modern biology; a terrifying power.' *Nation*. October 20, 1969. p. 406–409.

MULLER, H. J. 1961. 'The Human Future.' In *The Humanist Frame*. Edited by Julian Huxley. New York: Harper & Brothers.

MULLER, H. J. 1968. 'What genetic course will man steer?' *Bulletin of the Atomic Scientist*. March 1968. Pages 6–12.

RAMSEY, PAUL. 1970. *Fabricated Man. The Ethics of Genetic Control*. New Haven: Yale University Press.

ROSLANSKY, JOHN D. Editor. 1969. *Genetics and the Future of Man*. New York: Appleton-Century-Crofts.

SIMPSON, GEORGE GAYLORD. 1960. 'Man's evolutionary future.' *Zoologische Jahrbücher Abteilung Systematik, Ökologie und Geographie der Tiere 88*: 125–134.

SINSHEIMER, ROBERT L. 1969. 'The prospect for designed genetic change.' *American Scientist 57*: 134–142.

SONNEBORN, T. M. Editor. 1965. *The Control of Human Heredity and Evolution.* New York: Macmillan.

STRAUSS, BERNARD S. 1970. 'Man's future and the "secret of life." ' *Perspectives in Biology and Medicine 14*: 43–52.

TAYLOR, G. R. 1968. *The Biological Time Bomb.* New York: World.

WOLSTENHOLME, GORDON. Editor 1963. *Man and His Future.* London: Churchill.

9/Differentiation

Embryology, the science of the development of organisms, has been closely associated with genetics for more than a century. In 1898, Edmund B. Wilson entitled his monumental synthesis *The Cell in Development and Inheritance*. He and many others of that period saw inheritance and development as two aspects of the same fundamental biological phenomenon. Inheritance is the mechanism of transmitting biological information from one generation to another; development is the using of this information to produce a new individual. An embryo is provided with a complete set of hereditary instructions but, more often than not, with only the most meager supply of materials to carry them out. Its unfailing ability to do so has captured the imagination of generations of biologists. A fertilized egg, often too small to be seen with the unaided eye, divides itself into hundreds or thousands of smaller cells and then reorganizes these into parts of the older embryo and adult. The speed with which this is accomplished, the precision of the steps, and the miniaturization of the parts and processes can only evoke awe. But the most awesome problem of all is differentiation: a cell of one type, or its descendants, changes into a cell of a different type.

The first task of embryologists was to determine *what* happens during the course of development in a wide variety of animals and plants. They observed the sorts of events described for the amphibian egg in Chapters 10 and 11 of *Heredity and Development*.

EDMUND B. WILSON

In the closing years of the nineteenth century, an increasing number of embryologists became interested in *how* development is controlled. Many of their first experiments might be described as a geometry of development. Knowing what a whole embryo could do, they sought to determine the capabilities of the parts and their interrelations. This line of investigation led to the mosaic and regulative theories of development, which are useful ways of looking at development even today. The first selection, by Edmund B. Wilson, is devoted to this aspect of the history of ideas in development.

The Mosaic Theory of Development.
EDMUND B. WILSON

A remarkable awakening of interest and change of opinion has of late taken place among working embryologists in regard to the cleavage of the ovum. So long as the study of embryology was dominated by the so-called biogenetic law, so long as the main motive of investigation was the search for phyletic relationships and the construction of systems of classification, the earlier stages of development were little heeded. The two-layered gastrula was for the most part taken as the real starting-point for research, and the segmentation stages were briefly dismissed as having little purport for the more serious problems involved in the investigation of later stages. The cleavage is equal or unequal, total or partial, regular or irregular; the diblastic condition attained by delamination, migration or invagination; the gastrulation embolic or epibolic:—such were the general conclusions announced regarding the præ-gastrular stages in a large proportion of the embryological papers published down to the time of Balfour and even later. The last decade has, however, witnessed so extraordinary a change of front on this subject that it will not be out of place to review briefly the three leading causes by which it has been brought to pass.

First, it has become more and more clear that the germ-layer theory is, to a certain extent, inadequate and misleading, and that even the primary layers of the 'gastrula' cannot be regarded as strictly homologous throughout the animal kingdom. To assume that they are so involves us in inextricable difficulties—such as those for instance encountered in the comparison of the annelid gastrula with that of the chordates, or the comparison of the sexual and asexual modes of development in tunicates, bryozoa, worms and cœlenterates. This consideration led some morphologists to insist on the need of a more precise investigation of the præ-gastrular stages, and the desirability of taking as a starting-point not the two-layered gastrula but the undivided ovum. 'The "gastrula" cannot be taken as a starting-point for the investigation of com-

From *Biological Lectures Delivered at the Marine Biological Laboratory of Wood's Holl in the Summer Session of 1893*. Ginn & Co., Boston. 1894.

parative organogeny unless we are certain that the two layers are everywhere homologous. Simply to assume this homology is simply to beg the question. *The relationship of the inner and outer layers in the various forms of gastrulas must be investigated not only by determining their relationship to the adult body, but also by tracing out the cell-lineage or cytogeny of the individual blastomeres from the beginning of development.'*

The second of the causes referred to was the discovery of the so-called pro-morphological relations of the segmenting ovum. It is now just ten years since Roux and Pflüger independently announced the discovery that the first plane of cleavage in the frog's egg coincides with the median plane of the adult body (a fact announced many years earlier by Newport, whose observation fell, however, into oblivion). The same result was soon afterwards reached in the case of the cephalopod (Watase) and tunicate (Van Benden and Julin), and for a time it seemed not improbable that a general law had been determined. Later researches disappointed this expectation; for it was demonstrated that the first cleavage plane may be transverse to the body (annelids, gasteropods, urodeles), or even in some cases show a purely variable and inconstant relation (teleosts). The fact remained, however, that in the greater number of known cases definite relations of symmetry can be made out between the early cleavage stages and the adult body; and this fact invested these stages with a new and captivating interest.

The third and most important cause lay in the new and startling results attained by the application of experimental methods to embryological

study, and especially to the investigation of cleavage. The initial impulse in this direction was given in 1883 by the investigations of Pflüger upon the influence of gravity and mechanical pressure upon the segmenting ova of the frog. These pioneer studies formed the starting-point for a series of remarkable researches by Roux, Driesch, Born and others, that have absorbed a large share of interest on the part of morphologists and physiologists alike; and it is perhaps not too much to say that at the present day the questions raised by these experimental researches on cleavage stand foremost in the arena of biological discussion, and have for the time being thrown into the background many problems which were but yesterday generally regarded as the burning questions of the time. It is the purpose of this lecture to consider, briefly, the most central and fundamental subject of the current controversy.

It is an interesting illustration of how even scientific history repeats itself that the leading issue of to-day has many points of similarity to that raised two hundred years ago between the præ-formationists and the epigenesists. Many leading biological thinkers now find themselves compelled to accept a view that has somewhat in common with the theory of præ-formation, though differing radically from its early form as held by Bonnet and other evolutionists of the eighteenth century. No one would now maintain the archaic view that the embryo præexists *as such* in the ovum. Every one of its hereditary characters is, however, believed to be represented by definite structural units in the idioplasm of the germ-cell, which is therefore conceived as a kind of microcosm, not similar to, but a perfect symbol of, the macrocosm to which it gives rise (Hertwig). In its modern form this doctrine was first clearly set forth by Darwin in the theory of Pangenesis ('68). Twenty years later ('89) it was remodeled and given new life by Hugo de Vries, in a profoundly interesting treatise entitled *Intracellular Pangenesis* and in its new form was accepted by Oscar Hertwig, and pushed to its uttermost logical limit by Weismann. Kindred theories have been maintained by many other leading naturalists.

The considerations which have led to the rehabilitation of the theory of pangenesis are based upon the facts of what Galton has called *particulate inheritance.* The phenomena of atavism, the characters of hybrids, the facts of spontaneous variation, all show that even the most minute characteristics may independently appear or disappear, may independently vary, and may independently be inherited from either parent without in any way disturbing the equilibrium of the organism, or showing any correlation with other variations. These facts, it is argued, compel the belief that hereditary characteristics are represented in the idioplasm by distinct and definite germs ('pangens,' 'idioblasts,' 'biophores,' *etc.*), which may vary, appear or disappear, become active or latent, without affecting the general architecture of the substance of which they form a part. Under any other theory we must suppose variations to be caused by changes in the molecular composition of the idioplasm as a whole, and no writer has shown, even in the most approximate manner, how particulate inheritance can thus be conceived.

Based upon this conception two radically different theories of development have recently been propounded. The first of these—the so-called mosaic theory of Roux and Weismann, which forms the subject of this lecture—is based upon the assumption that the cause of differentiation lies in the nature of cell-division. Karyokinesis is conceived as qualitative in character in such wise that the idioplasmic germs are sifted apart, and cells of different prospective values receive their appropriate specific germs at the moment of their formation. The idioplasm therefore becomes progressively simpler as the ontogeny goes forward, except in the case of the germ-cells; these retain a store of the original mixture ('germ-plasm' of Weismann). Every cell must therefore possess an independent power of self-determination inherent in the specific structure of its idioplasm, and the entire ontogeny is aptly compared by Roux to a mosaic-work; it is essentially a whole arising from a number of independent self-determining parts, though Roux qualifies this conception by the admission that the self-determining power of the cell is capable in some measure, of modification, through interaction with its fellows ('correlative differentiation').

In the hands of Weismann this theory attains truly colossal proportions. The primary germs or units (which he calls 'biophores') are aggregated to form 'determinants,' the determinants to form 'ids,' and the ids to form 'idants,' which are identified with the chromosomes of the ordinary karyokinetic figure. Upon this basis is reared a stately group of theories relating to reproduction, variation, inheritance and regeneration, which are boldly pushed to their utmost logical limit. These theories await the judgment of the future. Brilliantly elaborated and persuasively presented as they are, they do not at present, I believe, carry conviction to the minds of most naturalists, but arouse a feeling of scepticism and uncertainty; for the fine-spun thread of theory leads us little by little into an unknown region, so remote from the *terra firma* of observed fact that verification and disproof are alike impossible.

In its original form the mosaic theory has, I believe, received its death-blow from the facts of experimental embryology, though both Roux and Weismann still endeavor to maintain their position. It is rather curious that the very line of research struck out by Roux, by which he was led to the mosaic theory, should in later years have ended in a view diametrically opposed to his own. In 1888 Roux succeeded in killing (by puncture with a heated needle) one of the first two blastomeres of the segmenting frog's egg. The uninjured blastomere continued its development as if still forming a part of an entire embryo, giving rise successively to a half-blastula, half-gastrula, and half-tadpole embryo, with a single medullary fold. Analogous results were reached by operation upon four-celled stages. It was this result that led Roux to compare the development to a mosaic-work, asserting that 'the development of the frog-gastrula, and of the embryo immediately derived from it is, from the second cleavage onward, a mosaic-work, consisting of at least four vertical independently developing pieces.' Roux himself, however, showed that in later stages the missing half (or

fourth) is perfectly restored by a process of 'post-generation,' which begins about the time of the formation of the medullary folds—a result which, in itself, really contradicts the mosaic hypothesis; for the course of events in the uninjured blastomere, or its products, is radically altered by changes on the other side of the embryo.

A more decisive result was reached in 1891 by Driesch, who succeeded, in the case of *Echinus,* in effecting a complete separation of the blastomeres by shaking them apart. A blastomere of the 2-celled stage, thus isolated, gave rise to a perfect but half-sized blastula, gastrula, and Pluteus larva; an isolated blastomere of the 4-celled stage produced a perfect dwarf gastrula one-fourth the normal size. Even in this case, however, the earliest stages of development (cleavage) showed traces of the normal development, the isolated blastomere segmenting, as if it were a half-embryo, and only becoming a perfect whole in the blastula stage. In the following year, however, the writer repeated Driesch's experiments in the case of *Amphioxus* (the egg of which is extremely favorable for experiment), and found that in this case there is, as a rule, no preliminary half-development whatever. The isolated blastomere behaves from the beginning like an entire ovum of one-half or one-fourth the normal size.

It is quite clear that in *Amphioxus* the first two divisions of the ovum are not qualitative, as the mosaic theory assumes, but purely quantitative; for the fact that each of the two or four blastomeres may give rise to a perfect gastrula proves that all contain the same materials. Nevertheless, in the normal development, these cells give rise to different structures—*i.e.,* they have a different prospective value—from which it follows that, in this case at least, differentiation is not caused by qualitative cell-division, but by the conditions under which the cell develops.

These facts are obviously a serious blow to the mosaic theory, and the efforts of Roux and Weismann to sustain their hypothesis in the face of such evidence only serve to emphasize the weakness of their case. In order to explain the facts of post-generation— *i.e.,* the capability of isolated blastomeres to produce complete embryos— both Roux and Weismann are compelled to set up a subsidiary hypothesis, assuming that during cell-division each cell may receive, in addition to its specific form of idioplasm, a portion of unmodified idioplasm afforded by purely quantitative division. This unmodified idioplasm ('accessory idioplasm' of Weismann, or in some cases 'germ-plasm'; 'post-generation or regeneration idioplasm' of Roux) remains latent in normal development which is controlled by the active specific idioplasm. Injury to the ovum— *e.g.,* mechanical separation of the blastomeres—acts as a stimulus to the latent idioplasm, which thereupon becomes active, and causes a repetition of the original development. By assuming a variable latent period following the stimulus, Roux is able to explain the fact that regeneration takes place at different periods in different animals.

Considered as a purely formal explanation this subsidiary hypothesis is perfectly logical and complete. A little reflection will show, however, that it really abandons the entire mosaic position, by rendering the assumption of qualitative division superfluous; and, aside from this, its forced and artificial character, places a strain upon the mosaic theory under which it breaks down. Both of the two fundamental postulates of the modified theory—*viz.,* qualitative nuclear division, and accessory latent idioplasm— are purely imaginary. They are complicated assumptions in regard to phenomena of which we are really quite ignorant, and they lie at present beyond the reach of investigation. The 'explanation' is, therefore, unreal; it carries no conviction, and no real explanation will be possible until we possess more certain knowledge regarding the seat of the idioplasm (which is entirely an open question), and its internal composition and mode of action (which is wholly unknown). In the meantime we certainly are not bound to accept an artificial explanation like that of Roux, however logical and complete, unless it can be shown that the phenomena are not conceivable in any other way.

We turn now to a brief consideration of opposing views, among which I ask attention especially to those of Driesch and Hertwig. In common with Kölliker and many other eminent authorities, these authors insist that cell-division is not qualitative but quantitative only, and hence is not, *per se,* a cause of differentiation, for there is no sifting apart of the idioplasmic units, but an equal distribution of them to all the cells of the body. In other words, the cleavage of the ovum does not effect an analysis of the idioplasm into its constituent elements, but only breaks it up into a large number of similar masses. Differentiation follows upon cell-division, is caused by the interaction of the parts of the embryo, and the character of the individual cell is determined by its environment—*i.e.*, by its relation to the whole of which it forms a part. 'The egg,' says Hertwig, 'is an organism, which multiplies by division to form numerous organisms equivalent to itself, and it is through the interactions of all these elementary organisms, at every stage of the development, that the embryo, as a whole, undergoes progressive differentiation. The development of a living creature is therefore in no wise a mosaic work, but, on the contrary, all the individual parts develop in constant relation one to another, and the development of the part is always dependent on the development of the whole.' There is therefore no necessary relation between the individual blastomeres of the segmenting ovum and the parts of the adult body to which they give rise; this relation is purely fortuitous. The most extreme statement of this view appears in the writings of Pflüger and Driesch. 'I would accordingly conceive,' says Pflüger, 'that the fertilized egg has no more essential relation to the later organization of the animal than the snowflake has to the size and form of the avalanche which, under appropriate conditions, may develop out of it.' Driesch, writing ten years later ('89), is no less explicit. He regards the blastomeres of the *Echinus* embryo, as "composed of an indifferent material, so that they may be thrown about at will, like balls in a pile, without the least impairment of their power of development." The ultimate fate of any particular blastomere is determined by its relative position in the mass; that is (to quote his own striking aphorism), 'their prospective value (*Bedeutung*) is a function of their location' (cf. His).

We shall presently return to these more extreme views, but I will here point out one all-important point which is definitely established by the work of Driesch and other experimentalists, and which is accepted by all opponents of the mosaic theory, namely, that the cell cannot be regarded as an isolated and independent unit. The only real unity is that of the entire organism, and as long as its cells remain in continuity they are to be regarded, not as morphological individuals, but as specialized centres of action into which the living body resolves itself, and by means of which the physiological division of labor is effected. This view, at which a number of embryologists have independently arrived, has been most ably urged by Whitman, in one of the lectures of this volume, though in connection with a general conception of development peculiarly his own.

It is important not to lose sight of the fact that Hertwig, no less than Roux and Weismann, conceives the idioplasm (which he would locate in the cell-nucleus) as an aggregate of units ('idioblasts') which severally correspond to the hereditary qualities of the organism; and since cell-division is not qualitative, every cell must contain the sum total of the hereditary character of the species. Differentiation is conceived by Hertwig (following de Vries) as the result of physiological changes in the idioblasts, some of which remain latent, while others become active, and thus determine the specific character of the cell, according to the nature of the active idioblasts. In regeneration such of the latent idioblasts are called into action as are necessary to carry out the regenerative process.

We have found good reason for the conclusion that the mosaic theory cannot, in its extreme form, be maintained. It remains to inquire whether the extreme anti-mosaic conception rests upon a more secure foundation, and whether the mosaic hypothesis may not contain certain elements of truth. I have elsewhere more than once pointed out that the views of Hertwig and Driesch have received a strong bias, from the circumstance that the discussion has hitherto been confined mainly to the echinoderm egg, which shows no visible differentiation in the cells until a relatively late period (16-celled stage).

The whole question assumes a somewhat different aspect when we regard such highly differentiated types of cleavage as we find, for example, among the annelids; and I would ask attention for a moment to the case of *Nereis*, which is, at present, the best known form. Differentiation here begins at the very first cleavage (which is conspicuously unequal), and it becomes more pronounced with every succeeding division. The median plane is marked out at the second cleavage; at the third the entire ectoblast of the trochal and præ-trochal regions is formed; at the fourth the material for the entire 'ventral plate' (including the ventral nerve-cord and the seta-sacs) is segregated in a single cell, that for the stomodæum in three cells; the fifth cleavage completes the ectoblast, and by the 38-celled stage the germ-layers are completely segregated (the mesoblast in a single cell) and the architecture of the embryo is fully

outlined in the arrangement of the parent blastomeres, or protoblasts.

We do not know whether, in this case, the first two blastomeres are qualitatively different, though there may be some ground for holding that they are, from the fact that the larger of the two contains a relatively larger proportion of protoplasm than the smaller.[1] But in any case their difference in size renders it impossible that they should play interchangeable parts in the cleavage. The entire later development is, however, moulded upon the 2-celled stage, every blastomere having a definite relation to it and a definite morphological value. The development is a visible mosaic-work, not one ideally conceived by a mental projection of the adult characteristics back upon the cleavage stages. The principle of 'organbildende Keimbezirke' has here a real meaning and value, and this would remain true even if it should hereafter be shown that both of the first two blastomeres of *Nereis*, if isolated, could produce a perfect embryo.

It is clear, from such a case, that the more extreme views of Driesch and Hertwig cannot be accepted without considerable modification. It seems to me, however, that they may be modified in such a way as, without sacrificing the principle of epigenesis for which they contend, to recognize certain elements of truth in the mosaic hypothesis; and I will attempt to indicate this modification by a comparison between *Amphioxus* and *Nereis*. In the case of *Amphioxus* we have the clearest evidence that differ-

---

[1] All my attempts to separate these blastomeres by shaking have thus far been unsuccessful.

entiation is, in a measure, dependent upon the relation of the cell to the whole of which it forms a part. The first visible differentiation in this case is at the third cleavage, which consists in an unequal division of each of the four blastomeres, so as to give rise to four micromeres and four macromeres, the former giving rise to ectoblast only, while the latter give rise to entoblast and mesoblast as well *(Diagram* I). If, however, the blastomeres of the 4-celled stage be separated (shaken apart) the course of events is entirely changed; for in this case each divides equally, not unequally, and ultimately gives rise to a complete quarter-sized dwarf, instead of one-quarter of a normal embryo, as it would have done under ordinary circumstances. The character of the fourth cleavage is here directly or indirectly determined in each cell by the relation of the cell to its fellows; and if this is true of any one stage of the ontogeny, a very strong presumption is created that it is true of all—that, in the process of progressive differentiation occurring in the course of every animal ontogeny, the character of each step is determined by the condition of the entire organism. The ontogeny is, in other words, a connected series of interactions between the various parts of the embryo, in which

each step establishes new relations, through which the following step is determined. The character of the series, as a whole, depends upon the first step, and this in turn upon the constitution of the original ovum. In *Amphioxus* differentiation proceeds slowly, the earlier blastomeres show no appreciable divergence, and the first stages show no trace of a mosaic work. In *Nereis,* on the other hand, a mosaic-like character appears from the beginning, because of the inequality of the first cleavage, which conditions the entire subsequent development through the peculiar relations established by it. The cause of the inequality must lie in the undivided ovum, and a study of the first cleavage-spindle shows that the inequality is unmistakably foreshadowed before the least outward sign of division appears; for the asters at the spindle-poles are conspicuously unequal in size, the larger aster corresponding with the future larger cell, (Diagram II). This difference is not connected with any determinable mechanical conditions; for the centrosomes lie nearly equidistant from the membrane (the egg is spherical), and the deutoplasm shows no perceptible inequality in horizontal distribution. The conclusion seems unavoidable that the differentiation in size is caused by a specific form of activity in the cytoplasm (or archoplasm), occurring prior to cell-division. But if a differentiation in size may have such an origin, we may fairly argue that other differentiations may likewise precede cell-division, and that in such cases the division may be, in a sense, qualitative.

It seems to me, that in these considerations we may find, in some

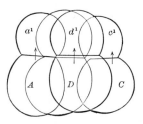

Diagram I

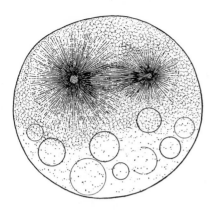

Diagram II.

measure, a reconciliation between the extremes of both the rival theories under discussion—that we may consistently hold with Driesch that the prospective value of a cell may be a function of its location, and at the same time hold with Roux that the cell has, in some measure, an independent power of self-determination due to its inherent specific structure. Such a view is only possible, however, if we regard the specific structure of the cell to have arisen not through the segregation and isolation within its boundaries of special idioblasts or germ-substances, that have been sifted out by qualitative division, but through a physiological specialization (as de Vries and Hertwig insist) that may have taken place before, during, or after cell-division, according to circumstances. If differentiation precedes or accompanies division, the latter process may be in a sense qualitative. If it follows, division will be purely quantitative, and in such a case we may rightly speak of differentiation as a result of cellular interaction. The segmentation of the egg presents more or less of a mosaic-like character, ac-

cording to the period at which differentiation appears, and the rate at which it proceeds, as expressed in limitations of the power of development in the individual blastomeres, and the differences in size and structure.

The general interpretation of development which I have thus endeavored to sketch will be found to differ widely in some respects from that set forth in one of the subsequent lectures of this volume, from which, through Professor Whitman's courtesy, I am enabled to quote. Whitman argues that 'cell-orientation may enable us to infer organization, but to regard it as a measure of organization is a serious error.' 'The question as to the presence of organization,' he says, 'is not settled by the *form* of cleavage. Eggs that admit of complete orientation at the first or second cleavage, or even before cleavage begins, are commonly supposed to reflect *precociously* the later organization, while eggs in which such early orientation is impossible are supposed to be more or less completely isotropic and destitute of organization. When the region of apical growth is represented by conspicuous teloblasts, the fate of which is seen to be definitely fixed from the moment of their appearance, we find it impossible to doubt the evidence of organization, or "precocious differentiation" as it is conventionally called. When the same region is composed of more numerous cells, among which we are unable to distinguish special proliferating cells, we lapse into the irrational conviction that the absence of definitely orientable cells means just so much less organization.'

It would be manifestly out of place to enter here upon any of the interesting discussions suggested by the passage just quoted, and I will therefore only add that Professor Whitman's position seems to me to rest upon a special and peculiar use of the word 'organization,' and that his view leads to a denial of the principle of epigenesis. No one would maintain that the living egg is 'destitute of organization,' but neither can any one maintain that the egg-organization is identical with that of the adult. Development is essentially a transformation of one form of organization into another along the path of cell-division and cell-differentiation; and it is undeniable that the adult form of organization is thus expressed earlier in some cases than in others—for example, in the segregation of the germ-layers in the polyclade, as compared with the annelid or gasteropod. We are still profoundly ignorant of the nature and causes of differentiation, and of its precise relation to cell-formation; and the question is probably not yet ripe for discussion. It is, however, impossible to maintain that differentiation in the Metazoa is entirely independent of cell-formation, when we recall the multitude of cases in which the lines of differentiation coincide with cell-boundaries.

HANS SPEMANN

Much of the fundamental work in experimental embryology during the early years of the twentieth century was done on the amphibian egg. Its development is described in Chapters 10 to 12 of *Heredity and Development*. In a long series of experiments, Hans Spemann (1869–1941) and his fellow workers made a profound analysis of the causal factors in early development. The culminating discovery was that of the organizing action of the dorsal lip of the blastopore. This is described by Spemann in a Croonian Lecture to the Royal Society.

CROONIAN LECTURE:
Organizers in Animal Development.
By HANS SPEMANN,
Professor of Zoology, Freiburg i. B.
(Lecture delivered Nov. 3, 1927.)
I.—*Introduction.*

One of the most remarkable features of the phenomena of life is their uniformity in the most unlike living forms. This fact enables us to choose for our researches such forms as will be, for one reason or another, the most suitable for observation or experiment. The results thus reached must, however, by no means be generalized without further proof; but again and again observations made on favourable subjects have been confirmed by further study in less favourable ones. The fertilization of the egg, for instance, was observed first in

From *Proceedings* of the Royal Society B, *102*: 177–187. 1927. Reprinted by permission.

the eggs of the sea-urchin, that is, in a species as unlike as possible to any higher animal; yet it has been shown since that fertilization goes on in the same way in all animals. The beginnings of a human individual do not differ essentially from those of an echinoderm.

I hardly suppose that many of you have up to the present taken any particular interest in the early development of the common newt, my only subject of research for the greater part of my life; I hardly think it is to such interest that I owe the honour of delivering this lecture. I imagine it was rather the conviction that the laws of development established for this low vertebrate hold true for all vertebrates, nay, for all animals; that development even of man follows the same principles. But for the same reason I dare not presume that all of you have present in your mind the first steps of development of the newt's egg. It will, therefore, be my first task to recall them to your memory.

(a) Scheme of Normal Development of the Triton Egg.—From the fertilized egg, morphologically a single cell, there arises by typical cell-divisions the *blastula,* containing a cavity, the *blastocœl,* formed by cells of very different size, shape and quality. The thick floor of the blastula is composed of the large yolk-laden vegetative cells; its thin roof is formed by the much smaller animal cells, with little yolk; the cells between both of them are middle-sized, forming the so-called marginal zone.

In a predetermined point a little below this zone an invagination takes place, growing deeper and deeper and advancing to both sides; this is the *blastopore,* with the *blastopore-lips.*

By this invagination the whole material of the floor and the marginal zone of the blastula is turned inside, forming the *archenteron.* Its cavity, still communicating with the blastopore, widens in front, whilst at the same time the original cavity of the blastula is reduced to a mere slit. This stage of development is known as the *gastrula,* and the process of invagination by which it is reached is called *gastrulation.*

The cell-layer derived from the animal cells of the blastula is called the outer germ layer or *ectoderm;* it gives origin to the epidermis and the whole nervous system. The floor of the archenteron, the *entoderm,* formed by the large yolk cells of the blastula, is the rudiment of the intestine; in this early stage it is not yet a closed tube, but an open furrow, its sides bending upwards without closing together. The opening of the furrow is covered by a plate of middle-sized cells, called the *mesoderm.* They are derived from the cells of the marginal zone, and form, as it were, the roof of the open archenteron. Out of this mesodermal plate are formed primarily muscles and skeleton, kidneys and peritoneum.

Now, to conclude this description, a few words about the origin of the central nervous system. The ectoderm in front of the blastopore, on the dorsal side of the gastrula, thickens to form the *medullary plate,* broad in front, narrower at the hinder end, surrounded by the *medullary folds.* This plate changes first into a furrow, then into a tube, covered by the surrounding epidermis. It is the *medullary tube,* the rudiment of the brain and spinal cord from the fore-end of this tube the rudiments of the eyes are pushed out, the so-called *primary eyeballs;* a little farther back the *ear-vesicles* are formed, the rudiments of the labyrinth.

(b) Deduction of the Problem of Determination.—From the facts of normal development thus outlined may be derived the problem with which our experiments were dealing. By an ingenious method worked out by W. Vogt, we may mark a definite part of the young gastrula by staining it *intra vitam* with nile-blue, and may follow it through development. It may thus be subsequently identified, for instance, in the medullary plate and later on in the primary eyeball. This part of the gastrula would then have been *'presumptive medullary plate,' 'presumptive primary eyeball.'* Another part farther away from the blastopore would prove to be *presumptive epidermis.*

Now the question arises: when will these parts of the gastrula *be determined* to differentiate either to medullary plate or to epidermis? Are these organs perhaps already determined at the beginning of gastrulation? Further development would then be *self-differentiation,* according to Roux. Or will they be determined later on, perhaps during gastrulation?

II.—*Experimental Deduction of the Conception of 'Organizer.'*

(a) Exchange of Presumptive Medullary Plate and Epidermis.—This alternative may be decided by a simple experiment. Between two embryos of the same age, young gastrula stage, little pieces of presumptive medullary plate and presumptive epidermis are exchanged. Using embryos of different colour, either natural or stained, we may identify the transplanted pieces

for many days and follow them in their development. As the pieces have been exchanged, each of them serves as a mark for the original place of the other piece, showing exactly what it would have become.

In fact, each piece develops not according to its origin, but according to the new place it occupies, presumptive medullary plate becoming epidermis in epidermis, presumptive epidermis becoming medullary plate in medullary plate. Hence it may be concluded:—

(1) That presumptive medullary plate and epidermis are relatively indifferent at the beginning of gastrulation.

(2) That there is some factor acting in the new place that determines the definite fate of the transplanted pieces.

Small pieces of the young embryo may be exchanged as well between different species, *i.e.,* heteroplastically. A very favourable combination is that of *Triton cristatus* and *Triton tæniatus* or *alpestris,* the eggs of the former being almost free from pigment, whilst those of the latter are much darker, especially those of *alpestris,* and densely laden with minute pigment granules.

In this way from a young gastrula of *tæniatus* a piece of presumptive medullary plate was taken, from a gastrula of the same age of *cristatus* a piece of presumptive epidermis; these pieces were exchanged. Both embryos developed normally. In the *tæniatus* after two days the white implanted piece of *cristatus* was seen in the medullary plate, fore end left side. Later on, when the folds had closed and the brain with the eyeballs was formed, the light piece was still visible shining through the skin. On sections through the head it is found in the brain and primary eyeball of the left side forming part of their wall. It is somewhat thicker than its surroundings, its cells are larger and free from pigment. It had been presumptive epidermis of *cristatus* and has formed a piece of brain, but brain of *cristatus*, amidst the brain of *tæniatus*. The embryo of *cristatus*, from which this piece had been taken, showed the implanted piece of *tæniatus* on the right side in the skin, in the region of the gill rudiments. Though presumptive brain, it had become epidermis, amidst the epidermis of the host; but epidermis of *tæniatus* according to its origin, with all its tendencies of growth. This could be seen by the form of the gills, which was that of *tæniatus*. The gills were further developed on the operated side; they contained rudiments of blood vessels of which nothing was to be seen on the other side.

(b) Ectoderm and Mesoderm exchanged.—By the same method O. Mangold examined the potencies of ectoderm and mesoderm, exchanging pieces of them at the beginning of gastrulation. He found that a piece of presumptive ectoderm transplanted in the region of the blastopore may be invaginated with the surrounding material and give rise to mesodermal organs such as notochord, somites, kidneys.

(c) The Centre of Organization.—Not all parts of the embryo are as plastic as this. A piece taken from the upper lip of the blastopore and transplanted into an indifferent region of another gastrula behaves in a totally different way. It does not adapt itself to the development of its surroundings, but sticks to its own way and forces the surrounding parts to follow its own direction. There arises a second embryo composed partly of the transplanted material, partly of cells of the host that have been induced to adapt themselves to the development of those ruling cells.

This experiment was made for the first time on my suggestion by Hilde Mangold. In the most instructive case which she obtained, a secondary embryo was formed with a medullary tube bearing ear-vesicles on its fore end, with notochord and two rows of somites. It was composed of cells of *cristatus* derived from the implanted piece, and cells of *tæniatus*, the host, that had adapted themselves to their development. The boundaries of the implanted piece, though very sharp and clear, were not respected by the general form.

Such a transplanted piece 'organizes' its new surroundings; I therefore call it an 'organizer.' It induces a 'field of organization.' The region of the early gastrula where these organizers lie may be called for the present a 'centre of organization.'

III.—*Experimental Analysis of the Function of the Organizer.*

(a) Extent of the Centre of Organization.—This point being reached we proceeded to a methodical analysis of the organizing faculty. The next question presenting itself was that of the extent of the centre of organization. After some preliminary experiments by Otto and Hilde Mangold, this question has been definitely settled by H. Bautzmann. He examined the surroundings of the beginning blastopore by transplanting small trial-pieces into the blastocœl of another

gastrula. By gastrulation these pieces are brought under the ectoderm and may show their power of induction. The region of the gastrula so determined as 'centre of organization' occupies a semicircular area above and beside the upper lip of the blastopore. This area coincides with that part of the gastrula which had been shown by W. Vogt to invaginate during gastrulation and form notochord and mesoderm.

This result is in harmony with a statement made by another student of mine, A. Marx, that a medullary plate may be induced by a piece of the roof of the archenteron, that is, of the material for notochord and mesoderm. The experiments of Bautzmann show, in addition, that these cells have this power even before gastrulation, and may not only induce medullary plate in the overlying ectoderm, but also mesodermal organs in their new environment.

(b) Origin of the Centre of Organization.—In the young gastrula the centre of organization is already determined; when does it originate? Probably very early in development, together with that shifting of material of the unsegmented egg that follows fertilization and produces the well-known 'grey crescent,' that area of the egg where later on invagination begins. The facts which point in this direction are the following: If in the 2-cell stage the dorsal blastomere is eliminated, that is, that part of the egg which contains the grey crescent, the ventral blastomere, though going on developing, and even gastrulating, forms no axial organs, that is, no medullary plate, notochord, somites. The same holds true for the case in which the dorsal half of the *unsegmented egg* has been cut off.

An older experiment of M. Moszkowski teaches the same thing still more conclusively. In the unsegmented egg the grey crescent was destroyed by puncture with a hot needle; the egg developed; it even gastrulated, but formed no axial organs. The crucial evidence would be induction of a medullary plate by a transplanted piece of the grey crescent. Such a piece when containing a supernumerary spermatozoon develops in the blastocœl. This experiment has been performed by me many times, but has not yet yielded a clear and positive result.

(c) Intimate Structure of the Centre of Organization.—Several facts are known concerning the intimate structure of the centre of organization.

Longitudinal Structure.—It is a significant fact that the secondary embryo need not have the same orientation as the primary. If it had, this would have pointed to some intimate structure of the host, in accord, perhaps, with Child's idea of gradients. Now such gradients may exist in the host and may be effective; there are reasons to believe it. Yet their influence is surely not decisive. The induced secondary embryo may even lie at right angles to the primary one. This is most striking when ear-vesicles are formed; they lie symmetrically to the right and left of the secondary medullary tube to which they are attached, but in relation to the host one behind the other. Therefore the organizer must have some longitudinal structure, which is not lost by transplantation and which determines the direction of its invagination and consequently of the induction. To clear up this question, organizers of oblong shape should be implanted in definite directions. Such experiments have been begun by Dr. Geinitz some years ago and have now been taken up by myself.

Laterality.—There must also be some medio-lateral structure of the centre, 'laterality,' to use Streeter's term. This may be concluded from experiments carried out quite recently by W. Vogt and K. Görttler, the knowledge of which I owe to the kindness of the authors. W. Vogt experimenting on Triton retarded the development in limited parts of the egg, either by localized lack of oxygen or localized cooling. For the latter purpose a piece of thin silver-plate was perforated by a hole, just large enough to hold a salamander's egg, deprived of its shell. The plate was fixed in a chamber, dividing it in two parts, in which water of different temperatures was circulating. The communication between the two parts was then closed by a salamander's egg, as by a cork. In consequence the two halves of the egg developed at very different rate; an embryo was produced that was composed of two halves of very different 'age,' with a perfectly sharp limit between them. For example, in one case, where the limit was transverse, there resulted an embryo with a rather long tail, but instead of a brain it had only a medullary plate.

Of greater interest for the question we are dealing with are such cases where the limit is median, dividing two lateral halves of very different age. From this Vogt concludes that the two halves developed independently from each other, as Roux had concluded from his classical experiments on half embryos. Vogt further concludes that this development is based on laterality of the two halves already existent in the 2-cell stage.

By a strange coincidence Prof. Julian Huxley had been performing similar experiments at about the same time, which have just been published in the last few days in the jubilee volume for Hans Driesch. His method differed from that of Vogt in that the gradients of temperature he applied were much less abrupt, so that his embryos did not show such a sharp limit between the two halves.

K. Görttler cut out a lateral piece of the upper lip of the blastopore and replaced it by the corresponding piece of the other side of another embryo, so that the gastrula went on developing with two left lips. Then two left halves of the medullary plate were formed, with two left medullary folds. The left half of the lip, as we must conclude, had some intimate structure that made it 'left'; that is, it had laterality.

This experiment of Görttler reaches farther than that of Vogt. From the results of Vogt a half-structure could only be inferred for the embryo as a whole, whilst the results of Görttler prove a half-structure more especially for the centre of organization, that is, of the presumptive mesoderm, which is invaginated during gastrulation and induces the medullary plate. The ectoderm of which the latter is formed, the presumptive medullary plate, may have had a similar structure; but this very fact would show the dominance of the organizing cells which can reverse the laterality of the induced material.

Regional Structure. — There are some facts that indicate yet another kind of structure of the centre of organization. This might be called a *regional structure*, as it determines the regions both of the mesoderm and of the medullary plate induced by it.

In the first perfect case of induction described by Hilde Mangold the ear-vesicles lay at the tip of the medullary tube; which means that the greater part of the brain with the eyeballs was lacking. Now I produced an isolated brain, with well-developed eyeballs in front and two ear-vesicles at the tip of the free posterior end. These different results may be caused by a difference in the inducing organizers, which might have been taken from different regions of the organizing centre. That mesodermal material which would come to lie under the anterior part of the medullary plate, would induce the brain; the material underlying the posterior part of the plate would induce the spinal cord.

In order to test this a piece of the upper lip may be excised at different stages of gastrulation and transplanted to the same region of another gastrula. As the mesoderm is wandering round the upper lip, the material passing it first will be brought to the anterior end and lie under the brain, whilst the material invaginated later will occupy the more posterior parts of the embryo lying under the spinal cord.

This experiment has been performed by me this summer. It proved that the centre of organization really has some regional structure, but at the same time it showed that the question is more complicated. The organizer may induce more than it ought to do; it has the tendency to build up a whole, behaving in this like a harmonious equipotential system of Driesch. And, further, it seems to make a difference according to the region of the host into which the organizer is implanted. Therefore the experiment must be varied still in another sense; as *different organizers*

have been transplanted to the *same place,* so the *same organizers* must be transplanted to *different places.* I have made this experiment, too, last summer; it yielded the result just stated.

(d) Nature of the Inducing Agent.— Of still greater interest than extent, origin and structure of the organizing centre, is the nature of its action; but obviously here our knowledge is very scanty. On the one hand, it may be said that this action cannot be of quite indifferent nature, *e.g.,* contact-action; for a piece of presumptive epidermis does not act in the same way. The stimulus must be *specific* to a certain degree. On the other hand, the stimulus is clearly not specific *within narrow limits.* This is shown by experiments of B. Geinitz. He found that induction may take place between an organizer of a frog, or toad, and a newt; that is, not only between embryos of different species or genus, but even of different order.

So far we know hardly anything about the most elementary question, whether induction is performed by chemical or by dynamical means. If the latter were true, only the living organizer could be supposed to induce; whilst morphogenetic substances might remain effective even after destruction of the cells. This shows the way this problem may be attacked.

In this connection I wish to mention two facts of a very promising character stated quite recently. One of them has been found quite independently by O. Mangold and myself in 1926; it might be called 'homoiogenetic induction' or 'assimilating induction.' We transplanted a piece of medullary plate into the blastocœl of a gastrula in order to bring it under the epidermis from the very first. In this way we made the surprising discovery that medullary plate may induce its like, that is, medullary plate. Mangold transplanted heteroplastically; I did so with stained material. We followed the whole development in the living object and on sections of preserved material, so we are quite sure of the fact itself. To explain it, we may suppose that the inducing agent may be given off again by the medullary plate which it had formed, to induce in indifferent material a new medullary plate.

Last summer, Mangold made the further discovery, not less important, that this agent is retained in the medullary substance for a surprisingly long time. A piece of the brain of a free-swimming larva may still induce a medullary plate. This fact seems to involve the chemical nature of the inducing stimulus. The case is similar to another known for many years, the regeneration of the lens of Triton from the upper margin of the iris, as discovered by Colucci and G. Wolff. According to the experiments of H. Wachs the lens regenerates under an inducing influence of the retina, the same part of the eye that induces the lens during the normal development. It must have retained this faculty which was needed during only a short time.

This seems to be of more general occurrence. Last summer, H. Bautzmann found something very similar. A piece of the very young notochord cut out of a neurula and transplanted into the blastocœl induced in the ectoderm a medullary plate. The notochord must have retained this faculty which had become active a long time before, when inducing the original medullary plate of the neurula.

IV.—The Rôle of the Organizing Faculty in Normal Development.

What rôle does this organizing faculty play in normal development? It has been shown by numerous experiments that the rudiment of the eye, the eyeball, may induce the formation of a lens in strange epidermis. It is very probable, therefore, that the eyeball exerts the same influence in normal development. The eyeball may therefore be called the organizer of the lens. Now an eyeball may be produced experimentally from such material as in normal development would have become epidermis, material that was transplanted to its new place and had thus been induced to form an eyeball. It is again very probable that in normal development the material of the eyeball is determined in the same way, by an inducing influence of the underlying mesoderm. Induced itself, it goes on inducing; so it may be called an *organizer of second grade.*

Another organizer, of second grade, B. Geinitz and I have produced by the following experiment. As I said before, O. Mangold has produced mesoderm out of presumptive ectoderm by transplanting presumptive epidermis of a young gastrula among presumptive mesoderm of another gastrula. This transplanted piece took part in the invagination and the further development of its new surroundings, where it formed notochord, somites, kidneys. B. Geinitz and I repeated this experiment with the same effect. Now Marx had shown, as I said before, that mesoderm may induce medullary plate. This experiment, too, has been repeated many times with the same result.

The question arises whether this inducing faculty of the mesoderm is acquired even by a piece of ectoderm that has been turned (experimentally) into mesoderm. To test this we combined the two experiments; that is, we forced presumptive ectoderm by transplantation to become mesoderm, took it out again some time after invagination and transplanted it afresh into the blastocœl of a young gastrula. It this way it was brought under the epidermis, where, in fact, it induced a highly perfect medullary plate.

In this experiment, as in the former, two processes of development, each of which had been determined by induction, were linked together. It seems to be not improbable that the whole development, at least of the amphibian embryo, is composed of single processes connected by organizers. The special mode of combination would have to be stated in the separate concrete cases; but in principle the old problem would be solved strictly according to the theory of epigenesis.

But as a matter of fact matters are much more complicated. About this let me add a few words in conclusion.

The experience gained by studying the development of the lens should make us cautious. There, as we have seen, the eyeball may induce a lens in foreign epidermis; but other experiments showed, with not less evidence, that the normal lens cells may also develop independently, forming a lens without an eyeball. In one subject at least (Rana esculenta, the water frog), both faculties have been shown conclusively to coexist, though either by itself would seem sufficient to warrant normal development. Rhumbler named this what engineers call 'double assurance,' and Braus adopted this

term for a very clear similar case he had discovered.

The first rudiments of organs might be determined in the same way; that is, the ectodermal material which is going to form the medullary plate might come to meet the inducing action which the underlying mesoderm exerts; this might even go so far that the material would form a plate without mesoderm. K. Görttler, examining this question during the last few years, has recently attained a clear positive result. I will only quote his last experiment which I consider to be the most conclusive and important. Presumptive medullary plate of a young gastrula was transplanted into the epidermis of a neurula; that is, of a somewhat older stage. There it developed sometimes in accordance with its new place, that is, to epidermis instead of medullary plate, as I had always observed after transplantation into a host of the same age; but sometimes it formed medullary plate in accordance with its origin.

This difference in behaviour Görttler believes to be caused by different orientation of the implanted piece. It develops according to its original destiny, if the intended movements of cells be not hindered by the new environment. There is a shifting of cells connected with development both in the implanted piece and in its new surroundings. If they are in harmony, the piece will follow its original intention and form medullary plate; if both interfere, it gets disturbed, as it were, and will follow the development of its new surroundings. This would show that even at the beginning of gastrulation a certain region is determined, though not irrevocably, to form medullary plate; mesoderm is then invaginated and pushed under it and acts in the same sense. How this harmony arises, that is the great question, and new experiments alone can answer it.

V.—*Summary.*

To summarize: The conception of organizers in development has been derived from experiments on amphibian embryos in the earliest stages. The different regions of such an embryo have not the same value for development; most of them are relatively indifferent and do not carry their destiny wholly within themselves. This can be shown by transplantation of these parts into other regions of the embryo; they may follow the development of their new surroundings. But there is a certain region in the embryo, parts of which behave in an entirely different way. Transplanted into a more indifferent region of the embryo they do not adapt themselves to their new surroundings, but retain their own character, and force the others to follow them. Such pieces 'organize' a new embryo, which is built up partly by the transplanted cells, partly by the cells of the host. For that reason they were called 'organizers,' while the region where they lie in those early stages of development was called the 'centre of organization.'

Further experiments were directed to determine the extent of this centre, its origin, its intimate structure and the nature of the organizing influence. The organizing centre coincides with the presumptive mesoderm; transplanted pieces taken from this region not only induce medullary plate, but also determine their new mesodermal neighbourhood to form notochord and somites. This centre seems to be foreshadowed by the grey crescent; defect of grey crescent prevents formation of axial organs. The cells of the centre must have some longitudinal structure which determines, at least partly, the direction of invagination and the orientation of the secondary embryo. There may be a corresponding structure in the ectoderm, but it is less effective. The same holds true for a regional structure which determines the regions of the medulla. All these processes obey the principle of 'double assurance.' The means of induction seem to be rather material than dynamical.

What has been achieved is but the first step; we still stand in the presence of riddles, but not without hope of solving them. And riddles with the hope of solution—what more can a man of science desire?

BIBLIOGRAPHY

These references are primarily books, monographs, and review papers. They will lead you to the original sources.

ADELMANN, HOWARD B. 1942. *The Embryological Treatises of Hieronymus Fabricius of Aquapendente.* Two volumes. Ithaca, N.Y.: Cornell University Press.
ADELMANN, HOWARD B. 1966. *Marcello Malpighi and the Evolution of Embryology.* Five volumes. Ithaca: Cornell University Press.

ALSTON, RALPH E. 1967. *Cellular Continuity and Development*. Glenview, Ill.: Scott, Foresman.

AREY, L. B. 1965. *Development Anatomy. A Textbook and Laboratory Manual of Embryology*. Seventh Edition. Philadelphia: W. B. Saunders. Primarily human embryology.

AUSTIN, C. R. 1968. *Ultrastructure of Fertilization*. New York: Holt, Rinehart & Winston.

BALINSKY, B. I. 1970. *An Introduction to Embryology*. Third Edition. Philadelphia: W. B. Saunders.

BALLARD, W. W. 1964. *Comparative Anatomy and Embryology*. New York: Ronald Press.

BARTH, LESTER G. 1953. *Embryology*. Revised edition. New York: Dryden Press.

BARTH, LUCENA J. 1964. *Development. Selected Topics*. Reading, Mass.: Addison-Wesley.

BELL, EUGENE. Editor. 1967. *Molecular and Cellular Aspects of Development*. Revised Edition. New York: Harper and Row.

BERRILL, N.J. 1961. *Growth, Development, and Pattern*. San Francisco: W. H. Freeman.

BODEMER, C. 1968. *Modern Embryology*. New York: Holt, Rinehart & Winston.

BONNER, JAMES. 1965. *The Molecular Biology of Development*. New York: Oxford University Press.

BONNER, JOHN TYLER. 1963. *Morphogenesis: An Essay on Development*. New York: Atheneum.

BRACHET, JEAN. 1950. *Chemical Embryology*. New York: Interscience.

BRACHET, JEAN. 1957. *Biochemical Cytology*. New York: Academic Press.

BRACHET, JEAN, and ALFRED E. MIRSKY. Editors. 1959. *The Cell*. Volume 1. New York: Academic Press.

BRØNDSTED, H. V. 1969. *Planarian Regeneration*. New York: Pergamon.

BROOKHAVEN SYMPOSIA IN BIOLOGY. 1965. *Genetic Control of Differentiation*. Number 18. Upton, NY: Biology Department, Brookhaven National Laboratory.

CLAVERT, J. 1962. 'Symmetrization of the egg of vertebrates.' *Advances in Morphogenesis 2*: 27–60.

COHEN, JACK. 1967. *Living Embryos: An Introduction to the Study of Animal Development*. Second Edition. New York: Pergamon.

COLE, F. J. 1930. *Early Theories of Sexual Generation*. Oxford: Clarendon Press.

COSTELLO, D. P., and others. 1957. *Methods for Obtaining and Handling Marine Eggs and Embryos*. Woods Hole, Mass.: Marine Biological Laboratory.

DALCQ, ALBERT M. 1938 *Form and Causality in Early Development*. Cambridge: At the University Press.

DALCQ, A. M. 1957. *An Introduction to General Embryology*. London: Oxford University Press.

DAVENPORT, CHARLES BENEDICT. 1897. *Experimental Morphology*. New York: Macmillan.

DAVIDSON, ERIC H. 1968. *Gene Activity in Early Development* New York: Academic Press.

DE BEER, SIR GAVIN. 1958. *Embryos and Ancestors*. Third Edition. Oxford: At the Clarendon Press.

DEHAAN, ROBERT L., and HEINRICH URSPRUNG. Editors. *Organogenesis*. New York: Holt, Rinehart & Winston.

DENIS, HERMAN. 1968. 'Role of messenger ribonucleic acid in embryonic development.' *Advances in Morphogenesis 7*: 115–150.

DETWILER, SAMUEL R. 1936. *Neuroembryology. An Experimental Study.* New York: Macmillan.

DURKEN, BERNHARD. 1932. *Experimental Analysis of Development.* New York: W. W. Norton.

EBERT, JAMES D. 1965. *Interacting Systems in Development.* New York: Holt, Rinehart & Winston.

ETKIN, WILLIAM, and LAWRENCE I. GILBERT. Editors. 1968. *Metamorphosis. A Problem in Developmental Biology.* New York: Appleton-Century-Crofts.

FLICKINGER, R. A. 1966. *Developmental Biology. A Book of Readings.* Dubuque, Iowa: Wm. C. Brown.

GALSTON, ARTHUR W., and PETER J. DAVIES. 1970. *Control Mechanisms in Plant Development.* Englewood Cliffs, N.J.: Prentice-Hall.

GASKING, ELIZABETH B. 1967. *Investigations into Generation 1651–1828.* Baltimore: Johns Hopkins University Press.

GOSS, RICHARD T. 1964. *Adaptive Growth.* New York: Academic Press.

GOSS, RICHARD J. 1969. *Principles of Regeneration.* New York: Academic Press.

GURDON, J. B. 1967. 'Control of gene activity during the early development of *Xenopus laevis.*' In *Heritage from Mendel.* Edited by R. A. Brink. Madison: University of Wisconsin Press.

HAMBURGER, VIKTOR. 1960. *A Manual of Experimental Embryology.* Revised Edition. Chicago: University of Chicago Press.

HAY, ELIZABETH D. 1966. *Regeneration.* New York: Holt, Rinehart & Winston.

HUETTNER, ALFRED F. 1941. *Fundamentals of Comparative Embryology of the Vertebrates.* New York: Macmillan.

HUXLEY, JULIAN S., and G. R. DE BEER. 1934. *The Elements of Experimental Embryology.* Cambridge: At the University Press. Reprinted by Hafner, New York.

JAFFE, LIONEL F. 1968. 'Localization in the developing *Fucus* egg and the general role of localizing currents.' *Advances in Morphogenesis 7*: 295–328.

JENKINSON, J. W. 1909. *Experimental Embryology.* Oxford: At the Clarendon Press.

JENKINSON, J. W. 1913. *Vertebrate Embryology.* London: Oxford University Press.

KAFIANI, CONSTANTINE. 1970. 'Genome transcription in fish development.' *Advances in Morphogenesis 8*: 209–284.

KUHN, A. 1965. *Vorlesungen über Entwicklungsphysiologie.* Second Edition. New York: Springer.

KUMÉ, MATAZO, and KATSUMA DAN. 1968. *Invertebrate Embryology.* Washington: National Library of Medicine. Public Health Service.

LAETSCH, WATSON M. Editor. 1969. *The Biological Perspective. Introductory Readings.* Boston: Little, Brown. Numerous articles on development.

LILLIE, FRANK R. 1919. *The Development of the Chick.* New York: Henry Holt. Revised by Howard L. Hamilton, 1952.

LOOMIS, WILLIAM F. JR. 1970. *Papers on Regulation of Gene Activity During Development.* New York: Harper & Row.

LOPASHOV, G. V., and O. E. STROEVA. 1961. 'Morphogenesis of the vertebrate eye.' *Advances in Morphogenesis 1*: 331–377.

MCELROY, WILLIAM D., and BENTLEY GLASS. Editors. 1958. *A Symposium on the Chemical Basis of Development.* Baltimore: Johns Hopkins University Press.

MEYER, ARTHUR WILLIAM. 1939. *The Rise of Embryology.* Stanford: Stanford University Press.

MONROY, A. 1965. *Chemistry and Physiology of Fertilization.* New York: Holt, Rinehart & Winston.

MORGAN, THOMAS HUNT. 1897. *The Development of the Frog's Egg. An Introduction to Experimental Embryology.* New York: Macmillan.

MORGAN, THOMAS HUNT. 1901. *Regeneration.* New York: Macmillan.

MORGAN, THOMAS HUNT. 1927. *Experimental Embryology.* New York: Columbia University Press.

NEEDHAM, JOSEPH. 1931. *Chemical Embryology.* Cambridge: At the University Press.

NEEDHAM, JOSEPH. 1942. *Biochemistry and Morphogenesis.* Cambridge: At the University Press.

NEEDHAM, JOSEPH. 1959. *A History of Embryology.* Second Edition. New York: Abelard-Schuman.

NELSON, OLIN E. 1953. *Comparative Embryology of the Vertebrates.* New York: Blakiston.

NEW, D.A.T. 1966. *The Culture of Vertebrate Embryos.* New York: Academic Press.

NEW YORK HEART ASSOCIATION. 1964. *Differentiation and Development.* Boston: Little, Brown. Also in *Journal of Experimental Zoology 157.* Number 1.

NIEUWKOOP, P. D., and J. FABER. 1956. *Normal Table of Xenopus laevis (Daudin). A Systematical and Chronological Survey of the Development from the Fertilized Egg Till the End of Metamorphosis.* Amsterdam: North-Holland.

OPPENHEIMER, JANE M. 1967. *Essays in the History of Embryology and Biology.* Cambridge: M.I.T. Press.

OPPENHEIMER, JANE M. 1970. *Cells and Organizers* American Zoologist *10*: 75–88.

PASTEELS, JEAN J. 1964. 'The morphogenetic role of the cortex of the amphibian egg.' *Advances in Morphogenesis 3*: 363–388.

RAVEN, CHR. P. 1958. *Morphogenesis: The Analysis of Molluscan Development.* New York: Pergamon.

RAVEN, CHR. P. 1959. *An Outline of Developmental Physiology.* Third Edition. New York: Pergamon.

RAVEN, CHR. P. 1961. *Oogenesis: The Storage of Developmental Information.* New York: Pergamon and Macmillan.

REVERBERI, G. 1961. 'The embryology of ascidians.' *Advances in Morphogensis 1*: 55–101.

ROSE, S. MERYL. 1970. *Regeneration: Key to Understanding Normal and Abnormal Growth and Development.* New York: Appleton-Century-Crofts.

RUGH, ROBERTS. 1964. *Vertebrate Embryology. The Dynamics of Development.* New York: Harcourt, Brace & World.

RUSSELL, E. S. 1930. *The Interpretation of Development and Heredity.* Oxford: Clarendon Press.

SAUNDERS, JOHN W. JR. 1968. *Animal Morphogenesis.* New York: Macmillan.

SAXEN, LAURI, and SULO TOIVONEN. 1962. *Primary Embryonic Induction.* Englewood Cliffs, N.J.: Prentice-Hall.

SCHMIDT, ANTHONY J. 1968. *Cellular Biology of Vertebrate Regenerations and Repair.* Chicago: University of Chicago Press.

Society for Developmental Biology. Each year a symposium is published for the society by Academic Press, New York. Recent volumes are:

23. *The Role of Chromosomes in Development.* 1964.

24. *Reproduction: Molecular, Subcellular, and Cellular.* 1965.

25. *Major Problems in Developmental Biology.* 1966.

26. *Control Mechanisms in Developmental Processes.* 1967.

27. *The Emergence of Order in Developing Systems.* 1968.

28. *Communication in Development.* 1969.

SPEMANN, HANS. 1938. *Embryonic Development and Induction.* New Haven: Yale University Press. Republished 1962 by Stechert-Hafner, New York.

THORNTON, CHARLES S. 1968. 'Amphibian limb regeneration.' *Advances in Morphogenesis 7*: 205–249.

TORREY, JOHN G. 1967. *Development in Flowering Plants.* New York: Macmillan.

TORREY, THEODORE W. 1967. *Morphogenesis of the Vertebrates.* Second Edition. New York: Wiley.

TRINKAUS, J. P. 1969. *Cells Into Organs. The Forces that Shape the Embryo.* Englewood Cliffs, N.J.: Prentice-Hall.

TYLER, ALBERT. 1967. 'Masked messenger RNA and cytoplasmic DNA in relation to protein synthesis and processes of fertilization and determination in embryonic development.' *Developmental Biology Supplement 1*, 170–226 (1967).

TYLER, ALBERT, and BETTY S. TYLER. 1966. 'Physiology of fertilization and early development.' in *Physiology of Echinodermata.* Edited by R. A. Boolootian. New York: Interscience-Wiley.

WADDINGTON, C. H. 1952. *The Epigenetics of Birds.* Cambridge: At the University Press.

WADDINGTON, C. H. 1956. *Principles of Embryology.* London: George Allen and Unwin.

WADDINGTON, C. H. 1957. *The Strategy of the Genes. A Discussion of Some Aspects of Theoretical Biology.* London: George Allen and Unwin.

WADDINGTON, C. H. 1962. *New Patterns in Genetics and Development.* New York: Columbia University Press.

WADDINGTON, C. H. 1966. *Principles of Development and Differentiation.* New York: Macmillan.

WEBER, RUDOLF. Editor. 1965–1967. *The Biochemistry of Animal Development.* New York: Academic Press.

WEISS, PAUL. 1939. *Principles of Development.* New York: Henry Holt.

WEISS, PAUL A. 1968. *Dynamics of Development: Experiments and Inferences.* New York: Academic Press.

WIGGLESWORTH, V. B. 1959. *The Control of Growth and Form: A Study of the Epidermal Cell in an Insect.* Ithaca: Cornell University Press.

WILENS, SALLY. Editor. 1969. *Organization and Development of the Embryo. Ross Granville Harrison.* New Haven: Yale University Press.

WILLIER, BENJAMIN H., and JANE M. OPPENHEIMER. 1964. *Foundations of Experimental Embryology.* Englewood Cliffs, N.J.: Prentice-Hall.

WILLIER, BENJAMIN H., PAUL A. WEISS, and VIKTOR HAMBURGER. Editors 1955. *Analysis of Development.* Philadelphia: W. B. Saunders.

WILSON, EDMUND B. 1892. 'Cell-lineage of Nereis.' *Journal of Morphology 6*: 361–480.

WILSON, EDMUND B. 1893. 'Amphioxus and the mosaic theory of development.' *Journal of Morphology 8*: 579–638.

WILSON, EDMUND B. 1898. 'Considerations on cell-lineage and ancestral reminiscence . . . *Annals of the New York Academy of Sciences 11*: 1–27.

WILSON, EDMUND B. 1905. 'The problem of development.' *Science 21*: 281–294.

WILSON, EDMUND B. 1928. *The Cell in Development and Heredity.* New York: Macmillan. Especially chapters 13 and 14.

WILT, FRED H. 1967. 'The control of embryonic hemoglobin synthesis.' *Advances in Morphogenesis* 6: 89–125.

WILT, F. H. and N. K. WESSELLS. Editors. 1968. *Methods in Developmental Biology.* New York: Crowell.

WISCHNITZER, SAUL. 1966. 'The ultrastructure of the cytoplasm of the developing amphibian egg.'*Advances in Morphogenesis* 5: 131–179.

WISCHNITZER, SAUL. 1967. 'The ultrastructure of the nucleus of the developing amphibian egg.' *Advances in Morphogenesis* 6: 173–198.

WITSCHI, EMIL. 1956. *Development of Vertebrates.* Philadelphia: W. B. Saunders.

YAMADA, TUNEO. 1961. 'A chemical approach to the problem of the organizer.' *Advances in Morphogenesis* 1: 1–53.

Index